Palgrave Historical Studies in Witchcraft and Magic

Series Editors: **Jonathan Barry, Willem de Blécourt and Owen Davies**

Series Foreword

The history of European witchcraft and magic continues to fascinate and challenge students and scholars. There is certainly no shortage of books on the subject. Several general surveys of the witch trials and numerous regional and micro studies have been published for an English-speaking readership. While the quality of publications on witchcraft has been high, some regions and topics have received less attention over the years. The aim of this series is to help illuminate these lesser-known or little-studied aspects of the history of witchcraft and magic. It will also encourage the development of a broader corpus of work in other related areas of magic and the supernatural: such as angels, devils, spirits, ghosts, folk healing and divination. To help further our understanding and interest in this wider history of beliefs and practices, the series will include research that looks beyond the usual focus on Western Europe and that also explores their relevance and influence from the medieval to the modern period.

Titles include:

Jonathan Barry
WITCHCRAFT AND DEMONOLOGY IN SOUTH-WEST ENGLAND, 1640–1789

Jonathan Barry
RAISING SPIRITS
How a Conjuror's Tale Was Transmitted across the Enlightenment

Edward Bever
THE REALITIES OF WITCHCRAFT AND POPULAR MAGIC IN EARLY MODERN EUROPE
Culture, Cognition and Everyday Life

Willem de Blécourt (*editor*)
WEREWOLF HISTORIES

Ruth Bottigheimer
MAGIC TALES AND FAIRY TALE MAGIC
From Ancient Egypt to the Italian Renaissance

Alison Butler
VICTORIAN OCCULTISM AND THE MAKING OF MODERN MAGIC
Invoking Tradition

Johannes Dillinger
MAGICAL TREASURE HUNTING IN EUROPE AND NORTH AMERICA
A History

Julian Goodare (*editor*)
SCOTTISH WITCHES AND WITCH-HUNTERS

Julian Goodare, Lauren Martin and Joyce Miller (*editors*)
WITCHCRAFT AND BELIEF IN EARLY MODERN SCOTLAND

Louise Kallestrup
AGENTS OF WITCHCRAFT IN EARLY MODERN ITALY AND DENMARK

Jonathan Roper (*editor*)
CHARMS, CHARMERS AND CHARMING

Alison Rowlands (*editor*)
WITCHCRAFT AND MASCULINITIES IN EARLY MODERN EUROPE

Rolf Schulte
MAN AS WITCH
Male Witches in Central Europe

Andrew Sneddon
WITCHCRAFT AND MAGIC IN IRELAND, 1890–1940

Laura Stokes
DEMONS OF URBAN REFORM
Early European Witch Trials and Criminal Justice, 1430–1530

María Tausiet
URBAN MAGIC IN EARLY MODERN SPAIN
Abracadabra Omnipotens

Robert Ziegler
SATANISM, MAGIC AND MYSTICISM IN FIN-DE-SIÈCLE FRANCE

Forthcoming:

Lizanne Henderson
WITCHCRAFT AND FOLK BELIEF AT THE DAWN OF ENLIGHTENMENT

Liana Saif
ARABIC INFLUENCES ON EARLY MODERN OCCULT THOUGHT

Palgrave Historical Studies in Witchcraft and Magic
Series Standing Order ISBN 978-1403-99566-7 Hardback
978-1403-99567-4 Paperback
(*outside North America only*)

You can receive future titles in this series as they are published by placing a standing order. Please contact your bookseller or, in case of difficulty, write to us at the address below with your name and address, the title of the series and one of the ISBNs quoted above.

Customer Services Department, Macmillan Distribution Ltd, Houndmills, Basingstoke, Hampshire RG21 6XS, England

Werewolf Histories

Edited by

Willem de Blécourt

palgrave
macmillan

First published 2015 by
PALGRAVE MACMILLAN

Palgrave Macmillan in the UK is an imprint of Macmillan Publishers Limited, registered in England, company number 785998, of Houndmills, Basingstoke, Hampshire RG21 6XS.

Palgrave Macmillan in the US is a division of St Martin's Press LLC, 175 Fifth Avenue, New York, NY 10010.

Palgrave Macmillan is the global academic imprint of the above companies and has companies and representatives throughout the world.

Palgrave® and Macmillan® are registered trademarks in the United States, the United Kingdom, Europe and other countries.

ISBN: 978–1–137–52633–5

This book is printed on paper suitable for recycling and made from fully managed and sustained forest sources. Logging, pulping and manufacturing processes are expected to conform to the environmental regulations of the country of origin.

A catalogue record for this book is available from the British Library.

Library of Congress Cataloging-in-Publication Data

Werewolf Histories / [edited by] Willem De Blécourt, Honorary Research Fellow, Meertens Institute, Amsterdam, the Netherlands.
pages cm

1. Werewolves – History. I. Blécourt, Willem de.

GR830.W4W47 2015
398.24′54—dc23 2015029077

Contents

List of Illustrations

Acknowledgements

Some time ago an enterprising independent publisher suggested I write a book about the history of werewolves for the general reader. At the time, I only knew about Dutch werewolves, although they were quite numerous and dated from the late sixteenth to the twentieth century. Armed with the support of Alfons Roeck's unpublished thesis on Dutch and Flemish legends, which had a chapter about the rest of Europe,[1] and Elmar Lorey's incomparable list of European werewolf trials,[2] I set out on a gruesome but exciting path of further enlightenment. I also consulted films and television series and, for a brief period, werewolf novels, since the Anglo-Saxon public needed something to relate to, bereft as they had been of indigenous werewolves for hundreds of years. It soon turned out that this lack of local variants had led to an over-reliance on a handful of historical French werewolves, which meant the field was now taken over by dabbling amateurs. Serious academic attention was sorely needed. I wrote my book but it was never published. I had already received proofs, and even compiled an index. But the publisher was taken over just before publication and the book was shelved. The publicity had been vigorous and the title featured on several websites. Some (online) bookshops and library sites have retained it to this day.[3] This phantom book has even managed to creep into other authors' footnotes and reading lists. Most of the chapters of the originally planned book have since been published as articles in journals or chapters in edited volumes;[4] two further articles which elaborate themes that were previously only touched on will also reach the public this year.[5] A proper academic monograph (or two) is still in the pipeline.

Meanwhile, other historians and folklorists have taken an interest in werewolves too.[6] This present volume offers the first opportunity to bring them all together. My intention is that the various contributions will open up the field for future scholarly attention, especially since they were conceived as first rather than final words. I am pleased to have so many excellent colleagues from different countries on board since a Dutch focus is hardly sufficient when it comes to werewolves, who varied so widely across Europe.

I would like to thank all those who have been involved in the different phases of this book: Caroline Oates, with whom talking werewolves is always rewarding (and who will be the co-author of my next werewolf book); the members of the Belief Narrative Network Committee, in particular Ülo Valk and Mirjam Mencej; Louise Nyholm Kallestrup for some last-minute translation; and, last but not least, Derek Davies who managed to spruce up the English of some of the non-native authors. But above all, my gratitude goes out to my friends and fellow authors whose patience was sometimes sorely tested but who, I hope, will be finally rewarded.

Notes

1. F. Roeck, *De Nederlandse weerwolfsage in de negentiende en twintigste eeuw* (Leuven, 1967).
2. Elmar Lorey at: http://www.elmar-lorey.de/prozesse.htm.
3. Willem de Blécourt, *Werewolves* (London, 2005). ISBN 1852854022.
4. Willem de Blécourt, 'I Would Have Eaten You Too: Werewolf Legends in the Flemish, Dutch and German Area', *Folklore* 118 (2007), 23–43; idem, 'A Journey to Hell: Reconsidering the Livonian "Werewolf"', *Magic, Ritual, and Witchcraft* 2 (2007), 49–67; idem, 'The Werewolf, the Witch, and the Warlock: Aspects of Gender in the Early Modern Period', in: Alison Rowlands (ed.), *Witchcraft and Masculinities* (Basingstoke, 2009), 191–213; idem, 'Monstrous Theories: Werewolves and the Abuse of History', *Preternature: Critical and Historical Studies on the Preternatural* 2 (2013), 188–212.
5. Willem de Blécourt, 'The Case of the Cut-Off Hand: Angela Carter's Werewolves in Historical Perspective', in: Hannah Priest (ed.), *She Wolf: A Cultural History of Female Werewolves* (Manchester, 2015), 148–165; idem, 'Werwolfberichte. Der Fall Peter Stump und seine Folgen', in: Rita Voltmer (ed.), *Europäische Hexenforschung und Landesgeschichte – Methoden,Regionen, Vergleiche* (Trier, 2016), in press.
6. The volume *She-Wolf* is mostly concerned with literary and cinematic werewolves, although it includes chapters by Rolf Schulte and Merili Metsvahi; both authors who are also contributors to the present volume. See also Rolf Schulte, *Man as Witch: Male Witches in Central Europe* (Basingstoke, 2009), 1–35; Merili Metsvahi, 'The Woman as Wolf (AT 409): Some Interpretations of a Very Estonian Folk Tale', *Journal of Ethnology and Folkloristics* 7 (2013), 65–92.

Contributors

Willem de Blécourt is a historical anthropologist and an independent scholar; he holds an honorary research fellowship at the Meertens Institute, Amsterdam. Apart from werewolves, his main areas of interest are the history of witchcraft in Europe, from the late Middle Ages up to the twentieth century, and the history of fairy tales during the same period. His most recent publications include *Tales of Magic, Tales in Print. On the Genealogy of Fairy Tales and the Brothers Grimm* (2012) and two edited volumes (with Owen Davies) about witchcraft after the witch trials, *Beyond the Witch Trials* and *Witchcraft Continued* (2004). He is currently writing *The Cat and the Cauldron*, a history of witchcraft in the Low Countries.

Johannes Dillinger is Professor of Early Modern History at Oxford Brookes University and extraordinary Professor of Modern History and Regional History at Johannes Gutenberg Universität, Mainz. He studied history, Catholic theology and education at Tübingen, Norwich and Trier. He won the Friedrich Spee award for outstanding contributions to the history of witchcraft. His publications include English and German monographs on the political representation of the peasantry in the early modern period, on terrorism, treasure hunting, witchcraft and on concepts of alternate history. Dillinger's research interests include state building, constitutional history, the history of political crime, church history and the history of folk belief, magic and witchcraft.

Matteo Duni teaches the history of the Italian Renaissance and of the witch-hunts at both Syracuse University's and New York University's centres in Florence. He is the author of *Under the Devil's Spell: Witches, Sorcerers and the Inquisition in Renaissance Italy* (2007) and of *Tra religione e magia. Storia del prete modenese Guglielmo Campana (1460?–1541)* (1999). He has also edited the two conference proceedings: *On Medieval and Early-Modern Witchcraft* (with Dinora Corsi): *Non lasciar vivere la malefica. Le streghe nei trattati e nei processi* (2008), and on the Inquisition (with Guido Dall'Olio and Andrea Cicerchia): *Prescritto e proscritto. Religione e società nell'Italia moderna* (in press). His survey of Italian witchcraft is due in 2018.

Richard L. Gordon is Honorary Professor in the Department of Classical Religious Studies and corresponding member, Max-Weber-Kolleg, University of Erfurt. His main research interest is the social history of Graeco-Roman religion. Recent publications include: 'Superstitio': *Superstition and Religious Repression in the Late Roman Republic and Principate* (100 BC to AD 300), in: S. Smith and Alan Knight (eds), *The Religion of Fools: 'Superstition' in Historical and Comparative Perspective* (2008); with Francisco Marco Simón (eds), *Magical Practice in the Latin West* (2010); 'Memory and Authority in the Magical Papyri', in: Beate Dignas and R. R. R. Smith (eds), *Historical and Religious Memory in the Ancient World: Essays for Simon Price* (2012); '"Will My Child Have a Big Nose?": Uncertainty, Authority and Narrative in Katarchic Astrology', in: Veit Rosenberger (ed.), *Divination in the Ancient World: Religious Options and the Individual* (2013).

Merili Metsvahi is an senior researcher at the Department of Estonian and Comparative Folklore, University of Tartu, Estonia. Her M.A dissertation (2000) concerned the belief background of the werewolf trial against Thiess in 1692. In 2007 she published her doctoral dissertation as *Indiviid, mälu ja loovus: Ksenia Müürsepa mõttemaailm folkloristi pilgu läbi* (Individual, memory and creativity: Ksenia Müürsepp's mental world from a folklorist's perspective), based on her own fieldwork. She has also published more than 30 scientific articles. Her research interests are Estonian folktales, werewolf legends, Estonian family history and women in Estonian folklore. She has taught courses on legends, on Estonian folk belief, on the methods of folklore field work (for the folklore students), and on the ways of teaching folklore at schools (for the pedagogy students) in the course of the last 14 years at the University of Tartu.

Maja Pasarić is a research fellow/senior assistant working at the Institute of Ethnology and Folklore Research in Zagreb, Croatia. Her research interests cover themes related to folk beliefs, mythology, ritual and cultural animal studies. Focusing mainly on cultural animal studies, she is interested in human–animal relationships investigated through the study of folkloristic and ethnographic materials, animal burials and zoomorphic objects. She explores folk beliefs about animals in relation with death and the afterlife, the role of animals in mortuary practices, animal sacrifice and the use of animals in different traditional cultures' magical practices. Her articles have been published in scientific and professional journals and books as well as in popular journals such as *Narodna umjetnost, Kroatologija, Literary Animal: Cultural Bestiary 2,* Zarez.

Aleks Pluskowski is a lecturer in the Department of Archaeology at the University of Reading. One of his specialisms is medieval European zooarchaeology, which he combines with an interest in religious diversity and the conversion from paganism to Christianity, as well as the environmental impact of crusading at the frontiers of Europe and Venetian colonisation of the Adriatic. He has published widely on these subjects, among other places in the edited volumes *Breaking and Shaping Beastly Bodies: Animals as Material Culture in the Middle Ages* (2007) and *The Ritual Killing and Burial of Animals: European Perspectives* (2012) and in the monographs *Wolves and Wilderness in the Middle Ages* (2006) and *The Archaeology of the Prussian Crusade: Holy War and Colonisation* (2012).

Rolf Schulte is a fellow in the Department of History at the University of Kiel and specialises in the nineteenth- and twentieth-century histories of France and Germany, and more specifically in the (early modern) history of witchcraft and witch persecution. He edited (with Burghart Schmidt) a volume on witchcraft in modern Africa (2007). His monographs include a study of the witch trials in Schleswig-Holstein: *Hexenverfolgung in Schleswig-Holstein 16.-18. Jahrhundert* (2001) and a more general study on male witches, translated into English as: *Man as Witch: Male Witches in Central Europe* (2009).

Michèle Simonsen taught French, ethnology and folklore at the University of Copenhagen. Among her books are *Le conte populaire français* (Que sais-je) (1981) and *Le conte populaire* (1984). After her academic career, she became a prolific writer of children's books. Her latest book is a collection of narratives from Lapland: *Contes et légendes de Laponie* (2014).

Christa Agnes Tuczay teaches old German language and literature at the Institute of German Studies, University of Vienna. She collaborated on the *Motiv-Index der deutschsprächigen weltlichen Erzählliteratur von den Anfängen bis 1400* and subsequent projects. She is co-editor and contributor to the volumes *Poetische Wiedergänger: deutschsprächige Vampirismus-Diskurse vom Mittelalter bis zur Gegenwart* (2005), *Faszination des Okkulten: Diskurse zum Übersinnlichen* (2008) und *Tierverwandlungen: Codierungen und Diskurse* (2011). Her specialisms are narrative research, the history of mentalities and magic, witchcraft history and medieval mysticism, about which she has published many articles. Her *Habilitationsschrift* is: *Ekstase im Kontext: mittelalterliche und neuere Diskurse einer Entgrenzungserfahrung* (2009). Most recently she published a cultural history of medieval fortune-telling: *Kulturgeschichte der mittelalterlichen Wahrsagerei* (2013) and a book on spirits and demons: *Geister, Dämonen – Phantasmen: Eine Kulturgeschichte* (2015).

Rita Voltmer is a senior lecturer in Medieval and Early Modern History, University of Trier. From 2013 until 2015 she was Visiting Professor in History and Religious Studies at the Arctic University of Norway (Tromsø). Her publications include *Hexen. Wissen was stimmt* (2008); and (with Walter Rummel): *Hexen und Hexenverfolgung in der Frühen Neuzeit* (2012). She has edited several conference proceedings and has published more than 80 articles in journals and as chapters in edited collections, among them contributions to the *Encyclopedia of Witchcraft* and the volume *Witchcraft and Masculinities in Early Modern Europe*. Her current research interests include: European and transatlantic witch hunts; medieval and early modern urban history; the history of criminality and poverty; medieval and early modern communication media, especially sermons and pamphlets; the Jesuit order, and the history of Northern Europe (the British Isles, Scandinavia).

1
The Differentiated Werewolf: An Introduction to Cluster Methodology

Willem de Blécourt

There is no werewolf history. At the most there are histories of werewolves, but these are fragmented and discontinuous. The werewolf is not a biological species, nor an afflicted human (although it is often mistaken for one),[1] and one of the problems lies in translation; not all the 'werewolves' in different languages can translate into one and the same 'werewolf'. Werewolves are cultural concepts, and thus not absolute but relative entities that are defined by their contexts. Language is one of these contexts and it cannot be assumed that, for instance, the English *werewolf* is similar to the French *loup-garou*, the German *Werwolf*, the Danish *varulf*, or the Latvian *vilkatis*. Nor can it be assumed that werewolves are associated with *lycanthropy*, a Greek term that by the second century had already become the medical term for a particular affliction. There may be some overlap between different terms, depending on the cluster of expressions of which the individual terms are part; an early cluster does not necessarily represent the ancestry of a later one. I will not attempt here to juxtapose the werewolf with male and female witches, which is only viable in the context of witch trials.[2] Instead, I will restrict my analysis to werewolves and concentrate mainly on groups of texts in which the subject performs.

These clusters, or culturally specific werewolf genres, are not necessarily the same, even for speakers of the same language. On the other hand, clusters can sometimes transcend language boundaries. If those who speak different languages share the image of a human being who, during a full moon, changes into a wolf or something resembling a wolf, and who often acquires his characteristics after having been bitten by a another werewolf, then they may refer to a similar kind of werewolf concept. In that case they are all influenced by the same American and, to a lesser extent, British films.[3] Today's werewolves are mainly known from visual media aimed at an adolescent public: films such as the *Harry Potter* and the *Twilight* series; television shows like *Being Human, True Blood, Teen Wolf*, or *Wolfblood*.[4] One of the consequences of this visuality is that, especially in the United States of America, the subject of present-day werewolf narratives is above all to do with their sightings rather than physical attacks.[5] But cinematic werewolves have their own history, which is quite different from the werewolf histories in learned tracts or oral traditions, which are the subject of this book.

In the following I will identify several clusters in which werewolves occur and can be understood.[6] I approach them through the local terms, rather than in an etymological sense, as I am more interested in the use of language than their (sometimes speculative) origin. Moreover, 'werewolf' is usually explained as a combination of 'wolf' and 'wer', the latter meaning man (from the Latin *vir*).[7] On closer inspection this is unsatisfactory, because historically 'wer' was only prefixed to wolf, but never to other animals whose shape man was known to change into; and the combination of two words from different language groups appears fabricated. 'Wer' is better understood as deriving from the Anglo-Saxon 'warg' (Old Norse *vargr*), which led to the French *garou*. If 'wer' was related to *vir*, then one would expect it to exist in a romanic language like French. Italians use *lupu mannaro*, the last word deriving ostensibly from the German 'man'. Literarily meaning 'strangler', 'warg' indicated someone outside the 'world', a socially deviant outsider; more specifically, a criminal and an outcast.[8] In this way the werewolf is opposed to socially integrated wolves, men with names like Beowulf, Rudolf, Ulf, or Wolfgang.[9] In Christian thought and early Bible translations, the werewolf (*werewulf*) was associated with or synonymous with the devil.

There is no academic history of werewolves. The corpus of reliable studies containing either a general overview or specific details of the werewolf's history is still very small and not always available in English.[10] Earlier books that have been reprinted are often badly out-dated.[11] Other books are squarely within the popular realm and only of limited use for informing present-day opinion.[12] In recent decades monster studies have emerged and witchcraft history has proliferated. Both contain references to werewolves, albeit profoundly different werewolves. In the recent *Research Companion to Monsters*, for instance, werewolves figure mostly in enumerations; they are merely analysed in a discussion of hyper-masculinity, in which modern films are irresponsibly linked to medieval literature.[13] The *Handbook of Witchcraft* only mentions the prosecution of werewolves in passing: one in early-modern France and some in the Netherlands and Hungary.[14] Werewolves on their own have not yet become a genuine subject of academic historical study; this is a first attempt to do so. While considering the werewolves of Antiquity and the early Middle Ages, two specific fields are targeted in this book: werewolf trials and werewolf legends. These are particularly rich in material and share slightly more characteristics than, for instance, twentieth-century traditional werewolf legends and films.

This volume marks the transition from popular werewolf publications to academic historical studies. But at the same time this requires some level of engagement with the arguments of popular writers. For instance, it needs to be explained why the motif of the full moon is a modern invention, since historical sources do not mention it as an instigator of metamorphosis.[15] In the material considered in this book the werewolf is mostly a nocturnal creature, and there is only a very rare story in which the moon enables somebody to observe a werewolf.[16] The full 'werewolf moon' has entered twentieth-century consciousness primarily through werewolf films, although even the history of werewolf films is sometimes ambiguous: there is, for instance, no full moon in the influential *The Wolf Man* of 1941, but only in its 1943 sequel *Frankenstein meets the Wolf Man*. Nevertheless the full moon, together with the other cinematic motif of the contagious bite, was already featured in *The Werewolf of London* of 1935 and before that in werewolf fiction such as Gerald Biss's, *The Door of the Unreal* (1919). Because of its intertextuality, moon lore can be considered a special modern werewolf cluster. Another cluster relates to the exaggerated amount of werewolf trials, which is discussed below.

In addition to fictional werewolves in novels, films and television series, several other potential werewolf clusters are excluded from this book. Issues of mutual coherence, or the absence of it, are already complex enough with regard to the current presentations, which are organised around 'genuine', albeit largely ascribed, werewolves, in contrast to the imaginary ones of literature. When (especially early) literary texts do figure in this volume, it is to penetrate the historical concepts underneath. Werewolves in early-modern trials had some counter-presence in everyday life, although the precise degree is certainly open to discussion. Some of the later legends about werewolves were first and foremost narratives rather than related to experiences, such as the Werewolf Husband (or Lover), which was meant to be an admonition against improper liaisons.[17] This tale was about a woman who was attacked by a werewolf when her male companion was away; she later recognised him as her attacker by the threads of her clothing stuck between his teeth. These stories were, however, embedded in oral exchanges about experiences with werewolves and advice on how to deal with them. Above all, the werewolves of everyday life were part of the outlook of many people; they were the subject of many similar yet subtly varied texts, as opposed to the unique and singular texts of literature. Admittedly, while this may serve as an argument to largely ignore the medieval werewolves of the romances and lays,[18] it is less convincing when it comes to the lycanthropes of medical history.[19] These, however, are not only very diverse, but they have nothing more in common with the early-modern werewolves than that they were thought to be men who became (and sometimes stayed) wolves. They certainly constitute separate clusters.

In the remainder of this introduction, I will contextualise estimates of the number of werewolf trials and set them against the main overview of historical publications of werewolf trials from the fifteenth to the seventeenth century. I will then discuss nineteenth- and twentieth-century legend texts, illustrating them with the results of fieldwork in Belgium. This leaves the eighteenth century as a historiographical gap, yet to be filled. A very brief discussion of the following chapters in this book can be found at the end.

Mythic numbers

Amateur werewolf history has evolved around a small number of mostly French cases. In 1865 the folklorist Sabine Baring-Gould included in his *The Book of Were-Wolves* 'A Chapter of Horrors', in which he described the 1521 trial of Bourgot and Verdung, the 1573 trial of Garnier, and those trials which took place in late sixteenth-century Franche-Comté, as reported in Boguet's *Discours de Sorciers*. Baring-Gould closed his chapter with reports of the tailor of Châlons and the 1598 Angers case of Roulet.[20] This survey was based on printed sources rather than on trial records; in 1933 Montague Summers provided a more extensive description on the basis of similar source material, mainly by adding details to these cases.[21] The next chapter in Baring-Gould's book, dedicated to Jean Grenier, was based on the work of Pierre de Lancre.[22] Apart from the six trials mentioned above, French werewolf prosecutions were reported in Orléans (1583), Rennes (1598) and Grenoble (1603);[23] but they have not found much resonance in werewolf publications.

In the 1970s, when a previously unparalleled spate of popular werewolf publications hit the market, new trial figures began to circulate. The sociologists Bill and Claire Russell found 21 cases in Summers' book detailed enough for them to analyse.[24] They also suggested that 'some 200 human werewolves' were supposedly burned 'at Labout'. This last figure is copied from a 1950s witchcraft publication.[25] Psychiatric papers of the 1980s alleged the even higher number of 600. 'Collective psychoses', as their authors put it, resulted 'in the execution of 600 people suffering lycanthropic syndromes in France during the Middle Ages'. This was said to have been accomplished by one man: 'Boguet (...) is reported to have condemned 600 sufferers to death'.[26]

At about the same time the exorbitant figure of 30,000 werewolves entered the public domain. According to several authors 'some 30,000 cases' were recorded in France between 1520 and 1630. Although not a single new case was published in support of it, this sensational figure appears to have stuck in the popular imagination. In Patricia Cornwell's crime novel *The Last Precinct*, a chief medical examiner discusses a werewolf murderer with a public prosecutor. The latter asks: 'Was he influenced by the French serial killer Gilles Garnier, who killed little boys and ate them and bayed at the moon? There were a lot of so-called werewolves in France during the Middle Ages. Some thirty thousand people charged with it, can you imagine?'[27] Novelists do not have to reveal their sources; but in this case that would have made little difference, for these are highly dubious. It may be deduced from the use of the word 'charged', that Cornwell probably found her information in Brad

Steiger's *Werewolf Book*; the moon and the Middle Ages are part of modern American werewolf lore. The figure of 30,000 also pops up in Rosalyn Greene's *The Magic of Shapeshifting* (1990), a New Age book that teaches its readers how to use their so-called 'animal medicine' to embark on a spiritual journey. 'We know', she writes, 'that "30,000 cases of lycanthropy were reported to secular and church officials between 1520 and 1630" in Europe. Thirty thousand in just about a century!' She refers to a book about the moon.[28] The medical anthropologist Cecil Helman also accepted the figure uncritically and likewise gave the locality as 'Europe', English shorthand for France.[29] The sources used by both Cornwell and Greene are based on newspaper articles and Helman quotes a popular medical weekly. Historical research is eclipsed here by (sloppy) journalism.[30] The modification to the word 'charged' was relatively new, as in earlier encyclopaedic publications the verb 'recorded' was used.[31] In his entry on 'Explanations for werewolves' Steiger even goes beyond 'charged' to 'condemned', a hasty conclusion that can also be encountered in the preface of an anthology of werewolf short stories.[32] Thus, in an encyclopaedia it is stated that 'In France *more* than 30,000 people were *executed* between 1520 and 1630 as suspected werewolves'.[33] The earliest appearance of this figure dates from the early 1970s; it is not mentioned in any previous publication on werewolves or witch trials. Since there was no project in France around 1970 to draw up a register of all the werewolf trials, the figure is as fantastic as that of the 9,000,000 witches supposedly burnt in Europe;[34] which is actually a logistical impossibility.

Boguet's figure of 600 alluded to witches rather than werewolves. As with the earlier 200, the figure of 30,000 can be traced back to the work of Pierre de Lancre and it may indeed have been considered a correction of it. It can be found in de Lancre's *Tableau de l'inconstance des mauvais anges et démons* from 1611. However, the figure 30,000 does not directly refer to werewolves (and is in that sense not entered in the pre-1970 witchcraft historiography), but to the estimated number of 'souls' (inhabitants) of the Pays de Labourd. De Lancre followed this with: 'among these people there is hardly a family which does not have anything to do with witchcraft'.[35] For him every Labourdin may have been a witch and every witch may have been capable of shape-shifting. But, that is no reason to grossly inflate an official report that was accepted by the church and the state from the opinion of one man; or, for that matter, to magically magnify the small Labourd into the whole of France. Neither is there any reason to conclude that the shape-shifting witches were considered to be werewolves. The witches were primarily women (since most of the time men were fishing at sea) who could change into dogs and cats, into pigs and horses, or into a raven or a hare.[36] Their crime, as de Lancre perceived it, was being in league with the devil and holding witches' sabbats, rather than devouring children whose remains were found scattered over the countryside. Reconstructing the rationale behind the designated period 1520–1630 is slightly tricky. The first year was probably chosen because it preceded the first known French trial of 1521. But 1630 seems puzzling: if derived from de Lancre's book, it should have been 1610. As there was no major werewolf trial in 1630 (at least that was known in 1970), nor a new French werewolf treatise, it is possible that the author, as yet anonymous, based the date on 1631, the year de Lancre died. The number of werewolf trials in France (Burgundy included) has never been properly counted, but based on the current state of research, 50 would not be an unreasonable estimate.

The numbers displayed by demonologists such as Boguet and de Lancre should not be taken at face value, but be understood within their own contexts. They were part of the political argument in defence of prosecutions as well as the position of the author. Boguet, who was the 'grand judge' of St Claude in Franche Comté from 1596 to 1611, sentenced to death 28 (not 600) of the 35 people he had tried for witchcraft, of whom four more subsequently died in prison.[37] In 1612 the *Parlement* of Dole, which had previously upheld Boguet's judgements, changed its policy and overruled him. Because he aspired to become a member of the *Parlement* himself, in which aim he only succeeded in 1618, he withdrew his book *Discours exécrable* from the printer when it was due for another reprint. This was a clear indication that he was willing to adapt his publicly voiced opinion on witches and werewolves to the then dominant judicial climate in France. De Lancre was less flexible. As far as can be ascertained the number of his convictions for witchcraft was neither 600 nor 200, but just over 80. In 1609 he was commissioned by the king to investigate the threat of witchcraft to the state in the south-western Basque part of France, but the chairman of the *Parlement* was also appointed, apparently to rein him in.[38] As Gerhild Williams points out, de Lancre went into the Labourd as a devout Catholic and a representative of the state, but also as an obvious party in a political conflict ('local jealousies and battles for border control') and as a traveller fascinated by the strange and repugnant customs of people whose language he could not understand. That he denounced as a witch nearly everyone in the region he investigated was a gross overreaction. Thus, his advice was ignored instead of drawing attention to the witchcraft problem that had become of central importance in de Lancre's life. The futility of his efforts became clear the following year when new judges were appointed to deal with the dozens of witches, who, thanks to de Lancre, were overcrowding the prison in Bordeaux. Most of these women were sent home when new rules of evidence were applied.[39]

The book de Lancre wrote about his Labourd episode was a rearguard action, and as Alfred Soman concludes: its 'bickering style (...) shows him to have been hopelessly out of touch with dominant judicial attitudes'.[40] It is perhaps apt to suggest that de Lancre knew the prevailing opinions very well. In the case of werewolves, he published what turned out to be the two paradigmatic verdicts. In 1599 the *Parlement* of Paris, to which Roulet had appealed, ruled that the appallant should be sent to a madhouse. De Lancre himself was a member of the Bordeaux *Parlement* it was decided in 1603 that Jean Grenier should spend the rest of his life in a monastery.[41]

The Lorey list

The most complete list of European criminal trials of werewolves to date has been compiled by the German journalist and writer Elmar Lorey.[42] He based it mainly on disparate publications and ordered it chronologically; the last version contains about 280 names of accused. While his enumeration also lists related though not similar concepts – such as, wolf-riding and wolf-banning – Lorey drew a line at slander cases, formal complaints by people who were called 'werewolf' by others, and the registers of fines levied at slanderers.[43] The list far exceeds the number of previously known cases, especially for the seventeenth century. The peak year is 1630, with 15 cases, mostly from Nassau and Sauerland. But in the second half of the seventeenth century more than 60 people, men as well as women, were still prosecuted as werewolves in northern and mid-Germany, and some in Flanders, the Franche Comté, Switzerland and Austria. Lorey's list also unearths the occasional oddity, such as the story about the man who was given a bean by the devil with the advice to plant it in the head of a murdered person. The new bean made the man invisible. His devil had also equipped him with a belt which made him change into a wolf, in which shape he did much harm to people and cattle. The subsequent trial in Paderborn dates from 1598; both the reference to invisibility and the criminal's roasted head displayed above a wooden wolf,[44] fit into the reception of the Stump trial of 1589.

A list compiled from published material is likely to contain some debatable entries as the choice will always be in favour of inclusion. For 1532, for instance, two unnamed men in Besançon are mentioned, who were found in Joseph Hansen's survey of the witch trials of 1900 rather than in his source collection.[45] In all likelihood this is a mistake and the year should be 1521, since nobody else refers to the case; in 1901 Hansen included the trial of 1521 rather than that of 1533.[46] In 1581 a shepherd called Petronius from a place called 'Dalheim' was prosecuted, a case which is traced back to the Guazzo compilation *Compendium Maleficarum* but not to its source, Nicolas Rémy; which would have made it clear that a place in Lotharingen (Lorraine) was intended rather than Dalheim in Germany.[47] The entry for the year 1588 figures a case concerning a woman from Apchon in the Auvergne which is elsewhere dated at 1558; this is not so much a trial than a story about a cut-off paw which ends up on the pyre. According to the second edition of Boguet's *Discours* the year is indeed 1588.[48] In 1591 a rather imaginative pamphlet appeared that told about the prosecution of 300 witches as werewolves in Jülich (Germany); it features in the list between square brackets, as no independent sources have been located, but it can be better comprehended as a part of the media hype after the Stump trial. These minor adjustments are partly due to mistakes made by Summers.

In an extensive discussion of the available historical werewolf treatises, though not of the present-day scholarship, Lorey has attempted to explain how the werewolf became demonised and liable to prosecution.[49] The debatable entries in his list do not play a role here. This essay provides both the counterpoint to Lorey's trial list and the frame in which he wants the judicial procedures understood. Early occurrences during the Swiss trials led Lorey to conclude that a werewolf (*warou*) was taken as a matter of fact in everyday life, without much concern about the influence of the devil. Theorists needed a change of mindset to accept the reality of animal metamorphosis, and thereby the culpability of the perpetrators, where they had previously considered it as despicable 'superstition'. This development can be traced through the French demonologists Nicolaus Jacquier and Petrus Mamoris, before it was taken up by Heinrich Kramer (aka Institoris) in his *Malleus Malificarum* of the late fifteenth-century. It culminated a century later in Jean Bodin's argument about the responsible mind, to which the now physically deemed possible metamorphosis was subject. More importantly, however, Lorey fails to discuss why around 1600 French authors, such as Claude Prieur, Jean Beauvois de Chauvincourt and Jean de Nynauld, reverted to the medical approach of lycanthropy, while Bodin continued to be relevant in Germany.[50]

In order to be workable the list needs to be rearranged into geographical clusters. When transposed onto a map (and corrected), it becomes obvious how particular trials, such as the one concerning Grenier in 1573–1574 or the 1589 one of Peter Stump, spread the notion that werewolves could be prosecuted (Figure 7.2). Afterwards, other men were accused of cannibalism as well as grave sexual misconduct; while their metamorphosis seems not to have been a crime in itself it facilitated their other actions. (Or, conversely one could argue that the prosecution of a devil-induced change was less complex than a charge of sexual crimes.) Already in mid fifteenth-century Switzerland cannibalism was understood as the *lupo moro*. From the Alps these trials had spread to Burgundy.[51] In the late sixteenth century they crossed over into the Rhineland, from where they spread further into Germany.[52] They gained in number as part of the ideological rift between Reformation and Counter-Reformation. Werewolf accusations only partly followed the paradigm of the Stump case of late 1589: sometimes it was adapted; on other occasions the concept was reinvented. In the course of the seventeenth century the accusations often became incorporated into larger witch trials. It can be concluded that the trial of Stump, usually the only German werewolf known to an English public,[53] was pivotal yet far from typical. In combination with the intellectual arguments, it could also be said that the work of Bodin contributed to this development. Sources for slander, if available, will document the geographical distribution much better than criminal trials, as insults show the occurrence of 'werewolves' in daily speech, rather than in the rarer judicial cases.

The list also figures ten trials from Lorraine, mostly from the first two decades of the seventeenth century; these are excerpted from the work of Robin Briggs.[54] In comparison to Grenier or Stump, they show the other end of the range of trials. In early seventeenth-century Lorraine the werewolf concept was vague and featured among witnesses rather than judges. As Briggs observed: 'The werewolf theme is merely hinted at in most of the Lorraine trials where attacks by wolves on animals are mentioned'. While witnesses testified they had seen, or thought they had seen, the accused in the form of a wolf, several of the latter simply denied it or replied that it had been their 'master', meaning a personal devil. Around 1600, herdsmen in the east of Lorraine were also familiar with the motif.[55] Whether the 'master's' habit of providing ointment to affect the change had also filtered down to the popular level remains open to discussion.

Lorey's list makes it clear that previous authors, proceeding from what they found in the sixteenth- and early seventeenth-century theoretical

Figure 1.1 In Geneva a man killed 16 children when he had changed himself into a wolf; he was executed on 15 October 1580. Coloured pen drawing, Johann Jakob Wick, *Sammlung von Nachrichten zur Zeitgeschichte aus den Jahren 1560–1587*, ms. F 29, fol. 167v. With permission from the Zentral Bibliothek Zürich. The right hand side of the image is reminiscent of the earlier lycanthrope in the woodcut by Lucas Cranach; the execution on the left side anticipates the Stump pamphlet (see Figure 7.2)

works, only caught the edge of the werewolf trials and confined themselves primarily to France. They largely missed the fifteenth-century Alpine origin and its seventeenth-century German continuation. Research into the French trial reports of the regional courts, or *parlements*, reveals not only more trials, but also different levels of intensity of persecution; Normandy, with its excess of shepherd male witches, did not have a werewolf 'problem'; something that is difficult to incorporate into a list.[56] Elsewhere, for example in central Germany, a particular group among the werewolves consisted of shepherds, as when their attempts to magically ban wolves backfired.[57] The total number of prosecuted werewolves in Europe probably did not exceed several hundred.

Werewolf legends

A number of modern stories and films depict Eastern Europe as the werewolf's cradle. This image is more the result of a specific kind of American folklore, which conflates werewolves and vampires, than of a specific regional tradition. The folklorist Harry Senn indeed starts his Romanian field-work report with the remark that 'in the contemporary American mind' werewolves and vampires are placed in 'the Transylvania mountains and forests of Romania' but he does little to rectify this misconception.[58] The presence in his material of the wolf leader (*'strigoi de lup'*), who mostly predates the werewolf tradition and only rarely became part of it, nevertheless points to the relatively recent arrival of werewolf tales. What is more, a yield of almost 50 percent rather uniform Werewolf Husband versions within the total corpus of 'werewolf' texts either indicates superficial field work or, as in the Belgium province of Antwerp, a marginal werewolf tradition.[59] Other Romanian werewolf motifs are weak, mixed up with witchcraft themes, such as sucking milk from animals. Another witch theme is that *'Tricoli'* (men changed into pigs rather than wolves) had to be wounded with a pitchfork for the wound to be visible once they were back in human form. These findings, carefully gleaned from Senn's material but not explicitly stated by him, are supported by the research of Germanist Inge Sommer in the same area. According to her, the Romanian *'Prikulitsch'* (shape-changer) legends only multiplied in the twentieth century.[60] This Romanian equivalent of the werewolf is hardly aggressive and the transformation is a result of fate rather than volition; belts or skins do not figure.[61]

There is thus every reason to accept the Werewolf Husband, or indeed most of the Romanian werewolf tradition, as a weak import from Germany. Germans had after all settled in Transylvania, or 'Siebenbürgen', since the Middle Ages and at one point the legend must have reached them.[62] A similar process occurred in neighbouring countries. Hungarians 'adapted' the werewolf 'from the folk beliefs of their neighbors': that is, again, primarily from Germans.[63] Among the Serbs and Croats '*vukodlak*' may literally mean 'wolf's hair'; but the creature behaves more like a vampire than a werewolf, or else it is a vampire transformed into a wolf; there is no evidence that the word 'has ever meant werewolf in the English sense'.[64] At least this calls for a thorough re-examination of eastern European 'werewolf' texts as they seem to be on the periphery, rather than in the centre, of werewolf lore. More to the north, the folklorist Oskar Loorits found that werewolf motifs in Baltic Estonia took a subordinate position within the legend material. Under German influence they had become fashionable and indeed popular but never affected the indigenous belief system. Popular legends, such as the Gift to the Werewolf and the Wedding Guests on the other hand, were in Loorits' opinion taken from the Russians.[65]

In western Europe werewolf legends seem more indigenous. Traces of werewolf trials also turn up occasionally in nineteenth- and twentieth-century oral traditions. As told in the German Westeifel, a rare werewolf priest who had been attacking cattle was finally caught. 'They made a pyre, put the priest on it and burned him. That happened in Auw near Manderfeld. That really happened. The ancestors said so'.[66] In a story from Mecklenburg a werewolf is injured by an axe and that same evening a woodcutter's brother, who is found to be wounded, is handed over to the authorities and burned. The same narrator told a version of the legend of the Werewolf Husband, which ends in a trial after the werewolf is finally given away by the threads between his teeth: 'She denounced him to the judge and he made him burn'.[67] In another version from Luxembourg, involving a werewolf father revealed to his children by the fibres in his mouth, it is related: 'The shepherd was burned and his ashes were thrown to the wind'.[68] It is necessary, however, to be cautious about details of this kind, as convictions were unlikely on the basis of only a warning narrative. Furthermore, it cannot be assumed that stories were handed down unchanged over the centuries. Some indeed had disappeared altogether; of the sixteenth-century stories that were so eagerly quoted in learned publications, no trace remained after 250 years.[69] The story from Luxembourg is derived from a manuscript source and the author might well have added the reference to the trial. Nineteenth-century legend texts in particular were usually augmented and polished by collectors and editors. The prototype text of the werewolf legend, published by the brothers Grimm, concerns a man who devours a foal, already incorporates a reference to a werewolf trial, and is taken from a demonological source.[70] Early modern texts sometimes even ended up published in their entirety in legend collections. The seventeenth-century pamphlet about the 'werewolf' of Ansbach (who, strictly speaking, was not a werewolf) has been incorporated into twentieth-century legend collections, without any indication that the tale ever circulated orally.[71] If the other examples above represent the genuine reminiscences of people interviewed by folklorists, trials were only remembered by remarkably few. In the legends from Hesse, where trials occurred earlier, they are not mentioned. Only the devil appears there 'sporadically' as provider of the werewolf's skin or belt.[72]

In most cases werewolf legends can only be traced back one or two generations. In a tale from Waldeck the narrator's grandfather had met a werewolf. The grandfather was at a loss what to do and hurled his spade at the beast. He missed and the spade flew over the werewolf. By chance this was fortunate, for after the implement had passed over him, the werewolf was stripped naked and turned out to be a neighbour, who begged not to be revealed.[73] A story from Lower Saxony about a werewolf belt begins with the statement: 'Grandfather told me this when I was still a boy'. In another tale from the same region are the words: 'I remember the stories old people told me, when I was still a boy, about fifty years ago, that in their youth the werewolf still roamed about'.[74] An informant from the Eifel area said: 'I did not experience this myself, I've got it from the stories of my father'.[75] Earlier generations could, of course, have in their turn heard the tale from their parents or grandparents; indeed, in the first example from Lower Saxony about the werewolf's belt, the narrator says exactly that. Most of the time this is unlikely, though, since the reference is not to something a father or a grandfather heard, but to something they had experienced. Such glimpses into a continuing tradition are no proof of unchanged content. If these examples suggest anything, it is that tales were transmitted primarily from men to men. The problem is that folklorists did not always ask about the origin of a tale and if informants mentioned it of their own accord, it could have been a rhetorical device. This may well have been the case with the Lower Saxony story, that was set four generations in the past. It is more elaborate than most legends, and thus in all probability it was told by an experienced storyteller. Its content, about a king who puts on a werewolf belt and has to be freed from it, is more anecdotal than legendary and there are no known parallels.

The notion of obtaining a werewolf's belt or skin from a pact with the devil illustrates several of the issues in contention. In seventeenth-century trials the devil always plays a role, yet in the legends he hardly appears. In the 1960s, in the Dutch hamlet of Gendt, south-east of Arnhem, the collector Dinnissen found only one informant, who mentioned that a werewolf had 'made a promise to the devil' and then was thrown a werewolf's skin down the chimney, even though Dinnissen had interviewed everyone appropriate.[76] Engelbert Heupers, whose research covered the whole Dutch province of Utrecht and parts of neighbouring Guelders, found only a few people who said that werewolfing was 'something of the devil, worse than witching'.[77] The Flemish folklorist Roeck considered this a typical Catholic motif, since it was also prevalent in French-speaking areas.[78] As it turns out, the most detailed Dutch werewolf stories featuring a devil stem from a small area around Catholic Maastricht. Dutch museologist Coen Eggen heard the folowing from a schoolmaster:

> If one wanted to become werewolf, one had to go at night to a certain cross-road, with a black chicken that had to be slaughtered there. Then a black coach would appear, drawn by black horses, and out would step a man, again in black: the devil. The applicant obtained a belt, or sometimes a shirt.[79]

Other versions gave different instructions. The would-be werewolf had to perform the ritual at midnight, draw a circle, or lie on the road. Although a werewolf was convicted in Maastricht in the early seventeenth century, the story hardly points to continuity or a stable tradition in a much wider area. It is both too local and its motifs are too tinged with romantic detail; the black chicken and the crossroads do not appear in the trial accounts. The only point suggesting continuity is that the story provides another Dutch instance of fusing werewolves with male witches (who sometimes enter a pact with the devil in this way). The story may have been translated from the French and printed in a local almanac, or it may have been told in schools before it was retold to inquisitive folklorists.

Local legends: the example of Maasmechelen (Belgium)

In 1967 the Flemish folklorist Alfons Roeck based his (unpublished) dissertation on 1,170 short Flemish and Dutch werewolf texts. He was not able to include the then ongoing Dutch research and he could only take into account the first 27 Master theses from the University of Leuven and three from the University of Ghent. Since then the amount of material has at least doubled. As of now, the Flemish Folk Tale Bank contains 1,845 werewolf legends, more than half of them (977) from the province of Limburg.[80] Its Dutch equivalent yields some 290 legends when the keyword 'weerwolf' is typed in;[81] this last bank still does not contain a number of the 1960s collections, such as those by Heupers, Krosenbrink, Linssen and Eggen.[82] This enormous amount of material has only partly been analysed.[83] Here I shall give some examples from one Flemish collection from the province of Limburg just to provide a sense of the kind of material available.

In 1970 one of the Leuven students, Piet Knabben, in the course of his research into 'folk beliefs', interviewed 37 informants in the municipality of Maasmechelen about local werewolf concepts, which resulted in 67 legend texts.[84] I have chosen this collection for two reasons. First, it is from an area stretching from Belgian Limburg, through the Dutch province of Limburg into Germany, that has a rich werewolf tradition; and second, Knabben's texts are transcripts of recorded interviews, which is not always the case with the other Leuven material. These texts draw one straight into the local werewolf lore. According to a 73-year old woman, a certain boy (name and family relation supplied) had been a werewolf for a while. When his father realised this he almost strangled him; but the boy 'could not help it, though he was an asshole' (or pervert) (4).[85] Another informant, now talking about a victim, did not want to reveal names: 'there was a boy', he said, 'who had to carry the werewolf several times. When he passed a certain place he was jumped upon' (16). The same man also knew about someone else who had been less intimidated. He drew his knife and cut the werewolf (*weerewouf*), perching on his back (17). One could meet a werewolf anywhere, another man said, 'but some places had a special reputation, where the road narrowed, or where there was a corner with many brambles and willows, then it could happen that, when you passed there at night, you had to carry him till you arrived home'. He continued: 'It did not do anything else to you, it could happen to anyone. The old people were not afraid' (54).

An 86-year-old man told how men had been joking about the werewolf, pretending that he was sitting on them and calling for help, but when someone would come to the rescue, the werewolf was gone and they all laughed (69). 'What did he look like? ' asked Knabben. 'Oh, an ordinary human, a human who was tormented, he had to harass people. He covered himself with something so that nobody could recognise him'. But someone had a knife and the werewolf's disguise dropped off: 'Take care not to betray me, otherwise it's your life', the werewolf threatened. 'Those were people, just like us. But they were tormented', the man repeated, 'and they had to go out to torment other people, for instance when someone had to fetch the midwife or the doctor' (at night) (70). 'He wrapped it around himself?' 'Yes, he had a hide hanging around him, I have never seen it'. Someone managed to burn a werewolf's hide and then he was saved. 'And then they knew who tormented the people' (71). A woman had often heard from her father that he had to carry the werewolf; when he passed a certain vegetable garden the werewolf flew out of a big bush and he couldn't do anything to stop it (85). According to another informant, however, a man who did not want to carry the werewolf hit him with his fist on his head. Then the werewolf stood naked before him and he knew who it was, 'because it were ordinary people. I also know who it was, but can't tell you' (122). A further informant, also a man, of 72, confirmed this: 'The werewolf was a normal human being who draped himself in a sheet, like a ghost, and then jumped on your back and then you had to carry him'. He had not seen it himself, but his grandparents talked a lot about it (124). This informant also knew a story about a farm maid who was not afraid of anyone. The farmer wanted to disprove this and put a hide over his head to frighten her, but he was hit with a wooden hammer and died (133).[86] The last was a warning tale, also told about people who pretended to be ghosts. Knabben's informants did not relate their own experiences and their stories took place one or two generations in the past. They had not seen the werewolf, only heard others talking about it. Some had been afraid, others not; some passively underwent the assault, while others fought it. Other interviews convey the same message. An 86-year-old man, for instance, said: 'when you went somewhere then he used to jump on your neck, stayed there for a while and jumped off' (171).[87] No mention is made of specific activities of the back-riding werewolf, such as peeing or licking, as occurs in legends from neighbouring communities.[88] Only a minority of the legends from Maasmechelen relate to stories, as opposed to narrated (second- or third-hand) experiences: the Werewolf Lover is referred to three times: once only briefly but with all the elements present (5); once truncated and containing a red handkerchief to ward off a dog, without any woman but with the recognition motif (169); and once as a story the narrator acquired from his father-in-law, which also contained mainly the handkerchief and the recognition of the werewolf afterwards (638).[89] The Werewolf Lover was hardly current anymore. A second tale was a little better known. In it a farm hand goes out regularly as a werewolf but is relieved of his urge by burning his wolf skin in an oven. Here there was variation, too; one labourer was said to have exclaimed after his belt was burned: 'what have you done now? Now I am cursed forever' (32). In the two next versions, he was saved (51), and his identity discovered (71): 'afterwards, he thanked them and went to visit the friars and he confessed it all and then he was saved' (115). In the last version, the werewolf had simply gone away: 'Yes, a werewolf, that was more a human being, but they had a skin, they wore it'. 'Where did they get that from?', the interviewer asked. 'I don't know where they got that from. One inherited it from another' (165).[90]

The Limburg werewolf is characterised as a back-rider (*Huckauf, Aufhocker*), a feature which pervades most of the neighbouring werewolf traditions. The back-riding werewolf must also have surfaced in mid-nineteenth-century France, since it was drawn by the artist Maurice Sand (Figure 1.2), but there needs to be more research into nineteenth-century French werewolf traditions.[91] Until a better explanation arises, my guess is that the trials in Limburg and neighbouring jurisdictions have resulted in a substitution of local back-rider figures by the werewolf. Back-riders are much more widespread and are named after all kinds of men; only in the German-Dutch regions which experienced werewolf trials are they called 'werewolves'. In this I differ from Roeck, who interpreted the back-riding werewolf as a sign that the werewolf was disappearing;[92] I suggest that it shows the strength of the concept instead. But it also means that conclusions about the Limburg werewolf cannot be projected on werewolves elsewhere, or vice versa.

Figure 1.2 Loup-Garou by Maurice Sand. *Le Magasin Pittoresque* (1857), 373

Geographical and temporal clusters

Some medieval and sixteenth-century reports mention multitudes of werewolves; but it is debatable whether or not they can be considered one and the same cluster. A possible wolf cult in antique Arcadia should be first and foremost understood within the ancient Greek context.[93] As Richard Gordon illustrates in the next chapter of this book, the 'cult' is also very much a construct of historiography, and undermined when one starts looking at multiple texts and archaeological evidence. The Lycaon myth is nevertheless important for its effect on the later imagery of man–wolf hybrids (the Arcadian king did not change back), even on early-nineteenth-century French 'folklore'. Metamorphosing Neuroi, as reported by Herodotus, comply more to the notion of a temporary change, as well as to packs; but they cannot be placed in the same category as later Nordic warriors or Estonian brigands.[94] Early modern trials feature predominantly single individuals, or sometimes up to two or three people suspected of being werewolves; these did not form packs and operated in a profoundly Christian context.

The diversity of werewolf concepts is therefore also apparent in the, not necessarily mutually exclusive, approaches of the contributors to this volume. Whereas, for instance, Aleksander Pluskowski looks for (sometimes elusive) traces of the beast in medieval mummings and archaeological artefacts, and considers it foremost as a disguise, Christa Tuczay focuses on the Nordic ideas of the soul. Johannes Dillinger discusses the curious logic of a number of demonologists, who left the soul out of consideration and concentrated on the power of the devil to deceive people, while Rolf Schulte establishes different degrees of the werewolf's presence on a popular level, using the difficult but unfairly neglected source of slander trials. Rita Voltmer's contribution, with its focus on the Jesuits' role in the Counter-Reformation, bridges these two approaches and is more context-oriented; by contrast, Dillinger's is an exercise in text comparison. Other trials and treatises are still waiting to be analysed.

As far as the trial distribution is concerned, the rare Italian case discussed by Matteo Duni can be understood as another off-shoot of the Alpine epicentre. But the main Italian werewolf under observation can hardly be seen as the result of over-zealous werewolf hunters, as later transpired in Germany. This raises the question whether the werewolf trials in the eastern Baltic lands, which featured many female werewolves, had their own dynamics or whether they were primarily influenced by German examples.[95] Or whether a Livonian werewolf healer in the late seventeenth-century shared any conceptual similarities with the French and German werewolves of a century earlier.[96] At this stage of the research, answers to these questions must necessarily remain tentative. Matters of translation, communication and adaptation, as well as local contexts, must be taken into consideration. Werewolves in sixteenth-century demonological thinking, for instance, seem to have changed from those in medieval romances; but, if the latter are of little relevance to demonologists, this change is probably only a function of a present-day perspective.[97]

On the other hand, there may be some connection between the trials and legends within a particular area. Nineteenth- and twentieth-century werewolf legends have a different centre of gravity in different areas of Europe. In some places – such as, the British Isles, Ireland, the Iberian peninsula, Bavaria, the Swiss and Austrian Alps, and Greece – werewolf references are practically absent, or only have a cursory presence in the odd saying in coastal areas: such as, Frisia and Portugal, where a seventh son was considered a werewolf without much further elaboration.[98] In Northern Scandinavia there is some overlap with Sami notions about soul journeys. In Southern Sweden, however, werewolves are rampant.[99] In various regions of France werewolves are also a regular feature in folklore collections, but the French folklore research into werewolves is as yet too fragmentary to provide a coherent representation;[100] the exception being the Dauphiné.[101] Apart from Sweden, a strong presence in the legends coincides geographically with earlier trials, although the precise nature of the connection between a trial and a legend cluster may vary. Switzerland presents an intriguing case in this respect, as the sporadic trials from the fifteenth to the seventeenth centuries do not seem to have had much impact on the later narration of legends.[102]

In areas with a full set of werewolf concepts, the beast is depicted in narratives as well as in first-hand reports as a figure of its victims' experience and of advanced warnings, in stories or directly, while its characteristics and actions are described, and advice is given about counter-measures. Separating the various werewolf clusters is vital to see at what points they may be combined again and to arrive at a reliable history. Only then may it become possible to answer questions about meanings which will not be the same for every cluster. The variety of werewolf clusters is certainly apparent in the last chapters of this book, which concern nineteenth- and twentieth-century legends from Estonia, Denmark and Croatia, written by respectively Merili Metsvahi, Michèle Simonsen and Maja Pasarić. In general, legends are characterised by their multitude. How, then, to assess matters of quantity? The late nineteenth-century Danish ethnographer Evald Tang Kristensen is said to have collected 'twenty-five thousand legends',[103] but in his major collection *Danske sagn*, published in the last decade of the nineteenth century, he only presented the relatively small amount of 42 werewolf legends.[104] As he also collected about half of them himself, the first question to ask is, whether these can genuinely represent Danish werewolf lore? One of the answers could be that Danish werewolf concepts were in decline, another that Tang Kristensen and his correspondents only touched the surface of this sexually loaded concept. Whatever the answer, it should evoke more caution when dealing with earlier texts. In the chapter on Danish legends several stories are discussed to arrive at an integrated semantic cluster, ranging from the cause of a werewolf's curse to his redemption. Yet contributing parts of this cluster are also found elsewhere and the question thus remains whether or not it is only possible to attribute a special meaning on the basis of a full cluster. This becomes the more urgent as it appears that the Danish werewolf is more the subject of artistic expression than of everyday concern, an issue that also emerges in the discussion of the Estonian werewolf legends. Although the latter are substantially different, the distinction between the werewolf as an everyday experience and as the subject of a good story can be discerned in both areas.

Apart from difference in substance, the following contributions also champion difference in approach. Werewolves cannot be comprehended on their own, but need to be seen as part of the ever-changing relations between humans and animals, and indeed gods or devils. Werewolves are to be compared with other human-animal disguises and transformations. They can be seen from the position of the learned and the mighty (or at least those who strove to be) as well as within the tensions of everyday life. Werewolves provide unique perspectives on the soul. They can enliven the history of the body, just as this is enhanced by considering shape-shifters. There is no werewolf history; there are only histories of werewolves.

Notes

1. Willem de Blécourt, 'Mounstrous Theories. Werewolves and the Abuse of History', *Preternature: Critical and Historical Studies on the Preternatural* 2 (2013), 188–212.
2. Willem de Blécourt, 'The Werewolf, the Witch, and the Warlock: Aspects of Gender in the Early Modern Period', in: Alison Rowlands (ed.), *Witchcraft and Masculinities in Early Modern Europe* (Basingstoke, 2009), 191–213.
3. In as far as psychiatric patients enact werewolf movies, I categorise them within the cinematic cluster. Cf. Jan Dirk Blom, 'When Doctors Cry Wolf: A Systematic Review of the Literature on Clinical Lycanthropy', *History of Psychiatry* 25 (2014), 87–102.
4. Kimberley McMahon-Coleman and Roslyn Weaver, *Werewolves and Other Shapeshifters in Popular Culture. A Thematic Analysis of Recent Depictions* (London, 2012).
5. Linda S. Godfrey, *Hunting the American Werewolf: Beast Men in Wisconsin and Beyond* (Madison, 2006); idem, *Real Wolfmen: True Encounters in Modern America* (New York, 2012). To prevent misunderstanding: I consider these books as source material rather than academic studies.
6. I have incorporated parts of the English version of my entry 'Wolfsmenschen', *Enzyklopädie des Märchens* 14 (2013), col. 975–986.
7. Cf. Charlotte F. Otten, *A Lycanthropy Reader: Werewolves in Western Culture* (Syracuse, 1986), 5–8.
8. Michael Jacoby, *Wargus, vargr. 'Verbrecher' 'Wolf'. Eine sprach- und rechtsgeschichtliche Untersuchung* (Uppsala, 1974), 77–93; Klaus von See, *Europa und der Norden im Mittelalter* (Heidelberg, 1999), 117–122; cf. Mary R. Gerstein, 'Germanic Warg: The Outlaw as Werwolf', in: Gerald James Larson (ed.), *Myth in Indo-European Antiquity* (Berkeley, 1974), 131–156.
9. This line of reasoning may be confusing since *varg* the is the current Swedish term for wolf. It is, however, of relatively young date as since 'the 10th century, the Germanic word *vargr* designated the outlawed criminal, has been transferred to the wolf', Martin Rheinheimer, 'The Belief in Werewolves and the Extermination of Real Wolves in Schleswig-Holstein', *Scandinavian Journal of History* 20 (1995), 281–294, cit. 284.
10. For example: Gaël Milin, *Les chiens de Dieu: la représentation du loup-garou en Occident, xie-xxe siècles* (Brest 1993); Elmar M. Lorey, *Henrich der Werewolf: eine Geschichte aus der Zeit der Hexenprozesse; mit Dokumenten und Analysen* (Frankfurt a.M., 1998).
11. Wilhelm Hertz, *Der Werwolf. Beitrag zur Sagengeschichte* (Stuttgart, 1862); Sabine Baring-Gould, *The Book of Were-Wolves* (London, 1865); Montague Summers, *The Werewolf* (London, 1933).
12. Ranging from titles such as: Adam Douglas, *The Beast Within* (London, 1992), to: Nathan Robert Brown, *The Complete Idiot's Guide to Werewolves* (Pittsburg, PA, 2009), and Barb Karg, *The Girl's Guide to Werewolves* (Avon, MA, 2009).
13. Asa Simon and Peter J. Dendle (eds), *The Ashgate Research Companion to Monsters and the Monstrous* (Farnham, 2012); Stephen T. Asma, *On Monsters. An Unnatural History of Our Worst Fears* (Oxford, 2009), 347–350.
14. Brian Levack (ed.), *Witchcraft in Early Modern Europe and Colonial America* (Oxford, 2013), 230, 240, 351.
15. Keith Roberts, 'Eine Werwolf-Formel: eine kleine Kulturgeschichte des Werwolfs', in: Ulrich Müller and Werner Wunderlich (eds), *Dämonen, Monster, Fabelwesen* (St Gallen, 1999), 565–581, 577; cf. Adam Douglas, *The Beast Within*, 38–40.

16. Cf. Paul Sébillot, *Le folk-lore de France*, II (Paris, 1905), 205.
17. Reidar Th. Christiansen, *The Migratory Legends* (Helsinki, 1958), ML 4005.
18. Leslie A. Sconduto, *Metamorphoses of the Werewolf. A Literary Study from Antiquity through the Renaissance* (Jefferson and London, 2008). See also the review of Karin E. Olsen in *The Medieval Review* 2 (2010) and Sconduto's reply in the following issue of the same journal.
19. Nadine Metzger, *Wolfsmenschen und nächtliche Heimsuchungen. Zur kulturhistorischen Verortung vormoderner Konzepte von Lykanthropie und Ephialtes* (Remscheid, 2011); see also: Nadine Metzger, 'Batling Demons with Medical Authority: Werewolves, Physicians and Rationalization', *History of Psychiatry* 24 (2013), 341–355.
20. Baring-Gould, *The Book of Were-Wolves*, 69–84.
21. Summers, *The Werewolf*, 223–234.
22. Baring-Gould, *The Book of Were-Wolves*, 85–99.
23. Caroline Oates, 'Metamorphosis and Lycanthropy in Franche-Comté, 1521–1643', in: M. Feher (ed.), *Fragments for a History of the Human Body*, I (New York, 1989), 304–363, esp. 361.
24. W.M.S. Russel and Claire Russel, 'The Social Biology of Werewolves', in: J.R. Porter and W.M.S. Russell (eds), *Animals in Folklore* (Cambridge, 1978), 143–182, 260–269, 153.
25. Pennethorn Hughes, *Witchcraft* (Harmondsworth, 1965), 184. The figure is repeated in H. Sidky, *Witchcraft, Lycanthropy, Drugs, and Disease: An Anthropological Study of the European Witch-Hunts* (New York, 1997), 122.
26. Paul E. Keck, Harrison G. Pope, James I. Hudson, Susan L. McElroy and Aaron R. Kulick, 'Lycanthropy: Alive and Well in the Twentieth Century', *Psychological Medicine* 18 (1988), 113–120, 114; T.A. Fahy, 'Lycanthropy: A Review', *Journal of the Royal Society of Medicine* 82 (1989), 37; Peter M. Yellowlees, 'Werewolves Down Under – Where Are They Now?', *The Medical Journal of Australia* 151 (1989), 663–665, 664.
27. Patricia Cornwell, *The Last Precinct* (New York, 2001), 312.
28. Rosalyn Greene, *The Magic of Shapeshifting* (York Beach, 2000), 3; Paul Katzef, *Moon Madness* (London, 1990), 58.
29. Cecil Helman, 'The Premenstrual Werewolf', in: *Body Myths* (London, 1991), 58–80, 75.
30. The papers quoted are: *National Examiner*, 6 April 1997; *Milwaukee Journal*, 18 March 1978; *World Medicine*, 30 May 1973.
31. Stephen Jones, *The Illustrated Werewolf Movie Guide* (London, 1996), 10; Colin Wilson, 'Werewolves', in: Jack Sullivan (ed.), *The Penguin Encyclopedia of Horror and the Supernatural* (New York, 1986), 453; Ian Woodward, *The Werewolf Delusion* (New York and London, 1979), 19.
32. Brad Steiger, *The Werewolf Book: The Encyclopedia of Shape-Shifting Beings* (Detroit and London, 1999), 105; Peter Haining, *Werewolf: Horror Stories of the Man-Beast* (London, 1987), 3.
33. Carol Rose (ed.), *Giants, Monsters, and Dragons: An Encyclopedia of Folklore, Legend, and Myth* (Santa Barbara, 2000), 391, my italics.
34. Wolfgang Behringer, 'Neun Millionen Hexen. Entstehung, Tradition und Kritik eines populären Mythos', *Geschichte in Wissenschaft und Unterricht* 49 (1998), 664–685. The more accurate estimate is ca. 40,000 convictions; also: Wolfgang Behringer, *Witches and Witch-Hunts. A Global History* (Cambridge, 2004), 157.

35. Pierre de Lancre, *Tableau de l'inconstance des mauvais anges et démons, où il es amplement traité des sorciers et de la sorcellerie*, Jacques-Chaquin (ed.) (Paris, 1982), 80; English translation: *On the Inconstancy of Witches: Pierre de Lancre's Tableau de l'inconstance des mauvais anges et démons* (1612), Gerhild Scholz-Williams (transl.) (Tempe, 2006).
36. Cf. Gustav Henningsen, *The Witches' Advocate: Basque Witchcraft and the Spanish Inquisition (1609–1614)* (Reno, 1980), 66, 111, 159, 250.
37. Oates, 'Metamorphosis and Lycanthropy', 326; E. William Monter, *Witchcraft in France and Switzerland: The Borderlands During the Reformation* (Ithaca and London, 1976), 71.
38. Henningsen, *The Witches' Advocate*, 23–24, 130–131, 317, 497, n. 52.
39. Gerhild Scholz Williams, *Defining Dominion: The Discourses of Magic and Witchcraft in Early Modern France and Germany* (Ann Arbor, 1995), 89–119. See also: Jonathan L. Pearl, *The Crime of Crimes: Demonology and Politics in France 1560–1620* (Waterloo, Ontario, 1999), 127–147.
40. Alfred Soman, 'Decriminalizing Witchcraft: Does the French Experience Furnish a European Model?', in: Gábor Klaniczay and Éva Pocs (eds), *Witch Beliefs and Witch-Hunting in Central and Eastern Europe* (Budapest, 1992), 379–400, 398.
41. About the last case: Caroline Oates, 'The Trial of a Teenage Werewolf', *Criminal Justice History* 9 (1988), 1–29, see p. 24, note 9 about Roulet and p. 29, note 52 about the decision of the Parlement de Paris.
42. www.elmar-lorey.de/Prozesse.htm.
43. Karl-S. Kramer, 'Hohnsprake, Wrakworte, Nachschnack und Ungebühr. Ehrenhändel in holsteinische Quellen', *Kieler Blätter zur Volkskunde* 16 (1984), 49–85; Alfred Höck, 'Bemerkungen zum "Werwolf" nach hessischen Archivalien', *Hessische Blätter für Volks- und Kulturforschung* 18 (1985), 71–75; Willem de Blécourt, *Termen van toverij* (Nijmegen, 1990), 113. See also the chapter by Rolf Schulte in this volume.
44. Wilhelm Richter, *Geschichte der Stadt Paderborn*, II (Paderborn, 1903), 159–160.
45. Joseph Hansen, *Zauberwahn, Inquisition und Hexenprozess im Mittelalter und die Entstehung der grossen Hexenverfolgung* (Munich, 1900), 504: 1533 instead of 1532, referring to Jules Baissac, *Les grands jours de la sorcellerie* (Paris, 1890), 316.
46. Joseph Hansen, *Quellen und Untersuchungen zur Geschichte des Hexenwahns und der Hexenverfolgung im Mittelalter* (Bonn, 1901), 513; he called the later case an 'Inquisitionsprozess', thus it should have been mentioned on p. 515.
47. Nicolas Rémy, *La démonolâtrie*, Jean Boës (ed.) (Nancy, 1998), 210; the errenous year 1558 in: Summers, *The Werewolf*, 228, cf. Otten, *Lycanthropy Reader*, 80.
48. Henry Boguet, *Discours des sorciers*, second edition (Lyon, 1608), 341–342.
49. www.elmar-lorey.de/werwolf/genese.htm.
50. Nicole Jacques-Lefèvre, 'Such an Impure, Cruel, and Savage Beast... Images of the Werewolf in Demonological Works', in: Kathryn A. Edwards (ed.), *Werewolves, Witches, and Wandering Spirits. Traditional Belief and Folklore in Early Modern Europe* (Kirksville, 2002), 181–197; Stuart Clark, *Vanities of the Eye. Vision in Early Modern European Culture* (Oxford, 2007), 63, 138.
51. Oates, 'Metamorphosis and Lycanthropy'; Rolf Schulte, *Man as Witch. Male Witches in Central Europe* (Basingstoke, 2009), 8–35.
52. Johanna Koppenhöfer, *Die mitleidlose Gesellschaft. Studien zu Verdachtsgenese, Ausgrenzungsverhalten und Prozeßproblematik im frühneuzeitlichen Hexenprozeß in der alten Graftschaft Nassau unter Johann VI. und der späteren Teilgrafschaft Nassau-Dillenburg (1559–1687)* (Frankfurt am Main, 1995); Elmar M. Lorey, 'Das

Werwolfstereotyp als instabile Variante im Hexenprozeß', *Nassauische Annalen* 112 (2001), 135–176.
53. Summers, *The Werewolf*, 253–259.
54. Robin Briggs, 'Dangerous Spirits. Shapeshifting, Apparitions, and Fantasy in Lorrain Witchcraft Trials', in: Kathryn A. Edwards (ed.), *Werewolves, Witches, and Wandering Spirits. Traditional Belief & Folklore in Early Modern Europe* (Kirksville, 2002), 1–24; the passages are reprinted in: Robin Briggs, *The Witches of Lorraine* (Oxford 2007), 122–127.
55. Elisabeth Biesel, '*Die oberste sey ein kostlichs weyb wie ein edell frauw gewessen*. Imaginationen des Weiblichen und des Männlichen im Spiegel lothringischer und maximinischer Hexenprozeßakten', in: Rita Voltmer and Günter Gehl (eds), *Alltagsleben und Magie in Hexenprozessen* (Weimar, 2003), 47–55, 49.
56. E. William Monter, 'Toads and Eucharists: The Male Witches of Normandy, 1564–1660', *French Historical Studies* 20 (1997), 563–595, 581.
57. Elmar Lorey, 'Vom Wolfssegner zum Werwolf. Hexereiprozesse im Nassauer Land', in: Rita Voltmer and Günter Gehl (eds), *Alltagsleben und Magie in Hexenprozessen* (Weimar, 2003), 65–79.
58. Harry Senn, *Were-Wolf and Vampire in Romania* (Boulder, 1982).
59. Marcel van den Berg, *De volkssage in de provincie Antwerpen in de 19de en 20ste eeuw* (Gent, 1993), 1858–1859.
60. Inge Sommer, '"Krechintza zwischen den Zähnen". Aspekte der Siebenbürgisch–Saksischen Prikulitschsage', *Neue Literatur: Zeitschrift des Schriftstellerverbandes der SR Rumänien*, 39, 3 (1988), 51–55.
61. Cf. Felix Karlinger and Emanuel Turczynski, *Rumänische Sagen und Sagen aus Rumänien* (Berlin, 1982), 79–81.
62. The tale circulated in seventeenth-century Germany, see: Heinrich Rimphoff, *Drachen-König* [etc.] (Rinteln, 1647), 124. It can already be found in the late 15th-century *Évangelie des quenouilles*, see: Claude Lecouteux, *Elle courait le garou. Lycanthropes, hommes-ours, hommes-tigres: une anthologie* (Paris, 2008), 189–190.
63. Tekla Dömötör, *Hungarian Folk Beliefs* (Bloomington, 1982), 122.
64. Jan L. Perkowski, *The Darkling: A Treatise on Slavic Vampirism* (Waterloo, 1999), 51. Cf. Ute Dukova, *Die Beziehungen der Dämonen in Bulgarischen* (München, 1977), 54–56; Paul Barber, *Vampires, Burial and Death: Folklore and Reality* (New Haven and London, 1988), 93. see also the chapter by Maja Pasarić.
65. Oskar Loorits, *Grundzüge des Estnischen Volksglaubens* (Uppsala, 1949), 311. See for the Russian *koldun*: W.F. Ryan, *The Bathhouse at Midnight. An Historical Survey of Magic and Divination in Russia* (Stroud, 1999), 72–78.
66. Matthias Zender, *Sagen und Geschichten aus der Westeifel* (Bonn, 1980), 75.
67. Karl Bartsch, *Sagen, Märchen und Gebrauche aus Meklenburg* (Vienna, 1879), 148.
68. Nikolaus Gredt, *Sagenschtz des Luxemburger Landes*, I (Esch-Alzette, 1963 [orig. 1883]), no. 846.
69. Rainer Alsheimer, 'Katalog protestantischer Teufelserzählungen des 16. Jahrhunderts', in: Wolfgang Brückner (ed.), *Volkserzählung und Reformation: Ein Handbuch zur Tradierung und Funktion von Erzählstoffen und Erzählliteratur im Protestantismus* (Berlin, 1974), 417–519; cf. Johann Jacob Bräuner, *Physicalisch- und Historisch-Erörterte Curiositäten* (Frankfurt am Main, 1737), 246–256; Jacob and Wilhelm Grimm, *Deutsche Sagen* (Berlin, 1816), 296–297.
70. Grimm, *Deutsche Sagen*, 293–294.
71. Leander Petzoldt, *Schwäbische Sagen: Von Odenwald bis zum Bodensee, vom Schwarzwald bis zum Lech* (Frankfurt a.M, 1985), 47–48.

72. Paul Zaunert, *Hessen-Nassauische Sagen* (Jena, 1929), 263–267; Heinrich Franz, 'Der Werwolfglaube unter besonderer Berücksichtigung der hessischen Überlieferung', *Hessenland. Hessissches Heimatblad* 31 (1917), 255–260.
73. Gustav Grüner, *Waldeckische Volkserzählungen* (Marburg, 1964), 42, 196.
74. Will-Erich Peuckert, *Niedersächsische Sagen* (Göttingen, 1966), nos 1398II, 1409.
75. Zender, *Westeifel*, no. 1135.
76. M.H. Dinnissen, *Volksverhalen uit Gendt* (Amsterdam, 1993), no. 215.
77. Engelbert Heupers, *Volksverhalen* (Amsterdam, 1984), no. 2723.
78. Alfons Roeck, *De Nederlandse weerwolfsage in de negentiende en twintigste eeuw* (Leuven, 1967), 291, unpublished.
79. Archive Meertens Institute, Amsterdam. Collection Eggen, 19.4.
80. VVB: www.volksverhalenbank.be.
81. www.verhalenbank.nl.
82. The last two are not published; the third only partially: Henk Krosenbrink, *De oele röp. Achterhookse volksverhalen* (Aalten, 1968).
83. Willem de Blécourt, '"I would have eaten you too": Werewolf Legends in the Flemish, Dutch and German Area', *Folklore* 118 (2007), 23–43.
84. Piet Knabben, *Resultaten van het sagenonderzoek in het zuidelijk deel van de Belgisch-Limburgse Maasvallei* (Leuven 1970), unpublished.
85. Dutch: *smeerlap*.
86. The numbers behind the quotes refer to the story numbers in the thesis. The corresponding numbers in the VVB are: 4=6458, 16=6470, 17=6471, 54=6517, 69=6531, 70=6531, 71=6533, 85=6547, 122=6583, 124=6585.
87. VVB 6648.
88. De Blécourt, 'I Would Have Eaten You Too', 37.
89. VVB numbers: 5=6459, 169=6643, 638=7156.
90. VVB numbers: 32=6492, 51=6514, 71=6533, 115=6576, 165=6639.
91. Sand's *loup-garou* etching does not appear in: George Sand, *Légendes rustiques* (Paris 1858) but the previous year in *Le Magasin Pittoresque*, p. 373; cf. Sébillot, *Le Folk-lore de France*, II, 206.
92. Also in: Alfons Roeck, 'Der Werwolf als dämonisches Wesen in Zusammenhang mit den Plagegeistern', in: Lutz Röhrich (ed.), *Probleme der Sagenforschung* (Freiburg i.Br., 1973), 139–148.
93. Richard Buxton, 'Wolves and Werewolves in Greek Thought', in: Jan Bremmer (ed.), *Interpretations of Greek Mythology* (Beckenham, 1987), 60–79; revised with a postscript in Richard Buxton, *Myths and Tragedies in Their Ancient Greek Contexts* (Oxford, 2013), 33–51; cf. Madeleine Jost, 'Deux mythes de métamorphose en animal et leurs interprétations: Lykaon et Kallisto', *Kernos* 18 (2005), 347–370.
94. Tiina Vähi, 'Werwölfe – Viehdiebe und Räuber im Wolfspelz? Elemente des archaischen Gewohnheitsrechts in estnischen Werwolfvorstellungen', in: Willem de Blécourt and Christa Agnes Tuczay (eds), *Tierverwandlungen. Codierungen und Diskurse* (Tübingen, 2011), 135–156; cf. Stefan Donecker, 'The Werewolves of Livonia: Lycanthropy and Shape-Changing in Scholarly Texts, 1550–1720', *Preternature* 1 (2012), 289–322.
95. Merili Metsvahi, 'Werwolfprozesse in Estland und Livland im 17. Jahrhundert. Zusammenstöße zwischen der Realität von Richtern und von Bauern', in: Jürgen Beyer and Reet Hiiemäe (eds), *Folklore als Tatsachenbericht* (Tartu, 2001), 175–184; Tiina Vähi, 'Hexenprozesse und der Werwolfglaube in Estland', in: Manfred Dietrich and Tarmo Kulmar (eds), *Die Bedeutung der Religion für Gesellschaften in Vergangenheit und Gegenwart* (Münster, 2003), 215–238.

96. Willem de Blécourt, 'A Journey to Hell: Reconsidering the Livonian "Werewolf"', *Magic, Ritual, and Witchcraft* 2 (2007), 49–67.
97. Cf. Jan R. Veenstra, 'The Ever-Changing Nature of the Beast. Cultural Change, Lycanthropy and the Question of Substantial Transformation (from Petronius to Del Rio)', in: Jan N. Bremmer and Jan R. Veenstra (eds), *The Metamorphosis of Magic from Late Antiquity to the Early Modern Period* (Leuven, 2002), 133–166.
98. Jurjen van der Kooi, *De nachtmerje fan Rawier: fryske sêgen oer it boppenatuerlike* (Ljouwert, 2000), 150–151.
99. Ella Odstedt, *Varulven i Svensk Folktradition* (Uppsala/Kobenhavn, 1943); cf. Bengt af Klinktberg, *The Types of the Swedisch Folk Legend* (Helsinki, 2010), 306–310.
100. Lise Andries, 'Contes du loup', in: Jean de Nynauld, *De la lycanthropie, transformation et extase des sorciers 1615. Étude critique augmentée d'études sur les lycanthropes et les loups-garous*, Préaud and Jacquin (eds) (Paris, 1990), 197–217; cf. Gudrun Staudt and Will-Erich Peuckert, *Nordfranzösische Sagen* (Berlin, 1968), 32–35; Felix Karlinger and Inge Übleis, *Südfranzösische Sagen* (Berlin, 1974), 104–109.
101. Charles Joisten, Robert Chanaud and Alice Joisten, 'Les loups-garous en Savoie et Dauphiné', *Le monde Alpin et Rhodanien* 20 (1992); Alice Joisten and Christian Abry, *Êtres fantastiques des Alpes. Extraits de la collecte Charles Joisten (1936–1981)* (Paris, 1995), 171–182; Raymond Moyroud, 'Loups et loups-garous autour des verreriers en Bas-Dauphiné sous l'Ancien Régime', *Le monde Alpin et Rhodanien* 30 (2002), 125–134.
102. Ursula Brunold-Bigler, *Wolfsmensch und Bärenhexe. Tiere in Sagen und Märchen aus den Alpen* (Chur, 2010), 152–169.
103. Timothy R. Tangherlini, *Danish Folktales, Legends, and Other Stories* (Seattle and London, 2013), 36.
104. Evald Tang Kristensen, *Danske sagn*, II (Arhus, 1893), 227–240.

2

Good to Think: Wolves and Wolf-Men in the Graeco-Roman World

Richard Gordon

It is a commonplace that metaphors tend to cluster around core themes – a tendency rather grandly termed the 'systematicity of metaphor'.[1] We might well extend the point to other, still more routinised forms of linguistic expression: such as proverbs, sayings and catchphrases. In Graeco-Roman antiquity, where, depending on period, between 85 percent and 95 percent of the entire population lived on the land, such tendencies are particularly observable in relation to animals, both wild and domestic. A standard modern Graeco-Roman Bestiary, careful but by no means complete, contains entries for 181 mammalian species, Mediterranean and exotic, from the common-or-garden to the virtually mythical.[2] Moreover, at any rate in Greek perspective, animals, above all mammals, were just part of a much broader continuum of animate beings, whose main constituents are gods, demi-gods and heroes, nymphs, mythical creatures such as centaurs and Sirens, humans (Greeks and others), beasts – and in some *dispositifs*, such as 'magic', even plants and stones, though they were otherwise not considered animate in the required sense.[3] In other words, the classification system encompassed, not merely what we would call natural species, but also super- and para-natural entities of different degrees, so that quite as much interest focused on establishing the external relations of the different groups as on the definitions of individual species.[4] Myths, 'folk wisdom' and catchphrases are some of the media through which relations between the groups were represented. Ancient stories about transformation and metamorphosis derive their cultural legitimacy from this conception of a densely-populated continuum of animate beings.

Since the publication more than half a century ago of *La pensée sauvage* by Claude Lévi-Strauss, it has become trite to note the intense interest of the so-called *Naturvölker* in their natural environment, the complexity and otherness of their classification systems (the 'science of the concrete'), and the intimate relation between these classifications and the local social order.[5]

It was Lévi-Strauss' complementary text, devoted to dismantling the traditional concept of 'totemism', that introduced the idea that animal species – not 'animals' as such – are not so much good to eat as 'bonnes à penser', good to think.[6] Lévi-Strauss' insight was soon picked up by the 'Paris School' of Greek cultural studies dominated by Jean-Pierre Vernant, whose members drew upon an extraordinarily wide variety of literary sources to explore the way in which animals were employed in Greek culture in order to conceptualise human social and gender relations, as well as types of reasoning.[7] In this endeavour, they profited enormously from the tradition of German Altertumswissenschaft, which habitually supplemented ancient descriptions of what was understood as the natural history of each species – often based mainly on Aristotle's *Historia animalium*, which treats around 550 different species under numerous classificatory headings – with a collection of the mythological and 'folkloric' materials connected with it.[8]

Here and there: some basic classificatory schemes

At this distance, and given the biases of our ancient source material for such matters, it is impossible to undertake the kind of close analysis that Stanley Tambiah provided of rural Thai classifications and rules regarding animals.[9] Ancient Greek and Italian cities are not closely comparable to modern Thai villages. We can nevertheless make some preliminary points. Although the Greek scheme of animate beings was so broad (which helped Aristotle to set up his biological paradigm, in which humans are animals – albeit the 'highest' animal),[10] they drew relatively sharp distinctions between man (*anthrôpos*) and beast (*thêr*). Beasts in turn were classified as belonging to land, water and air, thus creating 'nodes' for producing anomalies – such as amphibians, bats, and moles – which were exploited in certain *dispositifs*, such as magic.[11] Broadly speaking, the Greeks regularly distinguished between: (1) the city proper, often itself sub-divided into the defensible kernel, usually elevated, as in the case of the Acropolis at Athens or the Acrocorinth at Corinth, and the remainder of the heavily settled area within the walls, where they existed; (2) the cultivated land surrounding the city, settled with villages; and (3) the largely uninhabited and prescriptively 'unproductive' area beyond this, usually hilly or mountainous, mainly used for pasturage and hunting but also for stone-quarrying, gathering wild herbs, amorous dalliance and encounters with the divine – this was the realm of the goat-god Pan, and the Nymphs, both of them associated with awe, terror, and madness.[12]

Wolves belonged to this sphere of the wild but posed an economic problem because their range coincided with that of human pasturage: in other words, they killed sheep and goats put out to graze in the hilly or mountainous, at any rate prescriptively wild, territory that separated the various independent city-states of central and southern Greece from one another.[13] We can have no idea how many wolves still existed in the more heavily settled parts of this area in the Classical and post-Classical period – probably not many (Xenophon, for instance, does not mention wolf-hunting in his booklet on the sport, written c.370–60 BC). But they certainly continued to exist in the mountainous areas of western and northern Greece. The symbolic value of the wolf, however, had little or nothing to do with actual presence, real danger or quantifiable economic loss – it was the thought that counted.

The Greeks likewise distinguished between domesticated and wild animals (in effect, mammals only). Both categories were complex, in the sense that they contained sub-divisions based on proximity, familiarity and cultural value. Within the domesticated category, the dog occupied a privileged position, since dogs were the only animals regarded as (very lowly) members of the household.[14] Here however, we need to distinguish between: 'noble' or 'manly' dogs – mainly hunting-, fighting- and sheep-dogs, bred to protect livestock, which were kept outside; half-starved, flea-bitten, city-street mongrels, likewise never admitted to the house; and lap-, house- or cuddly dogs, kept as pets by women and children.[15] As members of the extended household, dogs were considered prescriptively inedible (if, under certain circumstances, sacrificable). At the same time, their habits of fawning, incessant barking and yapping, thieving food, mangling corpses, eating excrement and indiscriminate (also incestuous) copulation meant that they excited hostility and sympathy in almost equal measure.[16] Overall, however, dogs are the most-noted item in the Graeco-Roman bestiary. No less privileged, because of its association with military elites, and likewise prescriptively inedible, was the horse.[17] All other domestic animals, the plough-ox, other cattle, mules and donkeys, sheep and goats, pigs and domestic fowl, including the fighting cock and the quail, have their own place in the scheme: not considered members of the household, all were prescriptively edible (even if only a personal enemy would have eaten birds bred for fighting).[18] Some of these animals, notably the pig, had corresponding wild forms, to be encountered in the notionally wild zone beyond the cultivated fields, and were edible just like the domestic variety.[19] The wolf, however – the wild correlate of the dog – was, like its canine relative, considered inedible.

Chatting about wolves

In his excellent account of Greek ideas about wolves, Richard Buxton pointed out that the tradition is highly diverse: not only are different aspects of wolf behaviour stressed in different contexts but different literary genres convey their own divergent perspectives – in fable, for example, wolves are stupid.[20] If, as Otto Keller declared, '[Der Wolf ist] das entschiedene Hauptraubtier der altclassichen Länder gewesen', this fact made it all the more interesting to think – and transform into kaleidoscopic variety.[21] One might therefore begin with Homer's epic similes, where wolves figure as ferocious hunters in marauding packs precisely because they are analogies to heroic warriors.[22] One might also insist on the wolf's reputation as a solitary, asocial creature that can, like a human butcher, cut its victims' throats and bleed them.[23] But the sheer range and diversity of Greek wolf-lore is better represented by a chatty writer such as Claudius Aelianus (Aelian), whose *De natura animalium* (*On animals*), written in Greek in the second century AD, presents a compendium of tales and claims drawn from all manner of earlier sources about animals, mainly of course 'interesting' animals. Here the wolf is by no means of prime concern – the Hellenistic and Roman leisured reading-class lived in cities and on landed estates, where such animals scarcely figured: the elephant, the dog, the lion, the horse, the goat, the eagle, the asp and the dolphin, all of them nice literary animals, get more space. Taking Aelian as a guide, we can discern roughly five strategies in ancient wolf-lore, all of them applied to other animals as well. All demonstrate the principle of turning animals into counters in an essentially human game.

Productive etymology

The assimilation by Classical philology of Brugschian historical linguistics in the mid-nineteenth century turned ancient etymologies into a welcome proof of intellectual progress, since all could be shown to be 'non-scientific' in Brugschian terms, and thus naïve or ridiculous. Since the rise of cultural studies, however, it is the philologists themselves who have become the animals in the zoo, for ancient etymologies are now understood as providing key insights into mentalities and cognitive procedures.[24] However that may be, the method certainly provided a major antique method of constructing meanings around the notion 'wolf'.

Wolves, Aelian tells us, were loved by the god Apollo because he was born, like his sister Artemis, after his mother Leto had changed herself into a she-wolf for fear of Hera, the jealous wife of Zeus, father of gods and men, who had made Leto pregnant. This mythic connection between Apollo and the wolf is commemorated by the existence of a bronze statue of a wolf at Delphi, Apollo's main oracular site in Greece (*Nat. anim.* 10.26).[25] Now, this association between god and animal-species, which is also found in several stories associated with Argos, is the product of a persistent and early folk-etymology of the epithets *Lykios*, *Lykêgenês* 'Lycian/born in Lycia', applied to Apollo in Homer: the similarity of the Greek word for wolf, λύκος, *lykos*, with its adjective λύκειος, *lykeios*, was simply too tempting to neglect.[26] Aeschylus provides a fine example in *Seven against Thebes*, where the chorus prays: καὶ σύ, Λύκει' ἄναξ, Λύκειος γενοῦ / στράτῳ δαίῳ στόνων ἀντίτας, 'And may you, Lycian lord, be Wolf-like over against the opposing army, and (cause them to utter) groan for groan' (145f.).

Modern etymologists may claim that the two words *lykios* and *lykeios* are unconnected; the Greeks thought it obvious that they were and enthusiastically composed narratives that explored the resulting possibilities.[27] The statue at Delphi might perhaps instead have commemorated a wolf that revealed the identity of a temple robber who had stolen the temple funds and buried them on Mt. Parnassus – a motif that refers to another strategy we shall glance at later, the intelligent, quasi-human wolf.[28] Another possibility was offered by Apollo's oracular function. Thus, Pausanias the periegete, arrived at Sikyon, in the northern Argolid, where in his day there was a ruined temple of Apollo *Lykios*, was told a story about how an oracle from Apollo had once put an end to the constant attacks of wolves on the local flocks of sheep. The Sikyonians were told to collect the bark of some 'dry wood' to be found in the sacred enclosure of the shrine, and mix it with the flesh of the dead sheep. And 'as soon as' they tasted the bark, the wolves expired – a wonder indeed. Sadly, though, the local tourist guides did not know which sort of tree it was – and so a sovereign remedy against wolves was lost to science.[29] A late testimony claims that it was Apollo who, in the shape of a wolf, killed the mysterious Telchines, mythic ironsmiths who commanded the powers of magical 'binding' and unloosing.[30]

This same etymological strategy will concern us again when we turn to the story of King Lykaon and the cult of Zeus Lykaios on Mt. Lykaion in Arcadia.

'Empirical' observation

Like etymology, natural history offered plenty of scope for appropriating the wolf. The point, though, is that in antiquity (outside Aristotle's biological works) there are no 'simple' empirical reports of animal behaviour: every observation has been passed, often repeatedly, through the cognitive mincer. With Alexander's conquest of the Achaemenid Empire, the geographical area from which such 'observations' could be drawn expanded massively – Aelian can report authoritatively on wolf behaviour in Armenia, and even as far east as the Sea of Azov.[31] She-wolves spend 12 days annually in parturition; and why? because Leto too took 12 days to travel from the land of the Hyperboreans to Delos, where she gave birth to Apollo and Artemis.[32] In this case it seems likely that hunters' lore, never reliable or fixed, has been elaborated and specified to '12 days' in the light of a mythical detail, itself a product of the etymological conviction. It may be true, for example, that wolves lie down to suckle their young, but the fact is worth recording only because it serves, like their dentition and their claws, to categorise wolves with other big, dangerous animals, such as lions, leopards and bears.[33] It may be true that wolves tend to suffer from ankylosis of the spine, but the idea that they could not bend their necks, and so had to turn round fully to look at something, was interesting only because of the prior belief that one could be bewitched (be unable to utter a word) if a wolf's gaze fell upon one before one saw it – the suppressed term here being the belief that many animals with large, sharp or intelligent eyes, such as the seal, the hyena and certain snakes, could cast the evil eye.[34]

Natural justice

As a fierce wild animal, dangerous to flocks and therefore human economic interests, the wolf was proverbially ever hungry.[35] Adult pederasts who pursued boys were colloquially known in Athens as 'wolves'.[36] There was even a verb in Greek, λυκόω, 'tear apart (from the point of view of a wolf)', found predictably almost exclusively in the passive mood.[37] And yet wolves themselves had their enemies and their fears, in keeping with Nature's plan of just dealing. In Egypt there was a plant, named like Apollo λυκοκτόνος, 'Wolf-slayer', which would kill a wolf by convulsions if it so much as stepped on it. A similar story circulated about the squill, which would stop a wolf in its tracks; foxes had thus learned to gather the plant and throw them into wolves' dens. The entire island of Crete was inimical to wolves, and there were places on Mt. Olympus, home of the gods, where they would not venture.[38]

Wolves intelligent, just and pious

Especially when considered as an example of sociality or co-operation, the wolf was considered to display intelligence in solving problems. Thus, when faced with a wide river, they would hang onto the tail of the wolf in front of them, and so cross in safety. Using a similar method, they could attack and kill a bull that had taken refuge in a pond. Familiar with weather-signs, they would head for human habitations when a storm approached. Single wolves, however, might also show intelligence, even kindly intelligence: it was a wolf that saved Gelon of Syracuse from being killed when the roof of his schoolroom collapsed.[39] The wolves who live by the Sea of Azov are just like house-dogs (οἰκουροῦντες), and, provided they are given a share of the catch, assist the fisherfolk in their labours. But, if their reward is denied them, they punish them instead. Wolves might also demonstrate a natural piety and respect for the human institution of temple-asylum: thus, at Aulê in Arcadia, they would not pursue prey if the animal took refuge in the shrine of Pan, god of the wild.[40]

The properties of wolf-parts

The most widespread medical specialists in antiquity were the 'root-cutters' (ῥιζοτόμοι, *herbarii*), who collected herbs and plants of medicinal value but also used a great variety of animal parts in their practice.[41] These ubiquitous specialists employed a range of collection-practices and were one source of claims about the marvellous properties of certain plants and animal parts, which were often considered effective only if collected under special conditions.[42] Thus, the wolf was claimed to possess a special tuft of hair on its body that had to be removed while it was alive if it were to be effective for medicinal purposes; dropping the tuft deliberately might then be considered an optional ruse by wolves to avoid capture.[43] Other wolf products, however, were considered straightforwardly useful: when the droppings dried and turned white, they could be prescribed as a remedy against colic; wolf-fat smeared on the joints helped patients with tetanus cramps; the dried liver helped against liver complaints. Such prescriptions were not unaffected by developments in school medicine: after Marcellus of Side in the early second century AD, no doubt relying on earlier authors, described a melancholic condition named 'lycanthropy' ('Wolf-man syndrome'), whose symptoms included 'behaving like wolves and dogs' (whatever that might mean) and skulking at dawn in cemeteries,[44] we find claims that dried wolf's liver helps the 'moonstruck' ('lunatics') and the mad, as does a wolf's tooth; the roasted heart consumed on an empty stomach helps lycanthropes.[45] These prescriptions are apparently linked to a condition diagnosed in school medicine; generally speaking, however, parts taken from wolves have no special status in rhizotomic medicine – parts from many other creatures, such as moles, cicadas or chameleons, were considered more powerful.

We have already touched on the interest that attached to wolves' gaze and so to their eyes. The detached eye might itself consequently be attributed a transferred 'innate' power: it was red (another sign of the ability to cast the evil eye); it could be used to ensure victory in a court case; as a medical amulet, it would ward off all kinds of fever, and indeed hallucinations and spirit visitations (φαντάσματα). The right eye, presumably dried, would protect the wearer from blindness caused by ophthalmia – perhaps the left one too; and together with the first vertebra of the tail could be used for 'important magical practices'....[46]

Even such a selective survey of (mainly Greek) wolf-lore – I have largely ignored fable and proverb – is sufficient to show that there were no 'automatic' or 'visceral' associations with wolves in antiquity.[47] Aelian's heterogeneous collection of animal snippets confirms that the idea of the wolf, worked on over centuries, might evoke an entire range of associations, stories and implications. Unless the intended meaning were contextualised, 'wolf' might mean almost anything.

Metamorphosis

Stories about shape changing were a standard narrative device in Greek mythography.[48] Such tales told of a world suffused with divine presence. As we have seen, Greek cosmology presumed the existence of a hierarchy of partly knowable animate beings, which had once been very differently organised. It proceeded from Zeus, father of gods and men, through gods named and unnamed, continued through spirits and more or less impersonal forces such as Necessity, Fate and Fortune, to mankind, and so to animals and plants. Stories about metamorphosis through any of these shapes or forms were thus just concrete instances of the workings of a universal, if opaque, cosmic principle.[49] At the same time, such stories, in their innumerable variants, were bracketed in historically-differentiated ways: this or that version might locally enjoy a special status, in others be quite unknown; each re-telling was calculated to give its own specific sense in that particular context.[50] The theme of shape-shifting might be used in many different ways in such stories: it might evoke the gulf between the immortals and mortals, the moral distance between then and now, the permeability of ontological boundaries, say between humans and animals, humans and plants, enrich a physical location with an indelible narrative association, open up any of a variety of metonymic associations. Because such narratives had no 'essence', no fixed content or meaning, they could constantly be invested with new senses, acquire new twists through elaboration.

One feature remains constant, however: all such stories attempted to evoke a sense of surprise, of wonder (θάμβος).[51] Stories of metamorphosis are stories of the marvellous, a category produced by the conceptual grids developed by each culture, and so historically specific to each.[52] Two forms of non-supernatural marvel come to mind here: the exotic and the natural. The type of the exotic marvellous is the tale of Odysseus' encounters with folk whose non-Hellenism is defined by way of various infringements of Greek cultural and moral rules.[53] One function of such stories is to naturalise the audience's own visual, social and moral world; another, to suggest that there are in truth no limits to the variety of possible things. The world 'out there' pullulates with objects, peoples and practices, which one can barely imagine. In that context, metamorphosis itself is almost to be expected.

The natural marvellous is akin to the exotic, for many accounts of strange animals and animal behaviour, marvellous rivers and mountains, redoubtable plants and stones purport to derive from far-off lands.[54] But many others derive from local Greek and Italian histories and thus, like the lore in Aristotle's *Historia animalium* Book 10, ultimately from indigenous hunters, rhizotomists and country-folk. The immediate value of these stories varies with their genre and context: Aelian, for example, finds in them an extension of social rules and ethics, such that the animals often represent moral values; whereas Plutarch's *De sollertia animalium*, which makes use of much the same sort of material, tests them to discover the location of the boundary between human and animal. In local histories, they instantiated an identity based on nature as well as on the past. But as a whole, the natural marvellous, like the exotic, represents a massive effort of noticing, of reflection upon norms, rules and regularities, of sorting and selecting on the basis of widely-different criteria, which over time threw up a fine variety of ill-assorted debris that in turn provoked interpretation, one form of which is the metamorphic narrative.

These forms of the non-supernatural marvellous have a reciprocal relationship to the 'constituted transcendent', the amalgam of powers whose existence and significance is socially affirmed by religious institutions. On the one hand, the constituted transcendent provides the pattern for the earthly marvellous: the Homeric heaven, for example, with its metamorphic inhabitants, constructed a horizon of expectation that sustained all other forms of shape-shifting.[55] On the other, the non-supernatural marvellous generated the concrete evidence that proves that the traffic between the divine world and this one is constant, and, *a posteriori*, that the divine world truly exists. As such, it is a major resource in maintaining the vitality of religious systems, like the Greek and Roman, perpetually tending to be subordinated to, and rationalised by, political interests. Aspects of the vision of divine alterity repeatedly invaded and renewed the modes of civic religiosity, the religion of everyday, through dream, prophecy, portents, disasters.[56] And in the context of 'abnormal' religion – mystery-cults, seers and prophets, wonder-workers, magic – specific locations and substances were conceived as, or claimed to be, endowed with exceptional, reproducible power to alter normality. In a recipe in one of the Demotic magical papyri, for example, the head and blood of a hoopoe are to be cooked and dried, then painted onto the eyes to enable one to see 'shadows (of gods)'.[57]

If we attempt to functionalise the marvellous in this way, it comes as no surprise to find that it is in the Hellenistic period, and particularly from the mid-third century BC, that stories about metamorphosis are collected and to a degree systematised, though virtually nothing of these early efforts survives.[58] It so happens, however, that an ancient commentator to the *Metamorphoses* of Antininus Liberalis, a prose compendium probably of the second century AD, has inserted a note about the Hellenistic source of these versions.[59] The two authors most frequently cited are Nikander of Kolophon and Boios. Insofar as one can generalise, Nikander's stories tend to offer aetiologies for existing practices, while Boios seems to have created a new genre, stories about transformations into birds.[60] The most important point, however, is that few of the stories belong to the (rather elastic) class of 'PanHellenic' myths, known from epic, tragedy or any other major Classical genre; nor are they known to have been picked up by any of the early mythographers of the Classical period.[61] They are, at least originally, obscure local stories that could only have become known to littérateurs such as Nikander and Boios because they had been recorded – and duly elaborated, embroidered, poeticised – by local historians writing for the new audience of bookish readers that emerged in the Hellenistic period and whose works were made more generally available in royal libraries, such as those of Macedon, Alexandria, Pergamon – and later, of course, Rome.[62] It is this type of material that Ovid drew upon for his own *Metamorphoses*, in which he gives his own characteristic twists to the stories, often highlighting the subjectivity of the transformative experience: a device comically elaborated by Apuleius in his *Metamorphoses* almost two centuries later, an entire Bildungsroman told from the point of view of a man transformed into a donkey – and back.[63]

If the Hellenistic period did not invent transformative stories, it certainly domesticated them, in the sense that it recognised them as a genre, collected numerous non-canonical examples, and ran them through a predictable series of interpretative grids. Once that had been done, it was no longer possible, at least in the literate class, to take such stories at face value, however much authority they may have retained in their original local contexts, about which we know precisely nothing. They had become a literary device, where what counted was the wit, originality and conviction with which the poet could induce the reader to take the motif, here, now, seriously. Wolves, however, are conspicuously absent from this poetic genre, with the sole – one is tempted to call it 'very minor' – exception of the story of Lykaon, king of Arcadia.

Lykaon[64]

The most famous version of the story of Lykaon is that of Ovid's *Metamorphoses*, written in the last decade before the turn of the eras.[65] One reason for its fame is that it occurs early on in Book 1, so one did not have to read very far to come across it; another that it forms part of the run-up to the Deluge, which, for any Christian readers, gave this part of the poem a reassuring similarity to *Genesis*.[66] The story is narrated in heaven by Jupiter, who describes how he changed into human shape in order to inspect the morals of mankind. Without further explanation, he finds himself traversing wild and dangerous mountain ranges in Arcadia to reach 'the cold pine-woods' of Mt. Lycaeum.[67] As night falls, he comes across the house of Lykaon, whose lawlessness is prefigured in the descriptor *tyrannus* (218). Although Jupiter has made clear to the other inhabitants – in this antediluvian period there could be no cities – that he is a god, and they worship him as they ought (*pia vota*, 221), Lycaon decides to test whether he really is one. To this end, he cuts the throat of a Molossian hostage staying in his house, chops him up, and boils some of his parts, while roasting others (226–229). All these rapidly sketched details imply what Ovid does not explicitly say, that this is an impious mockery of the normal sacrifice of cattle, which involved cutting the animal's throat, collecting the blood, and then roasting the 'noble innards' and boiling the red meat. As soon as this impious meal has been served, Jupiter destroys the house and its *Penates* (household/family divinities).[68] Lycaon however, escapes into the fields, howling, and starts attacking the sheep 'in his settled bloodlust' (*solitaeque cupidine caedis*, 234). At this point Ovid resorts to his usual technique of emphasising the 'moment of hybridity':

> His clothes become hair, his limbs legs: he becomes a wolf yet preserves traces of his former appearance (*veteris servat vestigia formae*) – the same grey hairs, the same violent face; the same eyes glitter, it's the same mask of ferocity (236–239).[69]

Figure 2.1 Jupiter changes king Lycaon into a wolf. Print by Crispijn van de Passe, published in: Ovidius, *Metamorphoseōn Quidianarum typi aliquot* (Cologne 1602, 1607), Book I, print no. 6. (Rijksmuseum, Amsterdam RP-P-0B-15–890) (H 852). This is a reduced copy in reverse of a print from the workshop of Hendrick Goltzius after his invention and part of a series of engravings made in 1589 (H 540)

There is, however, no attempt to enter into the subjectivity of the transformation, as Ovid often does elsewhere with more sympathetic characters, and the brief narrative stops abruptly at this point, as Jupiter continues to rage against human wickedness in justification of his decision to punish mankind with extermination. There is no follow-up; in this version, Lycaon is an impious monster whose punishment fits both his name and his crime. At the same time, there is an unmistakable hint of how to read this transformation: less a wonder, more an allegory of vice.

What is a myth? How many variants can 'a myth' support? How many details need survive in the re-tellings? In the Classical field, these questions are often dealt with in handbooks of Greek myth (or on Wikipedia) by re-telling one version, usually from a prominent Archaic or Classical source, say a fifth-century tragedy, from Ovid's *Metamorphoses*, or from Apollodorus' *Bibliotheca*, presented as normative. This is convenient but misleading. In fact, the more often a myth appears in our sources, the more difficult it is to say what 'it' is. The simplest formulation is to say that each re-telling adapts a version of a story to the interests of the narrator/performer and the actual, intended or implied audience. In the Graeco-Roman case, we have no access to the living, oral versions, with their own innumerable immediate constraints and invitations, only to literary ones, subject to their own interpretative pressures: such as, genre (epic, epinician, tragedy...), pseudo-historicisation, Euhemerism, catasterism and so on, occlusion, abbreviation, ignorance, invention.[70] 'The myth' also shifts in relation to modern attempts at explanation: a structuralist account, say, of the myth of Oedipus creates, through its accentuations, a very different story from that of the philologist working on Sophocles' stagecraft.

The point can easily be illustrated by taking two other significant ancient versions of the Lykaon story, those of the epitome of Apollodorus' *Bibliotheca*, and Pausanias.[71] Pseudo-Apollodorus follows a version attested in the earliest surviving account, by Nicolaus of Damascus, in the late first century BC,[72] according to which it was not Lykaon, son of Pelasgus, who transgressed, but his 50 impious sons, who were tested by Zeus in the guise of an itinerant labourer. At the instigation of the eldest, Mainalos, they killed a boy belonging to a local Arcadian family and served up his viscera mixed with ordinary sacrificial meat. Zeus spotted the ruse and upturned the table (τράπεζα) at a place still named Trapezus, and killed them all, including Lykaon, with his thunderbolt. Only the youngest, Nyktimos, survived to become king of Arcadia. Pierre Borgeaud has read this version as just one phase of a 'culture-crisis' that extends over three generations (Lykaon, Kallisto, Arkas) and is ultimately resolved by the establishment of civilisation in the form of agriculture and regular sacrifice.[73] The motif of the impious sacrifice would fit that model, but metamorphosis into a wolf is irrelevant to it, and can therefore be dropped.[74] Is the metamorphosis motif then essential to 'the myth'? Or does it just fit with some versions that chose to elaborate on the narrative possibilities so plainly offered by Zeus *Lyk*aios, Mt. Lykaion and King *Lyk*aon, but not with others?

We cannot know how much detail has been lost in composing the epitome, though it clearly draws upon Hellenistic local histories of Arcadia, interested in creating a 'thick' local topography and fictitious genealogies to support political claims and hoped-for alliances between cities, by no means only in Arcadia, legitimised as citations from Archaic Greek sources.[75] Pausanias draws upon the same sort of material, but, having more space at his disposal, can afford to be much more circumstantial.[76] His account begins with mythical data shuffled into a typically rationalistic account of the civilising sequence: Pelasgos, 'the Arcadians' say – i.e. the local historian(s) he is mainly following – was the first human being there; but, of course, he must have had subjects, adds Pausanias, because otherwise he could not have been a king, which we know he was. Pelasgos taught his people to make huts for shelter against the weather and primitive clothes from the hides of wild pigs. He also weaned them away from their previous diet of raw leaves and roots and induced them to eat beech-mast (8.1.5–6).[77] It was Lykaon, too, who introduced elements of more developed civilisation by founding the city of Lykosoura at the foot of Mt. Lykaion and inaugurating the Lykaia, the festival in honour of Zeus Lykaios – for the Greeks, cities and festivals involving competitive athletics, chariot-races and theatrical performances were essential markers of civilisation.

At this point, however, the civilising process is abruptly halted by a comparison between the sacrificial practices of the first king of Athens, Kekrops, and of Lykaon, whom Pausanias presents as contemporaries. Both negated the rules of normative Greek sacrificial practice, namely blood-sacrifice centred upon the three major domesticates, cattle, sheep/goats, pigs: Kekrops because he refused animal sacrifice and allowed only a kind of pastry (πέμματα, in Attic dialect πελάνοι) baked in the oven to be offered to Zeus Hypatos (the Highest); Lykaon because he sacrificed a new-born baby to Zeus Lykaios. Although Lykaon was sufficiently aware of the (later) normative rules for Olympian sacrifice to smear some of the blood on the sides of the altar, Zeus did not take kindly to his offering: 'they say (φασίν) he was at once, during the sacrifice, turned from a man into a wolf' (8.2.3).

All too aware of the incredulity with which this tale was met by the sophisticates who preferred Zeus to use his usual instrument, the thunderbolt, Pausanias then devotes much more space to telling his readers why he chooses to find the tale plausible than on the narrative itself.[78] Starting with a wonderfully self-excusatory phrase: ἐμέ γ' ὁ λόγος οὗτος πείθει, 'well, I at any rate find this story convincing' [lit.: me at any rate this story persuades], he launches into a comparison between those early days of human life, when numerous (mythical) humans became gods (or heroes – he does not distinguish), such as Aristaios, Britomartis,[79] Herakles…, and his own Imperial age, when the only people who become gods are emperors, and that simply because they have the political power to insist upon it; and when the gods have stopped imposing instant condign punishment, preferring to impose it in the afterlife, so that greed and impiety rule the world. And if we grant that those mythical figures could become gods, why should we not believe that Lykaon could become a wolf, or Niobe a rock (8.2.4–5)? This, he admits, is an argument simply from τὸ εἰκός, plausibility (generally considered in rhetorical theory a weak type of argument);[80] but the story is assuredly very old. This argument looks as though it had been dusted off from somewhere else: we may remember that the Deluge is supposed to have occurred around the same period – Zeus' punishment for human wickedness; which does not suggest much moral difference between Lykaon's time and the Roman Empire's.

As I have said, Pausanias gives the antiquity of the story as one of his main reasons for accepting it. And this has been a very general assumption, especially since the standard edition of Hesiod's fragments, by Merkelbach and West, confidently attributed the version in the *Katasterismoi* of pseudo-Eratosthenes (which claims indeed to be Hesiodic) to the *Ehoai* (*Catalogue of Women*).[81] This version begins with the story of Arkas (the mythical figure who gave his name to Arcadia, though the word itself means 'bear' in Greek), but suddenly embarks on the story of Lykaon, fusing several versions of the story together – the word βρέφος (baby), for example, only occurs otherwise in the extant versions in Pausanias, who does not mention the etymology of Trapezus, which is a feature of Apollodorus' version. The 'Hesiodic' version also combines the motif of the transformation (Lykaon) with that of the thunderbolt (his house), which otherwise occurs only in Ovid and Hyginus, *Fab.* 176. All this looks like a conscious splicing of well-known variants. Nevertheless, it has been supposed that some version of 'the myth' was presented as a tragedy at Athens in the mid-fifth century, though it is frankly a mere guess.[82] As Bonnechere pointed out, the main motor of the story, the impious Thyestean feast, is a common narrative motif in Greece, associated both with earliest times and with certain families.[83] No doubt it was this that attracted a few tragic poets to the topic – one of them certainly in the late fifth century BC, though nothing is known of how the plot was dealt with, or whether the transformation into a wolf featured in it.[84] In view of the thunderbolt version, it cannot be excluded that the transformation is a relatively late addition to an older 'Thyestean' story, itself linked to the metamorphic stories of Kallisto and Arkas, prompted by the seductive Λύκαιος, Λύκαιον, Λυκάων, λύκος chain of associations.

Greek wolf-men

If Pausanias is prepared to believe that Lykaon, in remotest antiquity, became a wolf, just as Niobe became a rock, there are other Arcadian stories he dismisses as falsehoods. These include the tale that Niobe, in rock form, sheds tears every summer on Mt. Sipylos (8.2.7). And in relation to Zeus Lykaios, 'they say that at the sacrifices offered to him others too over the years (ἀεί τις) have turned into wolves, but, if they abstain from human flesh, they turn into humans again after nine years (ἐπὶ δεκάτῳ); if they do consume such flesh, however, they remain wolves for the rest of their lives' (8.2.6). These stories are quite incredible: Pausanias compares them to the belief that griffons are spotted like leopards or that Tritons (sea-monsters) can talk like human beings, or make noises by blowing into seashells (8.2.7) – for him they are of a quite different order of plausibility from the story of Lykaon, being produced by those who 'build falsehoods onto truths' (οἱ τοῖς ἀληθέσιν ἐποικοδομοῦντες ἐψευσμένα). It is however precisely these tales, which are evidently variants upon diverse stories about Lykaon, in that the idea of transformation is linked to sacrifices on Mt. Lykaion, that are our proper concern here.

Unfortunately, Pausanias' view of these narratives was evidently shared by many other members of the literate elite. Unlike the story of Lykaon, which could be retold in various genres (including tragedy) precisely because it was classed as a myth – a status legitimated by its parallelism to other stories of impious sacrifice – no extant source considered these stories of wolf-men worth taking seriously; though, it is certainly possible that some writers of *mirabilia* did find a place for them.[85] The earliest references are found at the end of the first quarter of the fourth century (say 370 BC). In his account of the formation of the lawless tyrant, Plato makes Socrates refer to a story (μῦθος) circulating about what happens at the temple of Zeus Lykaios: people (men) are inescapably turned into wolves if they consume human viscera there, mixed with those of animals.[86] One implication, that human sacrifice was practised on Mt. Lykaion, becomes explicit in a contextless remark by Aristotle's most famous pupil Theophrastus, linking Carthage, in antiquity the place par excellence where (infant) human sacrifice to Melkart was practised, with the festival of Zeus Lykaios, the Lykaia.[87] Pierre Bonnechere plausibly argues that these are essentially appropriations of Lykaon motifs that circulated in Greece, in this case Athens, as exemplifications of Arcadia as an internal other.[88] At any rate, we happen to know of two stories – one about a famous athlete, victor at Olympia, the other about a single Arcadian family, the Anthidai – which closely resemble those that Pausanias found so incredible.

The first is a story ascribed by Pliny to a book on Olympic victors by a Greek historian whose name is uncertain.[89] Demaenetus of Parrhasia (i.e. Arcadia) tasted of the viscera of a sacrificed boy (*puer*), as was then customary (*etiamtum... faciebant*), to Zeus Lykaios, turned into a wolf and nine years later turned back into an athlete. He took part in the boxing competition at Olympia and won. A propos of a victor-statue in Olympia, Pausanias tells the same story of an athlete named Demarchos, son of Dinytas, of Parrhasia.[90] Both Pliny and Pausanias find this incredible, the latter doubting – despite what he was to write in Book 8 – that it could have circulated in Arcadia at all. In the present context it is only necessary to emphasise that, by the Hellenistic period (it is of no account when 'Demaenetus/Demarchos' may have lived), stories about wolf-men could be told about athletes, evidently as a way of imaging their exceptional prowess. In this case, there can be no doubt that it was the fact that he came from Parrhasia, a town close to Mt. Lykaion, that prompted the connection to the wolf-story.

The second report is ascribed to 'the Arcadians' by an otherwise unknown author, 'Euanthes'.[91] It was customary – but we are not told how frequently, or why – for a male member of the family 'of Anthos/-us' (otherwise completely unknown), chosen by lot, to be taken to a pool in the area (not further specified, but evidently in Arcadia). There he hung his clothes on an oak tree, swam across the pool, and went into the wild (*abire in deserta*), where he turned into a wolf, living with these animals for nine years. If in this time he had eaten no human flesh,[92] he returned to the pool, swam back across it, and re-acquired human shape, albeit nine years older; and re-assumed his old clothes, still evidently hanging out on the oak tree. This version is interesting, since the motif of the impious sacrifice has been occluded in favour of a family peculiarity, and only survives in the condition of re-instatement into human form.

For the educated elite, the point of recounting such stories was to illustrate the credulousness of others – for Pausanias, as a Greek, that of the Arcadians; for Pliny, as a Roman *eques*, that of 'the Greeks'. We may simply note that, whereas Lykaon, in the versions in which he is turned into a wolf, never regains human form, in two of the 'foolish' stories this possibility is open, on condition that the original crime is not repeated, while 'Demainetos/Demarchos' simply becomes human again without more ado. These 'foolish' stories are therefore genuine were-wolf stories, which advertise themselves as such by introducing a typical motif of the marvellous – the period of nine years. The Anthidai story repeats the signal by specifying that the man's clothes will still be hanging on the oak tree, but slyly introduces a counter-note of 'realism' by specifying that the man will nevertheless have aged nine years. At one level, they lay claim to a relationship of continuity with *Lyk*aon, son of Pelasgos, founder of the *Lyk*aia, celebrated on Mt *Lyk*aion, which can only be achieved by stepping into the world of the marvellous. At another, they acknowledge but also repudiate his crime, inasmuch as, in their lupine form, the mutants have to respect the human rule that interdicts cannibalism – even though, from a wolf's point of view, eating human flesh would merely be permissible allophagy. That is why the Anthidai story introduces the idea that the mutant is no solitary wolf, but runs with the pack: this is a real wolf, but he has a choice – he can accept wolf-rules or he can retain a human conscience.

In other words, these stories – which I take to be Hellenistic – have the same significance as single stones in a large mosaic, such as the Nile mosaic at Palestrina: individually they enjoy low investment, but, conjoined with thousands of others, they form an immanent design – the design of a living, divinely-ordained kosmos, composed of innumerable elements, some of which are understood and have a firm place, many others of which are quite inscrutable.[93] Both types, however, are needed to sustain the vision of a cognisable world suffused with and dependent upon another invisible one, the baroquely-ramified divine world of the polytheistic imagination. From that point of view, the Hellenistic stories are no different from Herodotus' account, in the fifth century BC, of the Neuroi, a trans-Scythian people, every man jack of whom was said to turn into a wolf for a few days every year and then back again.[94] Herodotus does not believe it, but allows that the Scythians, whose story it was, insist upon its truth. Herodotus himself collected a sufficient number of bizarre facts about the Scythians, a people or peoples in the remote parts of Asia, as about the civilised Egyptians, Persians and others, to relativise his personal scepticism.[95] At the same time, however, he makes an interesting suggestion about how such reports might be understood, by introducing the idea of magical practice: the Neuroi, he says, seem to be a nation of γόητες (*goêtes*), people endowed with, or who claim to be endowed with, strange, disputed powers.

Mount Lykaion and its cults

In my view, many such 'foolish' stories circulated in Arcadia, negotiating in their various ways the inheritance of Lykaon; and, insofar as they were appropriated beyond Arcadia, acquiring new emphases and instrumentalisations at each telling.[96] But, except for scraps I shall come to in the next and final section, they are lost. This paper is not intended as a research report, but it is worth looking briefly at what scholars have made of these narratives and their supposed context, the cult of Zeus on Mt. Lykaion.[97] The common starting-point is Pausanias' comment – oddly enough not in the context of the Lykaon story – about the altar: 'They sacrifice on (it) in secret (ἐν ἀπορρήτῳ). It is not agreeable to me to go into the details – let it be as it is, and as it was in the beginning' (8.38.7).[98] As one might expect, the different explanations represent numerous ingenious variations on a small number of main paradigms.

Until the rise of folklore studies in the mid-nineteenth century, the story of Lykaon was understood to be centred upon the killing of a boy and taken as just one example of the widespread practice of human sacrifice, generally to avert evil, in primitive times.[99] This strategy was also a means of integrating the medical notion of lycanthropy into the picture. Thus, Carl August Böttiger, writing in 1795, starts from the premise that human sacrifice here might be a response to cases of lycanthropy in the medical sense:

> Man gab also den einheimischen Nationalgottheiten, Zeus und Pan, eine besondere dahin abzielende Bennenung von den Wölfen, man nannte sie Λυκαίοι, die Wölfsgötter, und opferte ihnen das wirksamste Sühnopfer, das das rohe Alterthum in solchen Fällen nur darbringen konnte, einen unschuldigen Knaben. Den Stifter und Opferpriester dieser grausamen, aber in Alterthum wenig befremdenden Sühnungsfeier nennen die alten Volksagen Lykaon.[100]

Although the hypothesis of human sacrifice has continued to feature in many modern accounts,[101] Wilhelm Hertz' analysis in *Der Werwolf* of numerous wolf-man myths, 'sagas' and legends as expressions of primitive cult-practice signalled the creation of a new paradigm.[102] Under the influence of Hertz, and more indirectly of Enlightenment theories of fetishist and totemic practices, particularly in colonial West Africa and North America,[103] the attention of classicists shifted to the transformation theme. What did metamorphosis mean? The leopard-men of Banana suddenly became topical.[104] It became common to assume the existence of an Arcadian wolf-cult, even to imagine there may have been regular were-wolf associations ('confréries des loups-garous'): i.e., rituals in which people represented themselves as wolves, in the course of which they may even have conducted human sacrifices.[105] Henri Jeanmaire even imagined that the cult represented an attenuation of a still more horrible practice, attested in the cult of Dionysos, the *sparagmos* of a living victim, and the consumption of the raw flesh.[106] The decisive blow to both these arguments, at least over the longer term, were the excavations conducted in 1903–1909 by Konstantinos Kourouniotês, both at the summit of Mt Lykaion and on the terraces some 200m lower down, where the famous games, the Lykaia, were celebrated.[107] On investigating the altar on the summit, Kourouniotês discovered that it consisted of an incredible quantity of sacrificial ash and bones, 30m in diameter and 1.5m high. As far as he could establish, none of the bones were human. The earliest traces of cult that he could find were two bronze tripods of the late eighth/early seventh century BC, from which he concluded that the *temenos* (sacred area) just below the altar had only been laid out – it was never truly monumentalised – during the Archaic period. He also discovered numerous votive offerings, including statuettes of Zeus and miniature tripods, of the type found, for example, at Olympia (which is not far away). Mt Lykaion began to look disagreeably normal.

What might replace the idea of totemic practice among adults involving human sacrifice? Oddly enough it was one of Pliny's narratives, about the Anthidai, that seemed to many the most promising option. Some of these accounts, though they drew on currently fahionable models, failed to attract supporters. One example is A. B. Cook, an early professor of archaeology at Cambridge, perhaps over-familiar with the vegetation-god scenarios of his Cambridge colleagues, James Frazer and Jane Harrison.[108] Reverting to an older etymology of Zeus Lykaios, which linked the epithet to light and the sun, he turned 'Anthos' into the personification of vegetation; the Anthidai must therefore have been a priestly clan whose function it was to ensure the fruitfulness of the crops each year. Lykaon's human sacrifice was intended to ensure this fertility; the metamorphosis represents his exile from Arcadia.

Appeal to van Gennep's idea of regularly structured *rites de passage* proved more successful. As this model became familiar to historians of religions, particularly in Italy and France both before and after the Second World War, the idea of the initiation of young men into the class of full warriors seemed to many an attractive way of accounting for numerous odd features of Greek myth, including Lykaon.[109] Although it did not quite fit Van Gennep's model, solitary, marginal 'wolves' might represent the period of marginality that in his model mediated between the rituals of separation and those of re-inclusion. The example always cited is that of the *kryptoi* ('the hidden ones') at Sparta, who had to live for a year out in the mountains and fend for themselves, becoming tough and useful by murdering uppity helots; and their analogues on Crete.[110] Wolf-men thus became 'wolf-men', a temporary status between childhood and manhood. Arcadia was suitably backward for such an institution to have survived into historical times; though, curiously enough, no ancient source mentioned it. The value of the category 'myth' here was that it could enshrine a folk memory of a very ancient institution. The 'proof' lay in the story of the Anthidai – taking off one's clothes, swimming across the pool, the transformation into a wolf.... The only slight problem was the nine years – too long surely for a rite of passage, though this has not proved an insurmountable problem to some.[111] The motif of eating human flesh was quietly forgotten.

The initiatory hypothesis is probably even now the most generally accepted explanation for what we may call the Lykaon complex.[112] Alternatives however soon emerged. Karl Meuli, who anyway believed that all animal sacrifice evoked guilt even among hunters and gatherers, argued that the idea of human cannibalistic sacrifice was a myth invented to justify the exclusion of the 'guilty' sacrificant, i.e. Lykaon was simply a type.[113] In the most elaborate discussion so far, Giulia Piccaluga argues that Lykaon, as the son of Pelasgos, belonged to a pre-civilised world, before humans began to be defined as 'eaters of bread', so that improper sacrifice ('Thyestean feasts') was a natural concomitant of the same cultural horizon. The Deluge is crucial to the myth, in that it brought this phase of pre-civilisation to an end by establishing a hydrological regime suited to rain-fed agriculture.[114]

The spread of structuralist ideas in the 1970s tended to reinforce such an account: the central theme of 'the myth' is the end of commensality between gods and men (which had already been the theme of famous analyses of the Prometheus myth in Hesiod's *Theogony*) and the loss of innocence involved in the establishment of civilisation, i.e. agriculture, regular marriage, private property, high rates of mortality.[115] In these accounts the focus is on the transition from a pre-cultural world, the age of the Pelasgians, to 'our' world, ordered by Zeus, an ambiguous transition provoked: in this case, by Lykaios' human sacrifice. The key incident, on this view, is Zeus' overturning the offering-table/dinner-table (the word τράπεζα means both) with the impious flesh upon it. The fear, represented by the transformation, is of falling back into the bestial, pre-civilised world.

The great advantage of such accounts is that they remain at the thematic level of the narrative instead of trying to find a real-world correlate that it somehow describes (or disguises). More recently, however, as the fascination with structuralism has waned, the presumed cult practices on Mt. Lykaion have again come to play a part in a debate over the status of human sacrifice in Greece.[116] Since such stories abound in Greek mythology, they pose a problem for any non-structuralist approach. One of the few recent scholars to argue in favour of the literal truth of the stories about the cult of Zeus, at least as regards the reality of human sacrifice, is Madeleine Jost, a specialist in the cults of Arcadia, whose main target is in fact the initiation thesis. She argues that such rites, performed by small groups of initiates, did indeed take place on Mt. Lykaion, which in turn gave rise to the myths.[117]

In this context it is relevant to note that, just over a century after Kourouniotês' excavations, an American-Greek team, under the direction of Anna Karapagioutou, David G. Romano and Mary E. Voyatzis, has been re-exploring the archaeological remains on Mt. Lykaion every year since 2004.[118] I confine myself to their remarks relating to the trench they drove through the great ash-altar on the southern summit in 2007.[119] Against Kourouniotês' belief that the summit was first used extensively in the Archaic period, it is now clear that, as Pausanias claimed, the cult there is extremely ancient: extensive ceramic deposits indicate that it was a sacred site from the Early to Late Helladic periods (say 2,500–1,100 BC), with the intensity of use growing markedly in the Mycenaean (Late Helladic) period – at one point in the ash-altar quantities of Mycenaean *kylikes* and other vessels, together with several Mycenaean terracotta animal-figurines and fragments of at least one terracotta human figurine, were found just above the bedrock. Although the stratigraphy is in many cases confused, it seems clear that the summit was an important Mycenaean site, in direct continuity from earlier sacral use.[120] This is itself a rare phenomenon in Greece, though it is of course not the same as saying that Zeus had always been the object of worship there.[121] At higher (i.e. later) levels in the trench, the team found Geometric and Archaic pottery and a series of miniature bronze tripods, which probably date from the eighth century BC. Against Kourouniotês, it now seems clear that the summit continued to be a cult site well into the Hellenistic period, the time when, in my view, narratives about were-wolves were being created. But on one central question, Kourouniotês' view has been fully confirmed: taphonomic analysis of the considerable amount of animal bone discovered in all the levels of the trench has shown that virtually all (c. 98 percent), as at Olympia, are from sheep and goat. There are no identifiably human bones, and almost no other animal species. Moreover, the bones are recognisably patellas, femurs, and tailbones: exactly the type of remains we find in 'normal' Greek sacrifice. Again like Olympia, the fill of the altar is almost entirely composed of decayed bone.

From marvel to magic

I have already mentioned Herodotus' tentative classification of the trans-Scythian Neuroi as γόητες, persons endowed with, or who claim to be endowed with, strange, disputed powers.[122] Like other exotic phenomena, their short-term transformation into wolves could find a convenient place in the ragbag of the Hellenistic genre of *Mirabilia*, in this case ethnographic.[123] But short-term transformations could be appropriated into other discourses too, notably that of magic. I have elsewhere tried to show that we need to distinguish, not merely between 'magical' practice and the discourse about illicit religious practices in Antiquity, but also between an earlier phase, down to the mid-Hellenistic period, in which the discourse has no fixed stereotypical figures, and the later period, for which the earliest tangible evidence is to be found in the crisis of the late Roman Republic, but which develops most emphatically in the Principate.[124] Once political power was focused upon a single figure, the emperor, rather than being widely dispersed within a ruling elite, 'magic' came to be defined increasingly in terms of the interests of that centre, specifically with 'illicit' divination and manipulation of the power of the gods, or of substances, to the detriment of others.[125] The fantastic figure of the Night-Witch (male or female), the inversion of all civilised and human values, proved a convenient mythic Other over against the organised religious order, centred upon the pious emperor as chief sacrificant, whose essential function was the legitimation of the theodicy of good fortune enjoyed by the elite.

We do not know how widely the Arcadian were-wolf stories circulated; at any rate, hardly any have survived in the tradition available to us – Hellenistic literature has anyway disappeared virtually without trace.[126] The few stories in Latin may ultimately derive from variants of this tradition mediated into Latin literature, which drew heavily upon Hellenistic originals, or from genuine Italic stories. The evidence, such as it is, hardly allows one to decide.

The earliest item is a passage in Vergil's *Eclogue* 8 (say 37 BC), put into the mouth of an anonymous young woman trying to regain the love of a faithless man, Daphnis. In the course of a long, quite imaginary, ritual, she takes some herbs and sings her incantation as she burns them on a small altar. These herbs, she says, have been collected in 'Pontus' (i.e. the Caucasus) by a professional *herbarius*, named Moeris, to whom she ascribes three feats: turning into a wolf and slinking into the woods, calling up the dead from their tombs, and 'singing' away field-crops from one man's field to another's.[127] Now this last was a practice already forbidden in the oldest Roman law giving, the XII Tables, and there had been a famous case in the second century BC in which C. Furimus Cresimus had successfully defended himself against precisely such an accusation.[128] Necromancy, i.e. divining by means of the spirits of the dead, was a practice that became central to the later stereotype of the negative witch; yet Varro could quite casually mention in the 40s BC that 'if blood is used (in "bowl-divination") one may also consult those who dwell in the Underworld' – the blood he means is surely animal, not human.[129] But, with one exception, lists of witches' powers, which at the very end of the Republic and the beginning of the Principate become something of a literary topos, never otherwise include auto-transformation into a wolf – the detail failed to establish itself.[130] Moreover, the motif does not occur in Vergil's model, Theocritus' *Pharmakeutris/Pharmakeutriai* (Idyll 2), which happens to be extant. Since stealing crops by incantation is a specifically Italic/Roman motif, and ends the list, we should perhaps infer that Vergil (or possibly Catullus before him, who wrote a poem, now lost, on the same theme) found the motif in a Greek context, possibly in one of the Greek tragedies about Medea – or even about Lykaon – or in the Latin translations or adaptations of such tragedies that appeared during the Hellenising phase of the later Republic.[131] 'Moeris' occurs nowhere else and, as far as we know, cannot be linked to Arcadia.

Vergil clearly implies that, whatever his origin, Moeris could change himself back and forth at will – he is a true were-wolf. Propertius evidently made the same assumption about his procuress. This is one major development between the Arcadian stories, which assume a single lifetime experience, and these poetic ones. The other is that transformation into a wolf is now, as Herodotus had suggested 450 years earlier, more appropriately contextualised as magic than as marvel.[132] But our poets have no space to tell us what Moeris and the procuress actually did when they went off into the woods, and no doubt no interest in doing so.[133] This deficiency is supplied by the third extant passage, from Petronius' fragmentary picaresque 'novel', the *Satyricon*, written around the mid-first century AD.[134] It is narrated in faux naïf first-person style at a dinner-party by a former slave, Niceros, as part of a competition in story telling – when he is finished, the host congratulates him on making his hair stand on end.[135]

The scene is southern, i.e. Greek-speaking, Italy. The narrator sets off at twilight, under a bright-shining moon, with a muscular soldier;[136] when they reach a cemetery, the latter takes off all his clothes, urinates on them and at once changes into a wolf – don't you dare disbelieve me! – howls once or twice and disappears like Moeris into the woods. The cemetery at once recalls Marcellus of Side and lycanthropy: his melancholics hung about such places; but this is mere background. The story continues: When the narrator touches the clothes, he finds they have turned to stone.[137] Scared to death, he goes back to his mistress' country-house, where she greets him with the news that a wolf has broken into the farmyard and attacked the animals, but one of the slaves managed to wound him in the neck with a spear. The narrator's eyes widen in horror. And on returning to the cemetery, he finds nothing but blood in place of the petrified clothes. Meanwhile, back home, a doctor is attending to a wound in the soldier's neck. *Intellexi illum versipellem esse*: 'And then I realised he was a *versipellis*, a shape-shifter'. The blood and the wound are the intradiegetic confirmations that the narrator's inference is true.[138]

The imaginative lacuna left in the witchcraft stories – what do werewolves do when they are not in human shape? – has here been filled with the thought, 'They do what ordinary wolves do, and kill sheep'.[139] The motif of child-killing, common in mediaeval and early-modern werewolf stories, is completely absent. One possible explanation is that eating sheep- rather than human flesh supplies a legitimation for the reversion to human shape: in the Arcadian tales, only the wolves who have eaten no human flesh turn back into humans. If so, this would make Petronius' reliance on Greek stories still more likely.[140] It is at any rate less closely linked to the discourse about magic than the story of Thelyphron's nose in Apuleius' *Metamorphoses* (2, 21–30), where a witch in the shape of a weasel enters a sealed room and causes the narrator to fall into a deep sleep, in the course of which his nose and ears are nibbled off and substituted by wax replacements.[141]

Conclusion

So far as I know, there are no other were-wolf stories that are to be dated later than Petronius (granted that Pausanias is evidently citing accounts that Pliny had already picked up Hellenistic Greek sources).[142] I would thus conclude that the attempt to incorporate the motif into the negative discourse of magic failed; it was felt more appropriate to leave it where the 'foolish' stories left it, in the relatively neutral realm of the marvellous. That is not necessarily to say that no tales of wolf-men circulated at the 'folk'-level in the Graeco-Roman world; I think it likely that they did. But we can only speak of the world of stories that individual writers chose to circulate within the literate class, granted that the great majority of that body of written narrative is lost forever.[143] This in turn, I would say, is to be linked mainly to two factors: the relative indeterminacy of the wolf's value in the Hellenistic and Roman Bestiary, including the *Aesopica*, where the animal simply ceases to be terrifying; and the fact that the motif of transformation into a wolf, itself fuelled by etymological inference, was so strongly linked to a single Greek cult, that of the Zeus worshipped on Mt. Lykaion in Arcadia.

There are two further considerations to bear in mind. First, the Hellenistic creation of a genre of metamorphic stories, against the background of the great continuum of living things, meant that there was far too much accessible competition within the corpus of myth. Lykaon, though a truly local story, could hardly get his head above the parapet. Second, the major resource for the construction of a negative stereotype of the witch was the figure of the ῥιζοτόμος, the root-collector, male and female, who was a master of herb-lore and could work both for good and ill. The stereotypes of this figure that were created in Latin poetry at the very end of the Republic and the early Principate do incorporate 'night-witch' elements, but these focus on necromancy, precisely because 'illicit' divination was the major concern of the prominent cases that came to trial, and necromancy could easily be turned into the most lurid form of divination. Were-wolves were of no interest in this connection. There was therefore no motive within the literate class to elaborate that particular fantasy, though we can see from Lucan's Erictho, the nightmarish cannibalistic witch in the *Bellum civile*, what Roman writers could do with the raw materials they did value, once they put their minds to it.[144] In other words, Vergil's Moeris proved a dead end.

Let me finish by citing the little joke with which Carl August Böttiger ended his own long contribution to this discussion two centuries ago:

> It may well be high time for the wolf in the fable appearing to me too, and, what was once the inevitable consequence of its unexpected appearance, to order me to become quiet.[145]

Notes

1. George Lakoff and Mark Johnson, *Metaphors We Live by* (Chicago, 1980[1], 2003[2] and often reprinted). They distinguish between 'structural' and 'orientational' metaphors.
2. Otto Keller, *Die antike Tierwelt*. 3 vols. (Leipzig, 1909–1920, repr. Hildesheim, 1963; 1980). This was a much expanded version of an earlier work confined, except for the dolphin and four species of birds, to mammals: *Tiere des klassischen Altertums* (Innsbruck, 1887, repr. Hildesheim 2001). Later research has added precision and many additional details to Keller's information, esp. in the area of fishes and invertebrates; see e.g. D'Arcy Wentworth Thompson, *A Glossary of Greek Fishes* (London, 1947); Ian C. Beavis, *Insects and Invertebrates in Classical Antiquity* (Exeter, 1988).
3. This is a central affirmation of the so-called 'Paris School' now subsumed into the Centre Gustave Glotz; see e.g. Marcel Detienne, 'Entre bêtes et dieux', *Nouvelle Revue de Psychanalyse* 6 (1972), 231–246, republished under the title 'Ronger la tête de ses parents' in his *Dionysos mis à mort* (Paris, 1977), 135–160; tr. as 'Between Beasts and Gods', in: Richard L. Gordon (ed.), *Myth, Religion and Society: Structuralist Essays by M. Detienne, L. Gernet, J.-P. Vernant & P. Vidal-Naquet* (Cambridge, 1981), 215–228 (see also n. 7 below). Much less is known about Italic or Roman attitudes towards animals, since there has been no research comparable to that on Classical Greece, and I largely ignore Italo-Roman attitudes here. But, as we shall see, some of the few true ancient werewolf stories exist only in Latin.
4. G.E.R. Lloyd, *Science, Folklore and Ideology: Studies in the Life Sciences in Ancient Greece* (Cambridge, 1983), 7–12.
5. Claude Lévi-Strauss, *La pensée sauvage* (Paris, 1962, edited and annotated in the Pléiade ed. [Paris, 2008] by Frédéric Keck), 553–872. On the acrimonious debate surrounding the English tr. by Sybil Wolfram, see Keck, 'Le totémisme aujourd'hui/ La pensée sauvage: Notice', 1774–1810 (1799–1801).
6. *Le totémisme aujourdhui* (Paris, 1962); the citation is on p.533 of the Pléiade ed. Classic exemplifications of the point, combined with criticism of Lévi-Strauss on certain key issues, are S.J. Tambiah, 'Animals are Good to Think and Good to Prohibit', *Ethnology* 8.4 (1969), 424–459, partly repr. in: Mary Douglas (ed.), *Rules and Meanings* (Harmondsworth, 1973), 127–166, and Dan Sperber, 'Pourquoi les animaux parfaits, les hybrides et les monstres sont-ils bons à penser symboliquement?', *L'Homme* 15.2 (1975), 5–24. Mary Douglas, 'Animals in Lele Religious Symbolism', *Africa* 27 (1957), 46–51, repr. in her *Implicit Meanings* (London, 1975), 27–46, may be said to have anticipated Lévi-Strauss' basic idea.

7. For example, Jean-Pierre Vernant, 'Entre bêtes et dieux', in his *Mythe et société en Grèce ancienne* (Paris, 1982²), 141–176; Marcel Detienne and Jean-Pierre Vernant, *Les ruses de l'intelligence: La métis des Grecs* (Paris, 1974); Pierre Vidal-Naquet, 'Bêtes, hommes et dieux chez les Grecs', in: Léon Poliakov (ed.), *Hommes et bêtes. Entretiens sur le racisme* (Paris, 1975), 129–142; Nicole Loraux, 'Sur la race des femmes et quelques-unes de ses tribus', *Arethusa* 11 (1978), 43–87, repr. in her *Les enfants d'Athéna: idées athéniennes sur la citoyenneté et la division des sexes* (Paris, 1981), 75–117.
8. This is true not only of Otto Keller (see n. 2) but also of the numerous entries devoted to animal species in the standard encyclopedia of Altertumswissenschaft, *Paulys Realencylopädie der classischen Altertumswissenschaft* (Stuttgart etc., 1894–1980, 34 double volumes plus 15 Supplements; hereafter *RE*). The very first lemma of vol. 1 (1894) cols. 1–4 is 'Aal', by Eugen Oder, who edited with Karl Hoppe the complex of ancient texts devoted to horse-leeching. Among other snippets, we learn of the eels in the shrine of Zeus at Labraunda in Caria, who wore golden necklaces and 'earrings' (Aelian, *Nat. anim.* 12.30); and of Krisamis or Kissamis of Cos, who killed an 'eel' (i.e. a giant freshwater catfish, *Silurus glanis*) that was taking his sheep, left it unburied despite a dream in which the fish appeared, warning him not to do so, and who subsequently perished with all his family (Suda s.v. Κρίσαμις; Zenobios IV 64 in: E.L von Leutsch & F.G. Schneidewin [eds], *Corpus Parœmiographorum Graecorum* 1 [Göttingen, 1839], 102).
9. See n. 6 above; note also Roy [Geoffrey] Willis, *Man and Beast* (London, 1974); Willis later extended his scope to include archaeology: Roy Willis (ed.), *Signifying Animals: Human Meanings in the Natural World* (London, 1990).
10. On Aristotle's classifications, see e.g. G.E.R. Lloyd, 'The development of Aristotle's theory of the classification of animals', *Phronesis* 6 (1961), 59–80; Arnaud Zucker, *Aristote et les classifications zoologiques* (Louvain-la-Neuve, 2005).
11. Cf. Richard L. Gordon, 'Magian Lessons in Natural History: Unique Animals in Graeco-Roman Natural Magic', in: Jitse Dijkstra, Justin Kroesen and Yme Kuiper (eds), *Myths, Martyrs and Modernity: Studies in the History of Religions in honour of Jan N. Bremmer*. Numen Suppl. 127 (Leyden 2010) 249–269.
12. For the conceptual scheme, see e.g., Richard Buxton, 'Imaginary Greek Mountains', *Journal of Hellenic Studies* 112 (1992), 1–15; idem, *Imaginary Greece: The Contexts of Mythology* (Cambridge, 1994), 81–96. Pan: Philippe Borgeaud, *Recherches sur le dieu Pan*. Bibliotheca Helvetica Romana 17 (Rome); Nymphs: Jennifer Larson, *Greek Nymphs: Myth, Cult, Lore* (Oxford, 2001). A form of madness was termed in Greek νυμφολήψις, *nympholêpsis*, lit. 'nymph-struck', cf. Corinne Ondine Pache, *A Moment's Ornament: The Poetics of Nympholepsy in Ancient Greece* (Oxford, 2011).
13. Alexandre Marcinkowski, 'Le loup et les Grecs', *Ancient Society* 31 (2001), 1–26.
14. 'Men and dogs are the only animals that belch when they have eaten well' (Aelian, *Nat. anim.* 4.20).
15. Cf. Liliane Bodson, 'Place et fonction du chien dans le monde antique', *Ethnozootechnie* 25 (1980), 13–21; on ancient dogs as pets, see her 'Motivations for Pet-keeping in Ancient Greece and Rome: A Preliminary Survey', in: Anthony Podberscek et al. (eds), *Companion Animals and Us: Exploring the Relationships between People and Pets* (Cambridge, 2000), 27–41.
16. Both in Greek and in Latin adjectives meaning 'doggish/dog-like' (*kyneos, kyn~* ; *caninus*) were negatively marked, often connoting shamelessness; in Greek, the gender-specific term 'bitch' was a term of strong abuse of women; in Latin, *canis*

repeatedly occurs as a term marking strong disapprobation, cf. Carla Mainoldi, *L'image du loup et du chien dans la Grèce ancienne d'Homère à Platon* (Strasbourg, 1984), 143–186; F. Orth, s.v. 'Hund', *RE* 8 (1913), 2540–2582 (2568f.). The clearest evidence of the ambivalence of the dog's status, however, is provided by the Greek physiognomic tradition, cf. George Boys-Stones, 'Physiognomy and Ancient Psychological Theory', in: Simon Swain (ed.), *Seeing the Face, Seeing the Soul. Polemon's Physiognomy from Classical Antiquity to Medieval Islam* (New York, 2007), 19–124.

17. Pliny's discussion of native domestic animals, for example, starts with the dog and continues with the horse, only then continuing with donkeys and then cattle (*Hist. Nat.* 8.142–166). He justifies this prioritisation with reference to the extreme fidelity to man of these species. Some of his stories are illuminating: one tells of a stallion that committed suicide after discovering that a mare it had covered was its own mother (ibid. 156). See also Liliane Bodson, 'Le temoignage de Pline l'Ancien sur la conception romaine de l'animal', in: Barbara Cassin and Jean-Louis Labarrière (eds), *L'animal dans l'Antiquité* (Paris, 1997), 325–354.
18. Maecenas, the counsellor of Augustus, is said to have started a fashion for eating domestic donkey foals at at high-society banquets (Pliny, *Hist. nat.* 8.170).
19. In the Hellenistic period, and during the Empire, wealthy land-owners kept many kinds of wild animals, including rabbits and hares, wild pig, different varieties of deer, hedgehogs, dormice, snails etc. in enclosures (*leporaria* in Latin), where they were fattened up for the kitchen, thus forming an anomalous class of animals neither domestic nor wild, cf. e.g., Varro, *De re rustica* 3.3.1–4. They also kept aviaries and fishponds for the same purpose.
20. R.G.A. Buxton, 'Wolves and Werewolves in Greek Thought', in: Jan N. Bremmer (ed.), *Interpretations of Greek Mythology* (London, 1987), 60–79 at 67. Stupid: e.g., *Aesopica*, Perry (ed.), nos. 153–160, 261, 274.
21. Keller, *Tiere* (see n.2), 158.
22. E.g. Homer, *Iliad* 16.35235–7 (the Danäans attack the Trojans as violently as wolves attacking helpless sheep); 16.156–166 (the Myrmidons are like wolves who have killed a stag in the mountains and tear it apart). See Mainoldi, *L'image* (see n. 16), 97–104. On the possible Indo-European roots of a connection between wolves, dogs and warriors, see Kim McCone, 'Hund, Wolf und Krieger bei den Indogermanen', in: Wolfgang Meid (ed.), *Studien zum indogermanischen Wortschatz* (Innsbruck, 1987), 101–154. Markus Egetmayer, the professor of Greek linguistics at the Sorbonne, has noted words such as *epilyk-* on Crete and Cyprus that denoted political offices: '*Epilukos/Opilukos*: un titolo greco di origine indo-europeo', in: Anna Sacconi et al. (eds.), *Colloquium Romanum: Atti del XII Colloquio internazionale di micenologia, Roma 20–25 febr. 2006 = Pasiphaë* 1 (Pisa and Rome, 2008), 251–268 (263–264). I thank Jan N. Bremmer for sending me a copy of this article.
23. Aristotle, *Hist. anim.* 9.6, 612b1–2 (bleeding); 6.18, 571b27–30 (solitary), cf. Leutsch & Schneidewin, *Corpus Paroemiographorum* (see n.8 above) II 243.10–11, ἀγορὰ λύκιος. The *Iliad*, however, at any rate the *Doloneia* in Bk. 10, generally considered a late inclusion into what was anyway a living text, also knows the stealthy wolf: the Trojan Dolon disguises himself as a wolf for his night raid (10.333–336; cf. Euripides, *Rhesus* 208–215). It is this aspect, of deviousness and stealth, that is sometimes picked up in Greek tragedy., e.g., Aeschylus, *Agamemnon* 1258–1260 (where Clytemnestra is a 'two-footed lioness', and Aegisthus, her lover, a wolf). The asociality of wolves is a main point of M. Detienne and J. Svenbro, 'Les loups

au festin ou la cité impossible', in: Marcel Detienne and Jean-Pierre Vernant, *La cuisine du sacrifice* (Paris, 1979), 215–233.

24. See e.g., Robert Maltby, *A Lexicon of Ancient Latin Etymologies* (Leeds, 1990); Amdreas N. Michalopulos, *Ancient Etymologies in Ovid's Metamorphoses: A Commented Lexicon* (Leeds, 2001).
25. The statue was sufficiently well known in the fifth century BC for the Spartans to have an inscription engraved on it: Plutarch, *Life of Pericles* 21.3. Pausanias, *Perieg.* 10.14.7, who also notes its existence, infers that the wolf was worshipped at Delphi. There was a cult of Artemis *Lykeia* at Troezene, which Pausanias likewise connects to wolves (2.31.4).
26. Markus Egetmayer, 'Lumière sur les loups d'Apollon', in: *Les sanctuaires syro-anatoliens de la préhistorie au début de l'ère chrétienne = Res Antiquae* [Brussels] 4 (2007) 205–219, has however attempted to show a connection between wolves and the stem λυκ- in relation to the name Lycia.
27. Cf. Mainoldi, *L'image* (see n. 16), 22–28. Pierre Bonnechere, *Le sacrifice humain en Grèce ancienne*. Kernos Suppl. 3 (Liège, 1994), 89 briefly discusses the modern etymologies.
28. Aelian, *Nat. anim.* 10.26; cf. 12.40; Pausanias, *Perieg.* 10.14.7.
29. Pausanias, *Perieg.* 2.9.7. Apollo also acquired the epithet λυκοκτόνος, 'Wolf-slayer'.
30. Servius, ad Vergil., *Aen.* 4.377. On the Telchines and their analogues, the Idaean Dactyls, see now Sandra Blakely, *Myth, Ritual and Metallurgy in Ancient Greece and Recent Africa* (Cambridge, 2006).
31. Aelian, *Nat. anim.* 17.31 (Armenia); 6.65 (Sea of Azov).
32. Aelian, *Nat. anim.* 4.4. The story is attested as early as Aristotle's *Hist. anim.* 6.35, 580a14–19.
33. Aelian, *Nat. anim.* 1.31; 5.50 (ii); 11.37 (this list includes the dog); 17.31 (in Armenia, wolves are like lions and leopards). This animal-lore accumulated *pari passu* with plant-lore – Theophrastus, in the later fourth century, already knew of a considerable number of Asiatic plants.
34. Ankylosis: Buxton, 'Wolves' (see n. 20), 62. Wolves cannot turn their neck: Aelian, *Nat. anim.* 10.26. Sharp sight, even at night, without a moon: ibid. (the Greek word λυκόφως, *lykophôs*, lit. 'wolf-light', meant twilight, gloaming). The idea that one could be bewitched if a wolf saw one first was a favourite literary topos, e.g. Pliny, *Hist. nat.* 8.80; Solinus, *Mirab.* 2.35; *Geoponica* 15.1.8, citing Plato, *Rep.* 1, 336d; Servius, ad Vergil., *Eclog.* 9.54; cf. W. Richter, s.v. Wolf, *RE Suppl.* 15 (1978) 960–987 at 971.
35. Aelian, *Nat. anim.* 7.1. 'Very fierce', ibid. 7.20.
36. Plato, *Phaedr.* 241d.
37. E.g. Xenophon, *Cyrop.* 8.3.41, λελυκωμένα πρόβατα, 'sheep mangled by wolves'. There was a kind of wolf called ἅρπαξ, 'Snatcher': Oppian, *Cyneg.* 3.304, cf. Lycophron, *Alex.* 1293 and 1309.
38. Resp. Aelian, *Nat. anim.* 9.18 (λυκοκτόνος); 1.36 (squill); 3.32 (Crete; Olympus). If taken in excess, the medicinal squill was believed to cause apparent or temporary death (Pliny, *Hist. nat.* 20.98). It was also used to break aggressive magical attack (ibid. 20.101).
39. Resp. Aelian, *Nat. anim.* 3.6; 8.14; 7.8; 13.1.
40. Aelian, *Nat. anim.* 6.65 (fishermen); 11.6 (Pan) – this story is unique to Aelian.
41. In general, see e.g. Jerry Stannard, 'Medicinal Plants and Folk Remedies in Pliny, *Historia Naturalis*', *History and Philosophy of the Life Sciences* 4 (1982), 3–23;

John Scarborough, 'The Pharmacology of Sacred Plants, Herbs and Roots', in: Christopher A. Faraone and Dirk Obbink (eds.), Magika Hiera: *Ancient Greek Magic and Religion* (New York, 1991), 138–174. Animal parts: Luis Gil, Therapeia. *La medicina popular en el mundo clásico* (Madrid, 1969), 188–194.

42. Cf. Richard L. Gordon, 'From Substances to Texts: Three Materialities of Magic in the Roman Period', in: Jan N. Bremmer (ed.), *The Materiality of Magic*, Acta of the International Conference held the Center for Advanced Studies, University of Cologne, May 2012 (Tübingen, 2015).
43. Pliny, *Hist. nat.* 8.83; ruse: Horapollo, *Hierogl.* 2.73.
44. Marcellus wrote a verse compendium of medical knowledge in 42 books, practically all of it lost except a prose synopsis of his account of lycanthropy (Suda, μ 205). Different versions of this are preserved in three late-antique/Byzantine medical textbooks, at greatest length by Aëtius, *Med.* 6.11 Olivieri, who links the condition both with wolves and dogs; this detail has disappeared from Oribasius, *Synops.* 8.9 Raeder, and Paul of Aegina, 3.16 (who depends directly on Oribasius). On the diagnosis, see Nadine Metzger, *Wolfsmenschen und nächtliche Heimsuchungen. Zur kulturhistorischen Verortung vormoderner Konzepte von Lykanthropie und Ephialtes* (Remscheid, 2011), 150–164.
45. These prescriptions are taken from the *Cyranides* Bk. 2, p.152 Kaimakis. Max Wellmann, *Marcellus von Side als Arzt und die Koiraniden des Hermes Trismegistos* (Leipzig, 1934) argued in favour of the idea that elements of Marcellus' poem are to be found in the *Cyranides*. Metzger, *Wolfsmenschen* (see previous n.), 248 rightly warns that this is speculative; but it seems to me more likely than not.
46. These properties ascribed to the wolf's eye are likewise taken from the *Cyranides* Bk. 2, p.152 Kaimakis. Wilhelm Hertz, *Der Werwolf: Über die Werwolfs-Verwandlung, Verwundbarkeit und Entzauberung* (Stuttgart, 1862; I have used the reprint, Leipzig 2013), 16 n.22, collected numerous analogous medico-magical uses for wolf-parts in European folk-lore.
47. Some proverbs using associations with the wolf can be found in: the standard collection, Leutsch and Schneidewin (see n. 8) I, 69, 269f., 279, 431; II 510f.
48. There is a useful compendium of such stories in: Paul M.C. Forbes Irving, *Metamorphosis in Greek Myths* (Oxford, 1990), 197–319.
49. Such stories form 'part of a complex network of belief and practice which, seen in its totality, is capable of conferring imaginative and emotional credibility upon even the "strangest" narrative details': R.G.A. Buxton, *Forms of Astonishment: Greek Myths of Metamorphosis* (Oxford, 2009), 230.
50. See e.g. Claude Calame, 'Mythe grecque et structures narratives: le mythe des Cyclopes dans l'*Odyssée*', *Živa Antika* 26 (1976), 311–328; idem, *Le récit en Grèce ancienne: énonciations et représentations de poètes* (Paris, 1986); idem (ed.), *Métamorphoses du mythe en Grèce antique*. Religions en perspective 4 (Geneva, 1988).
51. This is a repeated theme of Buxton's *Forms of Astonishment* (see n. 49).
52. I resume here some ideas first sketched in Richard L. Gordon, 'Imagining Greek and Roman Magic', in: Bengt Ankarloo and Stuart Clark (eds), *The Athlone History of Witchcraft and Magic in Europe*, 2: *Ancient Greece and Rome* (London, 1999), 159–275 (168–178).
53. E.g. Pierre Vidal-Naquet, *Le Chasseur noir* (Paris, 1981), 39–68.
54. For the writers of the paradoxographic tradition, cf. K. Ziegler, 'Paradoxographoi', *RE* 18.2c/Halbband 36.2 (1949), 1137–1166; the texts are reprinted in: Alessandro Giannini, *Paradoxographorum graecorum reliquiae* (Milan, 1965).

55. Albert Severyns, *Les dieux d'Homère* (Paris, 1966), 3–74; Jenny S. Clay, '*Demas* and *Audê*: The Nature of Divine Transformation in Homer', *Hermes* 102 (1974), 129–136; Hartmut Erbse, 'Homerische Götter in Vogelgestalt', *Hermes* 108 (1980), 259–274.
56. James Redfield, 'The Politics of Immortality', in: Philippe Borgeaud (ed.), *Orphisme et Orphée: en l'honneur de J. Rudhardt*. Recherches et Rencontres 3 (Geneva, 1991), 103–117 (at 103f.).
57. Hans Dieter Betz (ed.), *The Greek Magical Papyri in Translation* (Chicago, 1985), 201 (tr. Janet H. Johnson) = P.Leiden J383/P.Lond.10070, col. xiv 24.
58. The earliest such writers/collectors, Antigonos of Karystos, Didymarchos, Kallisthenes and Theodoros, are mere names to us, see e.g. K. Sara Myers, *Ovid's Causes: Cosmogony and Aetiology in Ovid's Metamorphoses* (Ann Arbor, 1994), 21–26; Forbes Irving, *Metamorphosis* (see n. 48), 19–21.
59. For the text, with an introduction, see Manolis Papathomopoulos, *Antoninus Liberalis, Les Métamorphoses* (Paris, 1968).
60. Nikander (second half of second cent. BC) is also known as the author of erudite didactic poems about poisonous animals (mainly snakes and scorpions), plants and minerals (*Theriaka* and *Alexipharmaka*); his transformational poems were called *Heteroioumena*, 'Shiftings' i.e. Metamporphoses. Boios (probably third century BC) is otherwise completely unknown; his poem was entitled *Ornithogonia*, 'The Coming-into-Being of Birds', and used as a source by Ovid.
61. Exceptions include Io (transformed into a cow), Niobe (turned into a stone), Hekabe (transformed into a bitch), Kadmos and Harmonia (turned into snakes). Apart from Io, none of these transformations is important to the narrative, they give merely a sense of a conventional ending. I ignore divine shape-shifting, since it is simply one among many other markers of difference between Here and There.
62. Mario Vegetti, 'La scienza ellenistica: problemi di epistemologia storica', in: Gabriele Giannantoni and Mario Vegetti (eds), *La scienza ellenistica* (Pavia, 1984), 427–470. An example of the style in mythography is the *Ἐρωτικὰ Παθήματα* by Parthenios of Nikaia (first century BC), whose gruesome 'Herippê' presupposes the invasion of Asia Minor by the Gauls in the third century BC, cf. Jane L. Lightfoot (ed.), *Parthenius of Nicaea: The Poetical Fragments and the Ἐρωτικὰ Παθήματα* (Oxford, 1999); Michèle Biraud (tr., comm.), *Passions d'amour: Parthénios de Nicée* (Grenoble, 2008).
63. Well described by Buxton, *Forms of Astonishment* (see n. 49), 112–114.
64. There are two excellent recent *mises au point* on the 'Lykaon complex': Jan N. Bremmer, 'Myth and Ritual in Greek Human Sacrifice: Lykaon, Polyxena and the Case of the Rhodian Criminal', in: idem (ed.), *The Strange World of Human Sacrifice*. Studies in the History and Anthropology of Religion, 1 (Leuven, 2007), 66–100; and Metzger, *Wolfsmenschen* (see n. 42), 180–197. The name Lykaon is found on Rhodes; for what it is worth, it has recently been claimed to derive from *Lukawanni, 'an inhabitant of Lukka', i.e., Lycia: René Lebrun, 'Syro-anatolica scripta minora, 1', *Le Muséon* 114 (2001) 245–251, cited by Bremmer, p.98 n.131.
65. Ovid, *Met.* 1.1.216–239; the Deluge is described at 262–292. The only other surviving version in verse is the five lines of Nonnus, *Dionysiaka* 18.20–24. All other versions are in prose. Many, but by no means all, of the sources were excerpted and discussed by Walter Immerwahr, *Die Kulte und Mythen Arkadiens, 1: Die arkadischen Kulte* (Leipzig, 1891) 1–24; more thoroughly by Giulia Piccaluga, *Lykaon. Un tema mitico* (Rome, 1968), 31–98.

66. Ovid's account, together with Pliny, *Hist. nat.* 8.81–82 and Petronius, *Sat.* 62, which I shall come to later, seems to have played some part in the late mediaeval Anglo-French stories about were-wolves: Willem de Blécourt, 'Wolfsmenschen', *Enzyklopädie des Märchens* 14 (2013), col. 975–986.
67. Two ranges are mentioned, Maenala, sacred to the god Pan, and Cyllene, the birthplace of Hermes. The text, however, conceals both of these associations; Maenala is simply 'bristling with the hidden lairs of wild beasts' (*latebris horrenda ferarum*, 216). The mountains are thus typological wildernesses, in no sense actual ranges, just as Lycaon's settlement is a *sedes*, a spot (218). The mountain now called Mainalon, near the modern village of Andritsina – itself within walking distance of the temple of Apollo at Bassae – is the highest in Arcadia.
68. The pious worshippers have now been forgotten.
69. Cf. Caroline Walker Bynum, 'Some Stories about Werewolves: Ovid's Lycaon', in her odd rag-bag, *Metamorphosis and Identity* (New York, 2001), 166–172.
70. See e.g., Claude Calame, *Greek Mythology: Poetics, Pragmatics and Fiction* (Cambridge, 2009) – worth persevering with, despite the very poor translation.
71. A convenient overview in tabular form of the differential occurrence of 9 motifs among the 18 known versions, 10 of which are late-antique and/or routine scholiasts, is provided by Metzger, *Wolfsmenschen* (see n. 42), 188 Table 3.
72. See Felix Jacoby, *Fragmente der griechischen Historiker* (Berlin and Leiden 1962), 90 F38. In this version, Lykaon is a pious man who claims that Zeus is present in some form or other at every sacrifice; the wicked sons decide to test this claim by mixing the viscera of a boy with animal viscera. Apollodorus is usually assigned to the second century AD. The original version of that text will thus have been roughly contemporary with Pausanias, writing in the second half of the second century.
73. Borgeaud, *Recherches* (see n.12), 44–66. Although I do not pursue the point below, Pausanias recounts a very similar three-generational narrative in 8.1.1–4.9.
74. The nature of the test is also different from Ovid's version: here it is not Lykaon who intends to test whether Zeus is really a god, but Zeus who wants to find out how impious the sons of Lykaon are. Not only is the victim a local child rather than an adult foreign hostage but the viscera (i.e., the heart, liver and lungs, which were normally roasted on spits) are now mixed with *animal* parts, i.e., there must have been ordinary animal victims. The opposition between roasting and boiling has disappeared.
75. The name *Mainalos*, for example, is taken from the range of hills named *Maenala* in Ovid (see n. 63), *Thesprotos* from an area in NW Greece. [Apollodorus'] list is however more than usually feeble; Mantineus, Kleitor, Stymphalos, Orchomenos are all place names in the Peloponnese. With six exceptions (Mainalos, Makareus, Manineus, Nyktinos, Orchomenos and Pallas), none of the 49 surviving names listed by the epitome is identical with any of the 23 names cited by Pausanias at 8.3.1–5.
76. Pausanias 8.1–5; the story of Lykaon's impious sacrifice is told very briefly at 8.2.3.
77. The Arcadians were generally regarded by other Greeks as backward and primitive, hence the derogatory term 'acorn-eaters' (βαλανηφάγοι, Herodotus, *Hist.* 1.66.2), i.e., pigs.
78. Vinciane Pirenne-Delforge in particular has emphasised the importance of Pausanias' wider pre-occupations in understanding his presentation of material; see esp. her *Retour à la source. Pausanias et la religion grecque.* Kernos Suppl.20

(Liège 2008) 67–72; 333–341, cf. already Claude Calame, 'Discours mythique et discours historique dans trois textes de Pausanias', *Degrès* (Brussels) 17 (1979), 1–30.

79. Aristaios was a culture-hero associated especially with Boeotia, Arcadia and the island of Ceos; Vergil recounts one version of a myth about him as the inventor of bee-keeping in *Georgics* 4.317–550. Britomartis was a minor deity worshipped mainly in northern Crete, whom Callimachus fused with the nymph Diktynna in his *Hymn to Artemis* (3), 180–200. She was later subsumed into the composite figure of Artemis.
80. Cf. *Rhet. ad Herennium* 2.2.3; Cicero, *De inv.* 2.5.16f.; Quintilian, *Inst.* 7.2.7f.
81. [Eratosth.], *Katast.* 1.8 Olivieri = frgs. 163, 165 M.-W. The *Katasterismoi* was a living text, constantly altered and adapted in antiquity.
82. A tragedy *Azanes* is listed among the plays of Achaios of Eretria, an important contemporary of Sophocles. Since the area of Azania lies on the border between Elis and Arcadia (Strabo, *Geogr.* 8.3.1), F. Meinecke suggested that it may have dealt with Lykaon (cf. Bruno Snell, *Tragicorum Graecorum Fragmenta*, 1 [Göttingen, 1971] 20 F2). The one surviving fragment does mention Zeus and his thunderbolt, but nothing whatever is known of the plot, so it remains a mere conjecture.
83. Bonnechere, *Sacrifice* (see n.26), 92.
84. A *Lykaon* was presented by the Athenian playwright Xenokles, together with *Bacchae*, *Oedipus* and the satyr-play *Athamas* in 415 BC (Aelian, *Var. hist.* 2.8, cf. Snell, *TrGrFrag* 1 [see n.82] 33 F1); another play of the same name was staged by the younger Astydamas, a great great-nephew of Aeschylus, in c.340 BC (*TrGrFrag* 1, 60 T60).
85. The only extant example, however, is the Anonymus Paradoxographus, *De transform.* p. 222.3–5 Westermann. The text can be found ap. Metzger, *Wolfsmenschen* (see n. 44), 194.
86. Plato, *Rep.* 565de. The passage is cited by Polybius, *Hist.* 7.13.7 in connection with the tyrannical Philip V of Macedon. At a guess, the likeliest source for Plato's remark is Xenokles' tragedy *Lykaon*, presented at the Dionysia in 415 BC (see n. 84 above). This would fit with the dramatic date of the *Republic*.
87. Ap. Porphyry, *De abstin.* 2.27.3 with 2.53.3, no doubt from Theophrastus' περὶ εὐσεβεῖας (*On piety*), itself no doubt taken from Plato, *Laws* 6, 782c. The same combination of Arcadia and Carthage in [Plato], *Minos* 315c (probably early third century BC). These passages make no mention of metamorphosis, since in the argumentative context it is irrelevant.
88. Bonnechere, *Sacrifice humain* (see n. 27), 86.
89. Pliny, *Hist. nat.* 8.82. The same narrative is quoted from Varro by Augustine, *Civ. Dei* 18.17, so we know that he is Pliny's immediate source, wherever Varro himself got it from. As for the name, the oldest mss. of Pliny read 'Euagropas', which C. Müller believed was a corruption of 'Euanoridas' of Elis, a former athlete who later compiled lists of Olympic victors and is cited by Pausanias at 6.8.1; the name appears in Pliny's list of authors for Bk 8 as 'Euanthe Agrippa' (Immerwahr, *Kulte* [see n. 65], 13f.). Another conjecture is 'Apollas'. Both Bremmer, 'Lycaon' (see n. 64), 84 and Metzger, *Wolfsmenschen* (see n.44), 205 take Jacoby's conjecture 'Skopas' for a fact (Jacoby, *FGrH* [see n. 72] 413 F1).
90. Paus. 6.8.2, without mention of a specific author. The confusion of names suggests that this was a free-floating story that could be attached to any Arcadian athlete.

91. Pliny, *Hist.nat.* 8.81; Augustine, *Civ. Dei* 18.17. Since this author is otherwise unknown, alternative suggestions abound. One is that it is an error for Neanthes (cf. Bremmer, 'Lycaon' [see n. 64), 85).
92. Pliny's text says simply *quo in tempore si homine se abstinuerit*, 'if he has in that time kept away from mankind'; Augustine's version suggests that this is a compressed way of saying 'if he has kept away from humans *and from human flesh*'; cf. Metzger, *Wolfsmenschen* (see n. 44), 206 n.37.
93. The last grand defence of such a view, albeit already blemished by the 'telluric imperative', is no doubt Alexander von Humboldt's *Kosmos. Entwurf einer physischen Weltbeschreibung* (Berlin, 1845; ed. by Otmar Ette and Oliver Lubrich, Frankfurt a.M. 2004).
94. Hdt., *Hist.* 4.105. The discussion by Metzger, *Wolfsmenschen* (see n. 44), 217–224 seems to me largely to miss the point here.
95. François Hartog, *Le miroir d'Hérodote. Essai sur la représentation de l'autre* (Paris, 1980), 23–127. Hartog's book, which argued a case that had then been perfectly orthodox in Paris for almost twenty years, attracted a good deal of misinformed criticism. I advised Cambridge University Press against publishing a translation precisely because of this Parisian cast, but they went ahead anyway.
96. Cf. Forbes Irving, *Metamorphosis* (see n.48), 55–56.
97. Useful accounts are to be found in Madeleine Jost, *Sanctuaires et cultes d'Arcadie*. Études péloponnésiens 9 (Paris, 1985), 263–267; and Metzger, *Wolfsmenschen* (see n. 44), 182–186.
98. Most scholars link this secret rite to other beliefs about the *temenos*, for example that any human who entered it would infallibly die within a year, and that living things, including animals, who entered it lost their shadows (8.38.6). On the importance not so much of 'secrets' as of secrecy viewed as a discursive framing of claims to power, for which Candomblé provides an excellent case-study, see Paul Christopher Johnson, *Secrets, Gossip, Gods: The Transformation of Brazilian Candomblé* (Oxford, 2002).
99. E.g. Jacob Bryant, *Von den Menschenopfern der Alten* (Göttingen, 1774) (an extract of 36pp. from his unreadable *Observations and Inquiries relating to various parts of Ancient History* [Cambridge, 1767]).
100. Böttiger, 'Älteste Spuren der Wolfswuth in der griechischen Mythologie', repr. in his *Kleine Schriften archäologischen und antiquarischen Inhalts, I* (Dresden and Leipzig 1837), 135–158 (150).
101. Martin Persson Nilsson, the dominant authority on Greek religion in the first half of the twentieth century, whose intellectual formation was complete by 1900, still maintained that human sacrifice was the 'core' of the Lykaon saga: *Geschichte der griechischen Religion*, 1 (Munich, 1955^2), 397–401.
102. 'Die heiligen Tiere waren der Gottheit liebste Diener; darum glaubte der Mensch, sich die Gottheit besonders günstig zu stimmen, wenn er ihr in Gestalt ihre Lieblinge entgegenträte': Hertz, *Der Werwolf* (see n. 46), 14. He treats the Lykaon myth as 'der älteste Werwolfsage' (p. 33).
103. The two words *fétiche* and *totem* both entered scientific vocabulary in the 1760s, though the first, as is well-known, actually derives from the Portuguese *fetiço*, a charm. Even Classical philologists rapidly became aware of such ideas: Böttiger, for example, calls Pan the fetish of Arcadian goat-herds: 'Wolfswuth' (see n. 100), 146.
104. James G. Frazer, *Pausanias' Description of Greece, Books 6–8* (London, 1913), 190.

105. E.g. WilliamRobertson Smith, *Lectures on the Religion of the Semites* (Edinburgh, 1889), 190; Lewis Richard Farnell, *Cults of the Greek States* (Oxford 1896–1909), 1: 41–42; Jean Przyluski, 'Les confréries de loups-garous', *Revue de l'Histoire des Religions* 121/122 (1940), 128–145; Richter, 'Wolf' (see n. 34), 973. As so often here, such ideas easily passed over into later theories, such as initiation; see e.g., Louis Gernet, 'Dolon le loup', in his *Anthropologie de la Grèce antique* (Paris, 1976 [orig. 1936), 154–71 (157f.); Liliane Bodson, Ἱερὰ ζῷα *Contribution à l'étude de la place de l'animal dans la religion grecque ancienne* (Brussels, 1978), 129.
106. Henri Jeanmaire, *Couroi et Courètes. Essai sur l'éducation spartiate et sur l'adolescence dans l'antiquité* (Lille, 1939), 560.
107. K. Kontopoulos had already reconnoitred the site in 1897. Kourouniotês' most important results were published as 'Ἀνασκαφαί Λυκαίου, in: *Αρχαιολογική Εφημερίς* [*Archaiologikê Ephemeris*] 1904, 153–214 with pl. 7–10. When he started the work, he was only 32.
108. Arthur Bernard Cook, *Zeus: A Study in Ancient Religion,* I (Cambridge, 1914) 63–89 (70–81).
109. E.g. L. Gernet, 'Dolon le loup' (see n.104); Jeanmaire, *Couroi et Courètes* (see n.105).
110. An influential early analysis of the *krypteia* as an initatory period was provided by Henri Jeanmaire, 'La cryptie lacédaimonienne', *Revue des Études Grecques* 26 (1913), 121–150. A brief account of the received views in: Anton Powell, 'Laughter in Spartan Society', in: idem (ed.), *Classical Sparta: Techniques behind her Success* (London and Norman, 1989), 5–19 (12–13).
111. Walter Burkert, for example, has this to say: 'Der Mythos spricht von einem "Knaben," der geopfert wurde, also einem Vertreter der Alterstufe, über die die Epheben sich gerade erheben: der Knabe muß sterben, wenn die Epheben in den Kreis der Männer treten, doch dem Einschluß folgt vorerst der Ausschluß. Das Leben als "Wölfe" in der Wildnis, das demnach die etwa 16–25 jährigen traf, ist dann ein Analogon der spartanischen Krypteia ... ': *Homo necans. Interpretationen altgriechischer Opferriten und Mythen*[2] (Berlin, 1997, ed.[1], 1972), 105. I for one see no great difference between this manner of argument and that of, say, Carl August Böttiger.
112. Recent supporters include Buxton, 'Wolf' (see n. 20), 70–72; Dennis D. Hughes, *Human Sacrifice in Ancient Greece* (London and New York, 1991), 103f.; Bonnechere, *Sacrifice* (see n. 27), 89–96; more cautiously, Bremmer, 'Lycaon' (see n. 64), 86–95. Forbes Irving, *Metamorphosis* (see n. 48), 55–56; Metzger, Wolfsmenschen (see n.44), 207–210 are two important recent voices against, with whom I fully agree.
113. Karl Meuli, 'Griechische Opferbräuche', in: Olaf Gigon e.a. (eds), *Phyllobolia; für Peter von der Mühll* (Basel, 1946) 185–188.
114. Piccaluga, *Lykaon* (see n.65), 99–146. She was of course extremely interested in Pausanias' report of the rain-making ritual performed by a priest at the altar of Zeus Lykaios, which involved an oak-branch being dipped in the water, which caused a vapour to rise, which then turned into a rain-cloud (8.38.4).
115. Detienne and Svenbro, 'Les loups au festin' (see n. 23); Marcel Detienne, 'At Lycaon's Table', in: idem, *The Writing of Orpheus* (Baltimore, 2002) 115–122 [= L'écriture d'Orphée (Paris 1989)]; Borgeaud, *Le dieu Pan* (see n. 12), 45–48.
116. The sceptics include Albert Henrichs, 'Human Sacrifice in Greek Religion: Three Case Studies', in: Jean Rudhardt and Olivier Reverdin (eds), *Le sacrifice dans l'antiquité*. Entretiens Fondation Hardt 27 (Vandœuvres, 1974), 195–235; Hughes, *Human Sacrifice* (see n.111); Bonnechere, *Sacrifice* (see n. 27).

117. In her thèse d'État (*Sanctuaires* [see n. 97], 258–267), Jost followed what had historically been the consensus, but was ultimately undecided about the best explanation; the attacks upon the reality of human sacrifice in Greece, particularly Bonnechere, seem to have pushed her towards a more explicit defence: see 'À propos des sacrifices humaines dans le sanctuaire de Zeus du mont Lycée', in: Robin Hägg (ed.), *Peloponnesian Sanctuaries and Cults*. Proceedings of the Ninth International Symposium held at the Swedish Instiute at Athens, 11–13 June 1994. Acta Atheniensia in-4° (Stockholm, 2002) 183–186; Madeleine Jost, 'Deux mythes de métamorphose en animal: Lykaon et Callisto', *Kernos* 18 (2005) 347–370.
118. Romano had conducted an earlier survey in 1996: David Gilman Romano, 'Topographical and Architectural Survey of the Sanctuary of Zeus on Mt. Lykaion, Arcadia', *American Journal of Archaeology* 101 (1997) 374.
119. See provisionally David Gilman Romano and Mary E. Voyatzis, 'Excavating at the Birthplace of Zeus', *Expedition* 52 (2010) 9–21 and online under Mt. Lykaion Excavation and Survey Project.
120. The preliminary report does not mention the possible effects of Kourouniotes' activities on the visible stratigraphy. Jost, *Sanctuaires* (see n. 97) already suspected a Mycenaean origin. The complete revaluation of Kourouniotês' dating, which Bremmer (see n. 64) could not have known, requires considerable amendment to his views. Purely coincidentally, the *Marmor Parium*, a Hellenistic chronography, dates the introduction of the Lykaia to before the year 1294 BC, so as to align it with the supposed dates of the Olympic Games and the Panathenaia at Athens: Jacoby, *FGrH* (see n. 72) 239 A17. Aristotle listed it in fourth place: frg. 637 Rose.
121. It can certainly not be taken as support for the thesis of Paul Wathelet, 'Homère, Lykaon et le rituel du mont Lycée', in: Julien Ries (ed.), *Les rites d'initiation. Actes du Colloque de Liège et Louvain-la-Neuve, 20–21 nov. 1984*. Homo Religiosus 13 (Louvain-la-Neuve, 1986), 285–297, that the Homeric Doloneia implies that Homer knew of the cult of Zeus Lykaios, though not of the impious sacrifice.
122. The best discussion of the range of associations connected with γόητες, in the fifth century by no means all negative, remains Walter Burkert, 'Γόης: zum griechischen "Schamanismus"', *Rheinisches Museum* 105 (1962), 36–55, repr. in his *Kleine Schriften III: Mystica, Orphica, Pythagorica* (ed. F. Graf) (Göttingen, 2006), 173–190.
123. Pomponius Mela, *Chorogr.* 2.14f.; Solinus, *Mir.* 15.2. The Neuri are followed immediately by the Anthropophagi.
124. Gordon, 'Imagining' (see n. 52), 204–210 (on the 'Night-Witch').
125. In referring to *veneficia*, the Roman law regarding magic, insofar as there was one, made no distinction between poison and other medico-magical substances. Thus, *philtra*, herbal concoctions, were thought to affect the mind and cause people to fall in love (again), cf. Gordon, 'Imagining' (see n. 52), 196–198, 244–247.
126. An Aesopian fable, 'The Thief and the Inn-keeper' (#419 Perry), uses the motif of a thief who threatens to turn himself into a wolf if the inn-keeper does not give him his expensive outfit. It seems to bear some relationship to that of Petronius, *Satyricon* 62, which I discuss below, though one cannot say which is earlier; cf. Eckard Lefèvre, 'Petrons Spuknovellen 61, 8–64,1' in: József Herman and Hannah Rosén (eds), Petroniana. *Gedenkschrift für Hubert Petersmann*. Bibl. Klass. Altertumswiss. NF, 2. Reihe 112 (Heidelberg, 2003), 147–157. The absence

of any connection to Arcadia might suggest, however, that the basic motif of were-wolfism was very familiar.

127. *Has herbas atque haec Ponto mihi lecta uenena /ipse dedit Moeris (nascuntur pluruma Ponto); / his ego saepe lupum fieri et se condere siluis / Moerim, saepe animas imis excire sepulcris, / atque satas alio uidi traducere messis* (95–99). This Eclogue contains the first surviving occurrence of the adj. *magicus* in Latin (*magicis... sacris*, 'with [or by] magical rites', l.66).
128. XII Tables 8.8a, b (451–0 BC) = Michael H. Crawford (ed.), *Roman Statutes*. 2 vols. (London, 1996) no.8.4. The account of Cresimus' trial, to be dated either to the 190s or 150s BC, was written up by the senatorial historian L. Calpurnius Piso Frugi in the late second century BC, and summarised by Pliny, *Hist.nat.* 18.41–43.
129. Varro, *De rerum deorum* 1, frg. IV Cardauns = Augustine, *Civ. Dei* 7.35.
130. The exception is a description of a drunken bawd by Propertius, *Eleg.* 4.5.13f.: *audax cantatae leges imponere Lunae / et sua nocturno fallere terga lupo*, '(she) dares to impose demands on the Moon by incantation and at night changes her shape into that of a wolf'. The bawd as a witch is another common feature of late Republican/Augustan poetry, as is the allusion to the Moon: everyone in antiquity knew that witches could 'call down the Moon' (cf. Gordon, 'Imagining' [see n. 52], 223f.). The rationalist explanation was that this was just a way of talking about eclipses; but Anne-Marie Tupet showed on the basis of magical practice in modern Algeria and Morocco that it was probably a technique of (mainly erotic) magic, in which the full moon is reflected in a vessel filled to the brim with water, which then 'boils'; this water can later be used as an ingredient in various procedures: *La magie dans la poésie latine, 1: des origines à la fin du règne d'Auguste* (Paris, 1976), 97–100. On Ovid, *Met.* 7.270f. see n.132 below.
131. Catullus' version is mentioned by Pliny, *Hist.nat.* 28.19. There were several Latin translations of plays about Medea, for example, in the Republican period, starting with Ennius: André Archellaschi, *Medée dans le théâtre latin d'Ennius à Sénèque* (Rome, 1990).
132. I take it that Ovid's inclusion of off-cuts from a wolf that can turn itself into a man among the ingredients of Medea's potion in *Met.* 7.264–271 is a sly inversion of Vergil's – or Catullus' – Moeris.
133. At any rate, eating human flesh has completely disappeared as a motif.
134. Petronius, *Sat.* 61.5–64.1. Jean Bouquet, 'Trois histoires fantastiques', *Annales Latini Montium Avernorum* 17 (1990) 17–35 rightly links this narrative to two other literary efforts in the proto-Gothic, both second-century AD, the ghost-story in Pliny, *Epist.* 7.27.5–11, and witch-stories in Apuleius, *Metam.* 1.11–19.
135. See the excellent account by Jürgen Blänsdorf, 'Die Werwolf-Geschichte des Niceros bei Petron als Beispiel literarischer Fiktion mündlichen Erzählens', in Gregor Vogt-Spira (ed.), *Strukturen der Mündlichkeit in der römischen Literatur*. ScriptOralia 19 (Tübingen, 1990), 193–217. The name 'Niceros' can mean either 'Overcome by erotic passion' or 'Eros wins out', though neither possibility seems very relevant here.
136. Alaberto Borghini, 'Lupo mannaro: il tempo della metamorfosi (Petr., *Sat.* LXII.3)', *Aufidus* 14 (1991), 29–32 points out that twilight recalls the Greek term λυκόφως (see n.34 above). The full moon is however surely more important.
137. In the Roman world, clothes were constantly being stolen while their owner – if too poor to afford a slave to watch over them – visited the public baths. Turning them to stone is thus a neatly comical way of preventing their theft. I take it

that the urination motif is intended to recall the territory-marking behaviour of dogs – the soldier is already almost a wolf.

138. Metzger, *Wolfsmenschen* (see n. 44), 237–242 argues that *versipellis* in the Petronian passage is intended just to imply that Niceros is naïve. I find this unconvincing. As Blänsdorf shows, the success of the narrative depends upon his being faux-naïf.
139. This again links the story to the Aesopica, which contain several 'fables' about shepherds, sheep and wolves (e.g., nos. 267, 210, 366, 451, 453, 477f. Perry). Particularly interesting are those that claim that a wolf will always kill sheep even if brought up in captivity (e.g., #209).
140. The alternative is, of course, that, under the impression of were-wolf stories, the narrator takes a perfectly natural occurrence (wolf in farmyard) as a marvellous one. To my mind, this reading is deliberately placed in doubt by the mention of blood where the clothes were, and the wound in the neck.
141. The word *versipellis* is used here too, so we cannot assume, as the Petronian passage might imply, that it is the Latin word for 'were-wolf'. On this point I agree with Metzger, *Wolfsmenschen* (see n. 44), 241f. The witches of Thessaly can transform themselves into almost any creature, including flies (2.22).
142. An exception may well be *Aesopica* #419 Perry (see n. 126), though the whole point is that the thief is only bluffing.
143. It is telling, I think, that the *only* Patristic citation is by Clement of Alexandria, *Protr.* 2.36.5 – the story was simply not considered sufficiently well-known to enter the Apologetic mainstream. No less than eleven (twelve if we include Nonnus, *Dionysiaka*) of the eighteen ancient texts that narrate 'the myth' in some form are late-Roman, even post-Roman, compilations.
144. Cf. Richard L. Gordon, 'Lucan's Erictho', in: Michael Whitby et al. (eds), *Homo Viator* (Studies for J. Bramble) (Bristol, 1987), 231–241.
145. Böttiger, 'Wolfswuth' (see n. 100), 158: Es mag also hohe Zeit sein, daß der Wolf in der Fabel auch mir erscheine und, was einst die unvermeidliche Folge seiner unerwarteten Erscheinung war, mir zu verstummen gebiete.

3

Into the Wild – Old Norse Stories of Animal Men

Christa Agnes Tuczay

The fascination with animal metamorphosis is reflected in stories and tales all over the world. The connotation of the shape-shifting ability is viewed ambivalently, often 'negatively and those with such powers are often sorcerers or witches', as Aðalheiður Guðmundsdottir correctly observed.[1] For this strand of medieval stories many scholars claim Celtic origin that combines metamorphosis with an evil spell. The idea behind the tales of the animal body as curse, burden and punishment provided a perfect foil for the ancient motif in the Christian context.

The early Northern tale-tradition provides a different interpretation of metamorphosis. In the saga-literature the famous battle-frenzied warriors the berserks change into so called were-animals: humans deliberately change into animal shape and transform into a stronger predatory animal because they have to do battle and combat. Besides the functional aspect for warriors in combat, the Northern tales of the animal transformation demonstrate a very interesting aspect that can be seen as a relic of an archaic belief-system in doubles and wandering souls. The Northern stories not only express the ancient desire to roam freely in the wild but the admiration and respect for the animal body that, according to Norse stories, seems to be hidden and asleep in the human body itself and can be awakened under certain circumstances.

The Old Norse concept of the soul

As the idea of the soul leaving the body has often been taken as the background to the belief in shape-shifting,[2] at this juncture it would seem useful to discuss the usage and meaning of the concept of the soul. While there are a good many studies of a concept of the soul that can be derived from Old Norse sources, little consensus has been reached on the subject. Defining the pre-Christian concept through traditions based on Christian and Antique dichotomies of body and soul is a dubious practice from the start; hence Heinrich Beck, for example, quite rightly insists that consideration be given to the individual context of each of the 'soul words'. The Christian concept of the soul, whose derivation from the loan word *sala* was recognised only belatedly, appears to have left no traces in the saga literature. Gustav Neckel thus attempted to banish the term from research, without offering an alternative.[3] The extensive discussion of these difficulties did not produce practicable results; even Claude Lecouteux's study on ghosts classifies the three central terms *hugr, hamr* and *fylgja* under the concept of the soul.[4] Compounding the issue is the fact that the words for the soul are not reflected in the texts directly, probably because they were considered 'obvious', and the tendency for different aspects of the terms to be accentuated depending on the narrative context. Nevertheless, we must consider that the creation of fixed definitions for these terms probably does little justice to actual heathen practice. Related to the terms above is *hamingja,* usually rendered as tutelary or guardian spirit, an entity that accompanies a person and that can go over to a relation after that person's death. A secondary meaning of *hamingja* is the strength of happiness which is bound up with a person's life and which is most significant in battle. In its visible form it is usually considered female and is also identified with the accompanying spirits or *fylgjur,* an interpretation which has, however, been rejected by Herbert Reier. The *fylgja* in animal form can also be as a tutelary when it appears as a woman; hence it would seem a certain contamination has taken place through literary usage. A connection usually exists between a person's character and the *fylgja* animal, which often appears in prophetic dreams. However, this is not limited to sleep, as in some exceptions the *fylgjur* can also be seen when awake – perhaps in a state of trance. A sign of being in a trance is also yawning, which signals the presence of *fylgjur.*[5]

It is rather difficult to provide a clear-cut answer to the question whether here 'reflections of archaic concepts of the soul are actually tangible: i.e., that the reader can directly participate in concrete forms of heathen belief not yet subject to the Christian perspective – or whether the effect of the *fylgja* and other motifs was not rather very much intended by the saga authors'.[6] The majority of scholars – and, given the material available for comparison, I include myself here – assume that the belief in *fylgjur* was authentic.[7] In his study on ghosts, Lecouteux postulates that this belief was not only inextricably bound up with the mentality of medieval man, but also that it is not just a literary or fictional motif.[8] Even if Böldl objects that 'historians of mentality' deal with literary and 'unformed' sources in opposition to one another, he regards the motif's frequency as evidence of its significance for the creators and recipients of the sagas, and further argues that its literary treatment need not point to the belief's 'retirement'.[9] However, he also stresses that there is no evidence of such a belief in earlier times, as the literary use subjects the motifs to a fundamental transformation. The soul leaves the body in animal form while the person sleeps, or the person wears the animal skin and in this way becomes the animal. The latter suggests the metamorphosis of the entire form, not merely the soul.[10] In his inaugural lecture as vice-chancellor at Kiel, Hugo Gering identified a connection between metamorphosis and the *fylgja*, the tutelary spirit: 'the entire concept is related to the belief in what is called the fylgja, the tutelary spirit watching over the human, his second ego that separated itself from him shortly before his death and became visible; both ideas feed into one another'.[11]

Eugen Mogk defines the berserker as the man dressed as a bear, the man in the bear skin, and regards the concept as a variant of the werewolf idea – a calamitous definition that continues to bring a great deal of confusion to the werewolf debate to this day.[12] 'Berserkers are in essence nothing more than werewolf myths'.[13] In a later article in Hoops *Reallexikon der germanischen Altertumskunde*, he attempts to unite two possibilities of metamorphosis in a single theory: 'As the name suggests, they are rooted in the belief that in ecstasy a human can send out his soul and take on the form of a bear or wolf'.[14] Mogk saw both the werewolf and the berserker as figures from folklore and mythology, with no connection to reality. His relation of the word *vargúlfr*, meaning 'dishonorable criminal', to the werewolf also caused much confusion and led to misinterpretations that continue to be repeated in many popular publications. Old Norse has no term specifically applicable to the werewolf; rather, it points to a neologism coined by the translator of Marie de France's lay. Like the verb *hamskiptask*, *vargúlftr* is a translation from *Bisclavret*. A variation on the belief in *ham* is the *hamför*, which also goes hand in hand with shape-shifting.

> König Harald gebot einem kundigen Mann im hamför, nach Island zu fahren und auszukundschaften, was er ihm sagen könne. Der schwamm in Gestalt eines Wales. Als er aber an Land kam, schwamm er ums Nordland herum nach Westen. Er sah, dass alle Felsen und Höhlen voller Landwichte waren.[15]

The English translation avoids mentioning *hamför* but names the man 'warlock'.

> King Harald bade a warlock to journey to Iceland and find out what he could tell him. He went in a whale's shape. And when he came to Iceland he proceeded west and north and around it. He saw that all mountains and hill were full of land-wights, some big and some small.[16]

He swims on and encounters dragons, snakes and a large bird, and later a large bull, and thus he cannot return to land. He reports all this to the king.

That these abilities are primarily to be attributed to people with heathen beliefs is suggested by the story of Thrand in the *Eyrbyggja saga* (ca. 1350). He is not described as being 'of one shape', but loses his ability after his conversion to Christianity. 'Thrand was extremely strong and exceedingly swift of foot. [...] they lost their troll nature when they were baptised'. Although he loses his ability for transformation some of his animal attributes remain, he is still sniffing like a dog or wolf.[17] Reier emphasises that the belief in strength specifically from conquest is expressed in the use of the terms *hamr* or *hamrammir.*[18]

The story of Odd Arngeirsson tells how one can leap from one shape to another. Odd's father and brother become the victims of a polar bear:

> Odd killed it (the bear) and brought it home, and the story goes that he ate the whole bear. He said he avenged his father by killing it and his brother by eating it. After that Odd turned out an evil man, very hard to deal with. He was such a great shape-charger that he set out one evening from Hraunhaven, and arrived in Thjorsriderdale the following morning to help his sister when the men of Thjorsriverdale wanted to stone her to death.[19]

The act of eating the bear transfers its strength to Odd, and this transfer is expressed not in a physical shift, but in the ability to cover incredibly large distances in an improbably short space of time. He willingly and actively penetrates the boundaries of time and space rather than being led (away) against his will as in the tales of rapture (see below). It is the potency of the ursine meal that enables him to achieve such high speeds.[20] Here the saga poet combines the earlier idea of transfer of strength with the ability to slip into a trance: it is well documented that Native American messengers could cover hundreds of kilometres in an astoundingly short space of time when they marched in the 'wolf trot'. One aspect that is in need of clarification is the connection to the idea of the tutelary mentioned above. The tutelary, the *fylgja*, appears as a large woman but also as the animal species that corresponds to a person's psyche. Conversely, a person's character corresponds to an animal: i.e., the person possesses the characteristics and customs of an animal, as Grönbech observes:

> The elements from which a soul creates the body (Old Norse: hamr, Old High German: hamo) are not recruited from outside, they are to be found within the soul and are likewise within the everyday body; whoever really appeared as a wolf, a bear, an ox, an eagle, always had the character traits of the wolf, the bear, the ox or the eagle within himself. His salvation was an inherent relation to an animal; even during the daylight hours he used its strength and its cunning, its courage, intuition and instinct within his own body.[21]

The berserker – bear or wolf-skinned?

The scald Þórbjörn Hornklofi provides the first account for Norse berserker in his *Hafnismal* where he seems to identify Berserker and *ulfheðnar* or wolf-skins. The famous stanzas of the poet are preserved by Snorri Sturluson:

> Savage berserker roaring mad,
> And champions fierce in wolf-skins clad,
> Howling like wolves; and clanking jar
> *Of many a mail-clad man of war.*[22]

In the ninth chapter of the *Vatnsdæla* saga, Harald Fairhair had berserks on his ship that he called *ulfheðnar*; they wore wolf-skins. Here berserker and *ulfheðnar* are identified together with berserkerdom and animal hides. The animal hide in question was not a bear's but a wolf's skin.

With the appearance of Sveinbjörn Egilson's dictionary, which posited a certain relationship of *ber* and 'bear', the berserker's shape-shifting received a new status. While Wilhelm Golther's *Handbuch der germanischen Mythologie* bracketed the berserker and the werewolf together, it made the distinction that the werewolf actually metamorphosed, whereas the berserker remained a (battle-frenzied) man. Jon Erichsen, the editor of the first source book of 1773, considered beserker a compound formed by the adjective *berr,* in the sense of 'bare' and *serkr,* shirt, seeing a parallel with *berbrynjaðr.*[23] Although he lists the *úlfheðnar* as a synonym, he gave no thought to the possibility that following *úlfheðinn, ber* might just as well be derived from the earlier form **beri,* bear.

In the *Ynglingasaga* the Christian poet Snorri Sturluson portrays the berserker as Odin's warriors:

> Odin could make his enemies in battle blind, or deaf, or terror-struck, and their weapons so blunt that they could do no more but than a willow wand; on the other hand, his men rushed forwards without armour, were as mad as dogs or wolves, bit their shields, and were strong as bears or wild bulls, and killed people at a blow, but neither fire nor iron told upon themselves. These were called berserker.[24]

Obviously Snorri's etymology does not indicate *ber* = 'bear' but berr = 'bare'.[25] Nevertheless many scholars argued for the bear etymology accounts and the depiction on Torslunda plate points to a wolf-skin masking of the depicted warriors (Figure 3.1). The helmet fittings prove the existence of *ulfheðnar* but not of bearskins. In Snorri's view and definition of the berserker the battle frenzy = *berserksgangr* is in the centre, not an assumed bear metamorphosis.

Figure 3.1 A helmet plate patrix depicting a dancing warriors from Björnhovda, Torslunda parish, Öland, Sweden. Picture 39, *Kongl. Vitterhets Historie och Antiqvitets Akademiens Månadsblad* (1872), 90. (wikimedia commons)

Detached documents like Böðvar Bjarki's bear clan prove that bear-metamorphosis was not unknown but the considerably consistent: dog and wolf motifs led scholars to identify berserker with werewolves.[26] In Chapter 35 of the *Egil saga,* presumably written by Snorri Sturluson, Skallagrim Kveldulfsson sets out to visit King Harald's court and chooses his companions carefully. He takes 12 of the strongest men, many of whom were capable of shape-shifting. The key passage can be found in Chapter 27, where Kveldúlf Skallagrim and his men attack the crew of King Harald's ships in a murderous rage. This famous story of Kveldulf = Eveningwolf, grandfather of Egil, narrates a family history of wolf-like creatures. Here the term *hamramr* is used synonymously with *berserksgangr,* the battle fury. Kveldulf is known for his sleepiness at evening that comes 'premonitory' whenever he changes his shape.[27] The *Egil saga* depicts how this power becomes diluted over the course of generations.[28] Kveldulf was a shape-shifter, a *hamramr,* and could also be taken by fury. His son Grim possessed supernatural powers after sundown (described as *hamask*), but was not able to shift. He used his raw strength for peaceful ends, for example by carrying a giant anvil when in a trance. Egil could be taken with rage, but was only described as *reiðr* (enraged), as this was not a family of berserkers but of noble leaders – who nevertheless possessed that ability.[29]

I find Blaney's interpretation persuasive when he identifies a clear shift of consciousness.[30] In most other instances *hamramr* remains unclear, as the narrators have not made the effort to 'define' the word and have provided no details. It is merely stated that the men concerned were capable of shape-shifting. It is only in the later *Sörla saga Sterka* from the sixteenth century that we encounter real bodily metamorphosis: Tófi is gripped with fury, suddenly appears extraordinarily ugly and transforms into a dragon.[31] A different example is provided by a man named Thoris's who has an initial experience of fury when he is attacked by a certain Gallti. But as Gallti also possesses this ability, the result is a fierce battle that lasts until Gallti submits. The *Göngu Hrólfs saga* relates:

> Jolgier flew into rage, raised the sword, and using both hands struck at Hrolf's waist. [...] Hrolf grabbed him by the shoulders, held him underwater, and kept him there till he drowned. Then Hrolf went ashore, utterly worn out. Every one of Jolgeir's men thanked him for what he'd done and said what a great man he was to have defeated such a dangerous berserk.[32]

In the same saga a Russian warrior enters a state of fury. Perhaps because of their ugliness, he and his brother are not only described as giants[33] and trolls, but the narrator further develops the analogy by having them use the staff, the traditional giant's weapon of the heroic epics.

> Scarcely any weapon could bite the coat he was wearing. When he grew angry, Rondolf would burst into a frenzy and howl like a troll.[34]

The onset of rage is usually displayed by howling, hyperactivity, reddening of the face, etc. Ljótr in the *Egil saga* begins to howl eerily and bites into his shield.[35] Egil bites his opponent's throat out and in his rage he also bites such a deep dent in his horn as to render it useless. Snækollr howls and bites into the edge of his shield,[36] as do Hauki,[37] Moldi and his comrades.[38]

The berserker Grim, having been abandoned as a child, is pictured as a feral child as wolf child: It was said he ate raw meat, drank the blood of people and animals, howled like a wolf and made people blind and deaf.[39] He could also change his shape faster than the eye could see. The analogy with the wolf is advanced further when it is asserted that berserkers bite their opponents' throats out.[40] The following story, from the *Göngu-Hrólfs saga* dating from the early fourteenth century, combines the three elements of berserkerdom – animal metamorphosis and magical heat, with its protagonist's magical traits – to great effect.

> He had a blood-brother called Grim Aegir, a powerful and thoroughly evil man. No one knew Grim's background or his family, for he'd been found on the beach at Laeso Island by the sorceress Groa. She was Thord's mother and reared Grim as her foster-son. She taught him so much about witchcraft that no one in Scandinavia could rival him, and his nature was utterly different from any other man's Some people think that Grim's mother must have been a sea ogress, for he could travel at will in both sea and fresh water: that's why he was called Grim Aegir (= Ocean). He used to eat raw meat and drink the blood of men and beasts. He would often change himself into the forms of various creatures and could do it so quickly the eye hardly saw it. His breath was so hot that even men in armour could feel it burning them. He could spew venom and fire at people, killing both them and their horses, and they were helpless against him. King Eirik put great trust in Grim and all his other men.[41]

Interestingly, the sagas contain far fewer specific references to 'genuine' bodily metamorphosis than we might expect. What they do entail is a wealth of physical change, particularly in the case of the berserkers. Where the terms *einhamr* and *hamramr* are used it is mostly impossible to determine whether it is a physical or psychic metamorphosis. In the *Book of Settlements* there are seven allusions to metamorphosis.[42] In three of them the poet does not go beyond merely mentioning it, providing no details which would indicate the kind of transformation. Only the archaic scene cited above in which Odd Arngeirsson eats the bear that killed his father and brother clearly describes physical metamorphosis.[43]

Reier considers this one of the most persistent surviving tales of ancient 'belief in magical strength', and their shape-shifting preserves all of their individual characteristics in a psychological portrayal. Their fight requires reserves of strength surpassing that which is considered normal; hence, the antagonists choose two of the strongest animals, the bear and the bull, to fight each other in a fury-like state. In 'mere' distorted human form such frenzied battles will later become *going berserk*. Judging by the chronological development of the concept suggested by the sources, animal metamorphosis for the purposes of gaining strength appears to precede fury.

Güntert was one of the first scholars to focus on the parallels between berserker stories and corresponding elements in Old Irish heroic epics. He assumed, however, that the Irish had borrowed the stories from the Norse and then embellished them with their penchant for exaggeration. For example, fury grips the hero Cú Chulainn considerably more strongly than in the Norse tradition. For purposes of illustration, let us cite the entire passage:

> Then his first distortion came upon Cú Chulainn so that he became horrible, many shaped, strange and unrecognisable. All the flesh in his body quivered. Her performed a wild feat of contortion with his body inside his skin. His feet and his shins and his knees slid so that they came behind him. His heels and his calves and his hams shifted so that they passed to the front. The muscles of his calves moved so that they came to the front of his shins so that each huge knot was the size of a soldier's balled fist. Then his face became hollow. He sucked one of his eyes into his head so that a wild crane could hardly have reached it to pluck it out from the back of his skull on to the middle of his cheek. The other eye sprang out onto his cheek. His mouth was twisted fearsomely. He drew the cheek back from the jawbone until his inner gullet was seen.[44]

In the aforementioned *Egils saga*, too, the distortion also borders on shape-shifting:[45]

> Egil was great of face, broad of forehead, with great eye-brows: the nose not long, but marvellously thick: that place with wide and long mustachios grow, the chin wonderfully broad and so all about the jaw. Thick-necked and great-shouldered beyond the measure of other men: hard-looking and grim-like whensoever he was wroth. He was of goodly growth and taller than any man else: his hair wolf-grey and close of growth, and became early bald.
>
> Now while he sat as is afore-writ, he kept a-twitching now one now another of his eye-brows down toward the cheek, and the other up toward the hair-rots. Egil was black-eyed and his eye-brows joined in the middle. Naught would he drink, though drink were borne to him but twitched his eye-brows. Now one now the other, down and up.[46]

The cooling-down of a heated berserker or ecstatic warrior can also be observed in the case of the Irish hero Cuchulainn[47] and in the Arthurian knight Kei.[48] Egil may fly into the rage of a berserker,[49] but he is not expressly seen as one.[50] Nevertheless, we cannot overlook his relationship with Odin, who has also made him a poet. This holds true for Starkaðr, with whom he shares a number of character traits.

Although these accounts represent extreme portrayals of ecstatic warriors and although these blatant paroxysms not only take on 'completely new dimensions',[51] but for that very reason can be considered fictional, some elements can be isolated as signs of a historical trance technique (which exists to this day).[52] One point that must be emphasised is that the terms *berserksgangr, eigi einhamr* and *hamask* were reserved for men, whereas *skipta hömum* or *hamleypa*, meaning 'shape walker', 'shape jumper', were a gender-specific term used for women. The use of *hamramr* is in any case limited to a few early sagas and the *Landnámabók*. In his study *Sejd*, which sadly remains untranslated, Dag Strömbäck claims that *hamramr* were reserved for family sagas,[53] but this is contradicted by the evidence in some of the heroic and adventure sagas, such as the *Göngu Hrolfrs saga*,[54] the *Sörla saga sterka*,[55] and all of the *Fornaldarsögur*. Notable evidence is, however, provided by the concentration in the earlier sagas, the *Landnámabók*, the *Heiðarvíga saga*[56] and the *Kormáks saga*[57] and in the sagas offering newer versions of older material, such as the *Harðar saga*,[58] the *Gull-Þóris saga*[59] and the *Flóamanna saga*.[60] The adjective *einhamr*, normally the character topos of a strong man who is, furthermore, not yet at ease with himself (Middle High German *ungehiure*), has a negative connotation and is also regarded as a hereditary trait.[61]

The equation of *eigi einhamr* to *berserk* in the *Eyrbyggja saga* (ca. 1350) suggests the physical fatigue that sets in after one had gone berserk, as two examples identifying *eigi einhamr* with physical shape-shifting illustrate. In *Orms þáttr Stórólfssonar*[62] (late thirteenth century) the reflexive *hamask* is mentioned in two instances: the prose passage between strophes six and seven in the twelfth-century helgi songs relate how Fránmar changed into an eagle.[63] In this context, the *Göngu-Hrolfr saga* examined above is an excellent example of the creative and playful use of the metamorphosis motif: Hrolf attacks Grim and attempts to strike him, but Grim escapes into the air in the form of a dragon and spits poison at his assailant before changing back to a man, his poison having killed nine men. Then, overcome with battle fury, Grim attacks Hrolf with such strength that he rams him into the ground. Here, the author uses the term *hamask* to describe the fury that has taken hold of Grim.[64]

The old word *hamask* is used in combination with the verb *bregða*, 'to wield, to throw', in a common term for the practice of shape-shifting, involving an abrupt, sudden movement, as in 'he threw himself into a bull shape'. In the Icelandic sagas the verb also appears occasionally, but not in the context of shape-shifting. It is only in the later *Fornaldarsagas* or in the heroic, fairytale or adventure sagas – and in the early Christian literature – that the word depicts the effect of demonic powers, be they used by friend or foe. The metonym 'to rage like a beserker', referring to human beings, is softened to the supernatural or fairy-tale 'to rage like a troll'.[65]

As discussed above, Skallagrim is taken by fury during an evening ball-game and nearly kills his own son. This saga combines all three terms, *hamramr, eigi einhamr* und *hamask*. In the first chapter we discover that Egil's grandfather, Ulf, lived in Norway and was Bjalfi's son (whose name, *Hallbera*, means 'to hide'). The sisters of Hammbjörn Hálftröll and the daughters of Ulf enn óargi (the fearless) are also related to him. All their names contain, in a different particle, variations of the word 'bear'. Hálftröll means half troll, or half giant, both of which are renowned for shape-shifting. Given his forebears are berserkers and half trolls, it is hardly surprising that Skallagrim possesses these characteristics. His appearance is described as black and ugly, a sign of a raging character. The evening wolf also dies from a fury that renders him so weak that he never recovers. The narrator remarks that this is why the enormous exertion of the fury can only be borne by young men.[66]

The raw strength of the berserker provides the saga poets with a wealth of variations; the motif is by no means related solely to belligerence, but is employed in connection with all manner of tasks. In the story of Styr, Halli, one of two berserkers living at his court seeks to marry his daughter. She is not too enamoured with the prospect, but sets each of them a difficult task. Both are successful, however, most likely owing to their raw strength falling into berserk fits:

> They both went into berserk fits and once they had worked themselves up into a frenzy they were not like human beings. They went mad like dogs and had no fear of either fire or iron.[...] The berserks went home that evening and were very tired, as is the nature of men who go berserk, that all their strength leaves them once their rage ebbs.[67]

The *Hrólfs saga kraka*, dating back to the fourteenth or fifteenth century, tells of the extraordinary Böðvar Bjarki,[68] in one of the few instances of a warrior's soul appearing not only in animal form, but in the animal form that corresponds to his character.[69] Interestingly enough, it should be emphasised that he is not known as berserker, but in opposition to them. Böðvar's father Björn had been cursed to roam as a bear by day, his son Böðvar is a shape-shifter, his ability innate:

> Hjorvard and his men saw a great bear advancing in front of King Hrolf's troop. The bear was always beside the king, and it killed more men with its paw than any five of the king's Champions did. Blows and missiles glanced off the animal, as it used its weight to crush King Hjorvard's men and their horses. Between its teeth, it tore everything within reach, causing a palpable fear to spread through the ranks of King Hjorvard army. Hjalti looked around for his companion Bodvar but did not see him. King Hrolf answered, "Bodvar will be where he serves us best, it is he who decides [...] Hjalti now ran back to the king's chamber, where he found Bodvar sitting idle. Hjalti spoke, "How long shall we wait for this most famous of men? It is a major disgrace that you are not on your feet. You should be testing the strength of your arms, which are as strong as any bear's [...] Then Bodvar stood up "But You Hjalti, by disturbing me here, have not been as helpful to the king as you think you have. It was nearly decided which side had gained the victory. You acted more out of

> ignorance than out of enmity to the king. In truth I can tell you that in many ways I can now offer the king far less support than before you woke me ... The bear was gone from King Hrolfs force.[70]

This later version of the story probably reflects a much older one, as the core element, the dialogue between Hjalti and Böðvar Bjarki, pre-dates it in the form of a poem and was, according to the *Ólafs saga helga* and the *Fóstbrœðra saga*, sung by the skald Þormóðr Kolbrúnarskáld to rouse the assembled army before the Battle of Stikastaðir in 1030.

Böðvar advises Hjalti to drink wolf's blood and Hjalti gains enormous strength. The concept of a soul that manifested itself in the form of an animal has not only been considered evidence of the saga's shamanic aspects,[71] but has also been used to establish its age. The passage cited above further develops the motif of metamorphosis. The enemy queen, possessed of magic powers, deploys a boar and revenants against the king's army. Although Böðvar, in a state of fury after Hjalti's rousing speech, piles up mountains of corpses, as his assailants are revenants, this has no effect on their numbers.[72]

Animal skins and animal bodies

Konrad Maurer's collection of 1855 and 1856 emphasises the motif of metamorphosis; in particular, that of the werewolf.[73] The words *hamramr, hamrammi, eigi einhamr* and *hamask*, which originally stood for metamorphosis, came to be used in the sagas to designate psychic as opposed to physical metamorphosis, which Maurer considered a diluted, indeed degenerate form of the motif.[74] As a further example he examines the expression 'to be gripped by fear' (*at verða at gjalti oder gjöltum*), which he took to have originally meant 'to become a wild boar'. Blaney questions the accuracy of this etymology, since *gjalti* is not related to boar but to *geilt*, meaning 'crazed', which in this context would also make more sense.[75] *Hamramr* too denoted not only psychic metamorphosis, as Maurer assumed, but also that of a physical kind. The *Landnámabók* (1275ff.) reports that Dufthak and Storolf Hængson became *hamramr* and fought one evening in the form of a bull and a bear:

> Dufthak of Dufthksholt was a freedman of those brothers; he was a man of exceeding great strength, as was also Storolf the son of Salmon, who then abode at Hvol. Between them there befell a dispute about grazing rights. A second-sighted man saw one evening nigh to nightfall, that a great bear went out from Hvol and a bull likewise out from Dufhaksholt, and they met on Storolfswold and set in anger on each other, and the bear got the best of it. In the morning it was seen that a dell was left where they had encountered each other, and the soil looked as if it had

> been turned inside out and there the place is now called Alda-grove; both of them were hurt.[76]

For Gering, *hamskiptask* means 'to change the outer shell'; this term was used just as much as *hamask*, 'to take on another shell', if not more frequently. Blaney objects that he has found only one instance of the usage of *hamskiptask*. Interestingly, he discovered it in the Old Norse translation of Marie de France's werewolf Lay *Bisclavret*.[77] Reier's considerations of the concept emphasise the 'encasing' connotation of the word *hamr*. He identifies a derivation from Gothic *ga-hamôn*, 'to dress oneself', and Anglo-Saxon *hama*, 'covering, suit' and especially from Old High German *hamo*, 'skin, veil, clothing'. Here he concurs with Alf Torp, who, like Gering before him, understands *ham* not only as covering but also as *fylgja*, tutelary, even if his argument is somewhat different:

> Hamr is essentially not the form of the human body, the human shape, but that of an animal; but not the whole body, but that which represents its external form, the skin, the fur. This only concerns us insofar as the word also expresses the animal's life force. It [the life force] can only be imagined as a bodily form where the human takes possession of the ham of a strong animal for that life force.[78]

The famous scene in the *Ynglinga saga* relates Odin's deliberate change of shape with an emphasis on the 'deformability' of skin:

> En þat bar til þess, at hann kunni þær þ´rottir, at hann skipti litum ol l´kjum á hverja lund, er hann vildi (chapter 6)
>
> [Odin had the ability and the art to change his skin and form it however he wanted.]:
>
> Óðinn skipti homum. Lá þá búkrinn sem sofinn eða daðr, en hann var þá fugl eða dýr, fisskr eða ormr ok for á einni svipstund á fjarlæg lond at sínum ørendum eða annarra manna.(chapter 7)
>
> [Odin could transform his shape: his body would lie as if dead, or asleep; but then he would be in shape of a fish, or worm, a bird or beast, and be off in a twinkling to distant lands upon his own or other people's business.] [79]

The Old Norse lexicon lists the particle *hamr* as meaning 'veil', 'skin', 'shape', derived from the root **kem*, 'to cover', 'to veil'. In modern Scandinavian languages *ham* still means 'skin', but also 'ghost'. In Old Norse *hamr* is often used in the sense of 'bird skin', for example the swan's dress of the valkyries *áltarhamir* in the introduction to *Volundarkviða*.[80] Freyja's feathered shirt in *Þrymskviða* v. 3, 5 and 9, to name just two examples. In view of this evidence, we can agree with Reier's assertion that *ham*, the animal skin, is worn because of the greater strength of the animal in question. *Hamr* also occurs in combination with the names of animals, especially with the wolf, i.e., *úlfr* and *vargr.*

In the *Volsunga saga* all elements of the berserksgang tradition, together with wolf metamorphosis, are to be found. Sigmund and Sinfjötli find two wolf-skins (*úlfahamir*) on the wall of a small hut, and when they pull them on, they become wolves:

> One time, they went again to the forest to get themselves some riches, and they found a house. Inside it were two sleeping men, with thick gold rings. A spell had been cast upon them: wolfskins hung over them in the house and only every tenth day could they shed the skins. They were the sons of kings. Sigmund and Sinfjotli put the skins on and could not get them off. And the weird power was there as before. They howled like wolves, both understanding the sounds. Now they set out into the forest, each, going his own way. They agreed then that they would risk a fight with as many as seven men, but not with more and the one being attacked by more would howl with his wolf's voice. [...]

The Young Sjinfjotli breaks this promise by killing 11 men without calling his uncle, who gets very angry about his nephew's disobedience:

> Sigmund leapt at him so fiercely that Sifjotli staggered and fell. Sigmund bit him in the windpipe. That day they were not able to come out of the wolfskins. Sigmund laid Sifjotli over his shoulder, carried him home to the hut and sat over him. He cursed the wolfskins, bidding the trolls to take them. [...]
>
> Then they went to the underground dwelling and stayed there until they were to take off the wolfskins. They took the skins and burned them in the fire, hoping that these objects would cause not further harm. Under that magic spell they had performed many feats in King Siggeir's kingdom.[81]

Before Sinfjötli's birth, Signy swaps shapes with a sorceress, hamr, translated here with 'shape'. Signy and a sorceress change shape – *skipta homum* – and Signy visits her brother Sigmund in his forest underground dwellings. As he does not recognise her he sleeps with her and fathers Sjinfotli:

> It is now told that once while Signy was sitting in her chamber, a sorceress, exceedingly skilled in the magic arts, came to her Signy said to her: 'I want the two of us to exchange shapes.' The sorceress answered: 'It shall be as you wish.' And she used her craft so they changed shapes. The sorceress now took Signy's place as Signy wished. She slept with the king that night and he did not notice that it was not Signy beside him.[82]

This tale has often been read as an initiation narrative.[83] If we examine the sequence of events more closely, a number of ruptures become apparent:[84] the young Sinfjötli may have got through some examinations, but his Uncle Sigmund considers him too young and inexperienced to be his companion. So Sigmund roams the woods with his nephew in order to 'kill people'. They attack and plunder farms. One day the story with the transformation into a wolf occurs. In the form of wolves they agree that each of them may fight by himself with up to seven men. If there are more than seven opponents, the other wolf should be called to help. They go their separate ways and a short time later the young Sinfjötli has to fight against 11 men but manages to kill them. When he is found seriously injured by Sigmund, he boasts to him of his heroic deed, whereupon Sigmund is overcome with rage and bites his throat out.[85] He takes the dead Sinfjötli to the hut where the metamorphosis began. There he observes how a weasel brings another one back from the dead with a life-giving herb, and is able to revive Sinfjötli. One important detail has to be mentioned: the modality of the metamorphosis seems to anticipate the later tales circling around the early modern werewolf concept. Transformation normally takes places with the means of a pelt or by magic ring.[86] Both modalities develop the story-tradition to a view of the animal state as curse or spell.

This unusual tale contains not only a ritual death and a resurrection, but also metamorphosis and fury as narrative elements. Another aspect has to be considered if Sinfötli's story indicates to Völsungs as wolf-clan as *úlfhéðnar* or wolf warrior clan 'that performs its deeds in the name of the wolf'. The described 'consecratory' ritual seems to bring themselves closer to and identify themselves with Odinn, as Guðmundsdóttir demands.[87] Although many studies[88] on lupine metamorphosis regard a wild life as part of a warrior's education, the difficult preparation undergone by Sinfjötli, consisting of no more than a single fight, seems somewhat excessive. If we do read his death as an initiation, its cause, the breaking of a taboo, at first glance seems too forced: its breaking too easily motivated, and the breaking of the promise too severe a punishment. Upon closer examination the parallels to professional berserker-warriors and other warrior bands become obvious: 'Die Zeit, die Sigmund und Sinfjötli "als Wölfe" im Wald verbringen, bezieht sich ursprünglich auf die Verwilderungsphase zweier Wolfskrieger, die einer Initiation vorausgeht – und zwar im speziellen Fall der Initiation in die elitäre Kriegertruppe der Berserker'.[89] The remarkable point older studies, and especially recently McCone and Speidel, have emphasised is the role of the outcast existence of the wolf-men, that has its parallel in the berserkers and adolescent warriors of the Chattes.[90] Although the hybrid character of the story assembles and crossfades the remnants of a ritual story with malediction, it nevertheless provides one of the most vivid descriptions of the protagonists' perceptions of a wild animal body, with all given advantages and without the wistfulness the early modern werewolves seem to suffer from.

Conclusion

Ultimately, three categories of animal metamorphosis can be identified in the sagas. First: The animal soul or free soul leaves the warrior Böðvar Bjarki's body[91] and appears in animal form to fight for him while he lies seemingly dead. This shows his efficiency as a warrior most spectacularly, as he actually appears as an animal. Böðvar's metamorphosis is described as *hamför*, and involves his human soul, subject to his will, leaving his body in the form of an animal and endowing him with its particular abilities or great strength. The concept of the soul denoted by *hamför* cannot be exchanged as easily as the *hamskiftning* type; rather, it is a type of 'internal organism' i.e., transported to the outside world through ritual or invocation. The soul is housed by the body and the shape-shifting takes place as the result of a conscious decision to avail oneself of the animal's abilities. The soul is sent on business or into battle. If it is injured, the body displays the same injuries.[92]

Second: Fury primarily means a form of psychic metamorphosis in which a warrior (usually a berserker) behaves like the animal in question. This also involves a willful intention, assisted by a rousing speech and other external factors that send the warrior into the state of trance that lends him super-human powers and renders him impervious to fire and weapons. And finally: The third form is genuine metamorphosis, of the kind which to this day is commonly associated with the werewolf; The Tale of Sinfjötli and his family is a good example.

Notes

1. Aðalheiður Guðmundsdóttir, 'The Werewolf in Medieval Icelandic Literature', *Journal of Englisch and Germanic Philology* 106 (2007), 277–303, here 277.
2. Cf. the unjustly neglected work by Reier on the Old Norse concepts of the soul: Herbert Reier, *Heilkunde im mittelalterlichen Skandinavien: Seelenvorstellungen im Altnordischen* (Kiel, 1976).
3. Gustav Neckel, *Walhall. Studien über germanischen Jenseitsglauben* (Dortmund, 1913), 111.
4. Claude Lecouteux, *Geschichte der Gespenster und Wiedergänger im Mittelalter* (Köln, 1987), passim.
5. Peter Buchholz, *Schamanistische Züge in der altisländischen Überlieferung* (Münster, 1968), 51f.
6. Klaus Böldl, *Eigi einhamr: Beiträge zum Weltbild der Eyrbyggja und andere Isländersagas* (Berlin, 2005), 113.
7. Lecouteux, *Geschichte der Gespenster*, 209.
8. Melissa Berman, in her dissertation on the Egil saga, draws up a catalogue of the motifs most prevalent in the adventure sagas, particularly the *Egil saga*, for instance: *Shamanistic Shape changers, Shape-changers with one alter ego or no apparent physical alteration; shield-biting, howling, etc. as berserksgangr* (in five sagas), *sudden surge of power through berserksgangr* (in ten sagas), *fits of anger, berserk fit during hólmganga* (five examples), *Sorcerers with special battle protection* (they are either explicitly called berserker or iron cast sorcerers), *throat-biting, killing many single-handedly* (thirteen parallels); Melissa A. Berman, *Fiction in 'Egils saga'*. Ph.D. (Stanford, 1983).

9. Böldl, *Eigi einhamr,* 113; a persuasive argument, also employed in connection with mystics' experience of ecstasy as being real.
10. Wolfgang Golther, *Handbuch der germanischen Mythologie* (Leipzig, 1895), 60ff. Walter Baetke considered these motifs borrowed from Christianity, Baetke, *Die Isländersaga* (Darmstadt, 1973), 347; Röhn and Simek both draw attention to the difficulty of reconstructing heathen beliefs on the basis of concepts of the soul alone, Hartmut Röhn, *Untersuchungen zur Zeitgestaltung und Komposition der Íslendingasögur* (Basel and Stuttgart 1976), 290; Rudolf Simek, *Lexikon der Germanischen Mythologie* (Stuttgart, 1995), 202.
11. Hugo Gering, *Über Weissagung und Zauber im nordischen Altertum* (Kiel, 1902), 13.
12. Guðmundsdóttir lists 14 indigenous Icelandic sources, 'The Werewolf', 278.
13. Eugen Mogk, *Germanische Mythologie* (Straßburg, 1898). 273. Cf. Michael Jacoby, *Wargus, vargr. 'Verbrecher', 'Wolf'. Eine sprach- und rechtsgeschichtliche Untersuchung* (Uppsala 1974), passim.
14. Eugen Mogk, 'Berserker', in: Johannes Hoops (ed), *Reallexikon der germanischen Altertumskunde,* I (Straßburg, 1911–1913), 260–261.
15. *Götterdichtung und Spruchdichtung,* Felix Genzmer (transl.) (Jena, 1922) (= *Thule: altnordische Dichtung und Prosa,* II, Felic Niedner (ed.))
16. Oddr Snorrason, *The saga of Olaf Tryggvason,* Theodore M. Anderson (transl.) (Ithaca and London, 2009), 173.
17. *Eyrbyggjas Saga,* Paul Schach (transl.) (Nebraska, 1959), 128. The *Eyrbyggja saga,* which reflects the time before and after the acceptance of Christianity, is particularly interesting because of the many motifs from the belief in revenants and souls. Its episodic structure has invited a variety of interpretations, from 'aimless' to 'interacting with listeners'. Cf. the latest, pioneering work by Böldl, *Eigi einhamr,* who analyses all the tendencies of the saga's reception and treatment by scholarship.
18. Reier, *Heilkunde im mittelalterlichen Skandinavien,* II, 25ff.
19. *The Book of Settlements: landnamabok,* Hermann Pálsson and Paul Edwards (eds) (Hatmondsworth, 1988), 109.
20. This motif many werewolf-stories have picked up. Cf. Wilhelm Hertz, *Der Werwolf. Beitrag zur Sagengeschichte* (Stuttgart, 1862), 48.
21. Vilhelm Grönbech, *Kultur und Religion der Germanen* (Hamburg, 1937), 273. English translation: *The Culture of the Teutons* (London, 1931).
22. Snorri Sturluson, *Heimskringla,* Erling Monsen and A.H. Smith (ed. and transl.) (Cambridge, 1932), 64.
23. 'De berserkis et furore berserkico'. Appendix to: Árni Magnússon, *Kristni Saga, sive Historia Religionis Christianæ in Islandiam introductæ: nec non þattr af Isleifi Biskupi, sive Narratio de Isleifo Episcopo* (Hafnia, 1773), 142–163.
24. Snorri Sturluson, *Heimskringla* (1932), 11.
25. Ludwig Rübekeil, *Diachrone Studien zur Kontaktzone zwischen Kelten und Germanen* (Wien, 2002), 111.
26. Rübekeil, *Diachrone Studien,* 118.
27. Cf. *Egil's saga,* E.R. Eddison (transl.) (New York, 1968 [orig. 1930]), 246.
28. *Die Geschichte vom Skalden Egil,* Felix Niedner (transl.) (Jena, 1914) (= *Thule* III).
29. Egil's family contains many figures that are characteristically wild and certainly possessed of the berserker's fury. Particularly Egil's unusually large head has caused Jesse Byock to presume Paget's syndrome, a genetic bone disorder, although recent scholarship disputes this, Jesse L. Byock, *Feud in the Icelandic Saga* (Berkeley, 1982).
30. Benjamin Blaney, *The 'Berserkr': His Origin and Development in Old Norse Literature.* Ph.D. (Boulder 1972).

31. *Fornaldar sögur Nordrlanda eptir Gömlum Hándritum*, C.C. Rafn (ed.) (Kaupmannahöfn, 1830), 423f. Cf. http://www.germanicmythology.com/FORNALDARSAGAS/SorliStrekaHardman.html.
32. *Göngu-Hrólfs saga,* Hermann Pálsson and Paul G. Edwards (transl.) (Edinburgh, 1980), ch. 6, 42f.
33. Giantesses were often called *brúðir berserkia* which Buchholz takes as evidence for the relation to ecstatic warriors. Cf. Peter Buchholz, *Vorzeitkunde. Mündliches Erzählen und Überliefern im mittelalterlichen Skandinavien nach dem Zeugnis von Fornaldarsaga und eddischer Dichtung* (Neumünster, 1980), 157 n. 391.
34. *Göngu-Hrólfs saga,* 94f.
35. *Egil's Saga,* 173.
36. Bandamanna saga, in: *Fünf Geschichten aus dem westlichen Nordland,* W.H. Vogt and Frank Fischer (transl.) (Jena, 1914) (= *Thule* X).
37. Hallfreðar saga, in: *Vier Skaldengeschichten,* Felix Niedner (transl.) (Jena, 1923) (= *Thule* IX).
38. *Fünf Geschichten aus dem westlichen Nordland* (Jena, 1914), 142.
39. Scholars agree that the phenomenon of feral children has little if nothing to do with werewolves, cf. Brian J. Frost, *The Essential Guide to Werewolf Literature* (London, 2003), 40.
40. Especially often in the *lygisögur* (lying sagas), and also in the Egil saga, *Egils's saga* (1968), 4. Cf. *Die Geschichte vom Skalden Egil* (Jena, 1914) (=*Thule* III), ch. 4, 15; ch. 65, 220; *Drei Lygisogur,* Åke Lagerholm (ed.) (Halle 1927) (= *ASB* XVII) ch. 4, 15; *Hávardar saga,* ch. 21, 54; in the *Völsunga saga,* see: *Isländische Heldenromane* (Jena, 1923) (=*Thule* XXI), ch. 8, 131, and *The Saga of the Volsungs,* Jesse Byock (transl.) (Berkeley, 1990), 44f; *Thorsteins saga Vikingssonar* FAS II ch. 23, 452; Thorstein þáttr uxafots X ch. 10, 363.
41. *Göngu-Hrolfs Saga,* 30f.
42. *Book of Settlements: landnamabok,* 60, 136, 171, 204, 207, 210f, 224.
43. The fairytale of the Bearskin, seldom encountered in popular tradition, also displays the traits of an initiation narrative. From the sixteenth century on an expression still in use today became common in German: 'auf der faulen Haut liegen'. That refers to someone lying lazily on a skin, a sloth-Bear. Originally it was conceived of as a bearskin. Grimmelshausen takes the pejorative term bearskin ='*Bärenhäuter*', which can be traced back to 1563, and lends it a positive connotation. Cf. Heinz Rölleke, 'Bärenhäuter', *Enzyklopädie des Märchens* 1 (1977), col. 1225–1232.
44. *Táin Bó Cúailnge,* O´Rahilly (transl.) (Dublin, 1967), 201–202. Cf. Helmut Birkhan, *Kelten: Versuch einer Gesamtdarstellung ihrer Kultur* (Wien, 1997), 968f; Rübekeil, *Diachrone Studien,* 135f.
45. About Egil's 'bizarre psyche' see: Anne Heinrichs, 'Gunnhild Ozurardóttir und Egil Skalla-Grímsson im Kampf um Leben und Tod', in: Heinrich Beck and Else Ebel (eds), *Studien zur Isländersaga* (Berlin, 2000), 72–108, 80f.
46. *Egils saga* (1968), 111. In *Vatnsdœla saga* Ljót contorts: 'She had cast her clothes up over her head and was walking backwards and had thrust her head back between her legs; the look in her eyes was ugly as hell as she glared at them like the witch she was', *The Vatnsdalers' Saga,* Gwyn Jones (transl.) (New York, 1944). Cf. *Fünf Geschichten aus dem westlichen Nordland* (Jena, 1914) (= *Thule* X), 60f.
47. Like Dietrich's of Bern characteristic hesitation, Cuchulainn tarries. This peculiar behaviour Campbell has isolated as an important element in a hero's life that helps to arouse his fury, see: Joseph Campbell, *The hero with a thousand faces* (Cleveland, 1967).

48. Kei was able to survive in water for days and could expand his size enormously and dry objects because of his body heat. Cf. Helmut Birkhan, *Keltische Erzàhlungen vom Kaiser Arthur* (Kettwig, 1989), 44, 48.
49. There is also a reference to a bear, as he pretends to show Önundr a bear in the night, and lures and kills it.
50. Baldur Hafstað, *Die Egils Saga und ihr Verhältnis zu anderen Werken des nordischen Mittelalters* (Reykjavik, 1995), Chapter 1, considers Egil as person with extra society origin.
51. Birkhan, *Kelten*, 971.
52. Michael P. McGlynn, 'Bears, Boars and Other Socially Constructed Bodies in *Hrófs saga kraka*', *Magic, Ritual and Witchcraft* 4 (2009), 152–175, 160f.
53. Dag Strömbäck, *Sejd: textstudier i nordisk religionshistoria* (Stockholm, 1935).
54. *Göngu Hrolfs Saga.*
55. Cf. A.G. van Hamel, 'The Saga of Sörli the Strong', *Acta Philologica Scandinavica* 10 (1936), 265–295.
56. *Fünf Geschichten von Ächtern und Blutrache* (Jena, 1922) (=*Thule* VIII).
57. *Vier Skaldengeschichten* (Jena, 1923) (=*Thule* IX).
58. *Fünf Geschichten von Ächtern und Blutrache* (Jena, 1922) (=*Thule* VIII).
59. Die *Gull-Þóris saga oder Þórskfiringa saga*, Konrad Maurer (ed.) (Leipzig, 1858).
60. Richard Perkins, *An Edition of Flóamanna Saga With a Study of its Sources and Analogues* (Oxford, 1972).
61. In the Harðar saga, in: *Fünf Geschichten von Ächtern und Blutrache* (Jena, 1922) (=*Thule* VIII).
62. *Norwegische Königsgeschichten* (Jena, 1928) (=*Thule* XVII).
63. The term *Hamask* denotes the already mentioned dragon metamorphosis of Tófi.
64. *Göngu Hrolfs Saga*, ch. 31, 98f.
65. Heinz Dehmer in his study *Primitives Erzählungsgut in den Íslendinga-Sägur* (Leipzig, 1927), 86f, discusses the role of the berserker in the folk-tale *Die vom Berserker bedrohte Jungfrau*, classifying it as a sub-category to the tale of the dragon bride. He lists 6 fundamental motifs that are rarely to be found in the same tale. The role of the berserker as villain is a reflex of his outlaw position in Icelandic society.
66. The physical strain that goes together with the out-of-body experience, the berserks share with magicians and other shape-shifters.
67. *Eyrbyggja Saga*, 56; also: *Die Geschichte vom Goden Snorri* (Jena, 1920) (=*Thule* VII). Here the berserkers have to pass a suitors' test, as in the later tales in which a giant or the devil erect a building.
68. Böðvar's remarkable ability is derived from his provenance. He is the son of Bera, that is, the she-bear, and Prince Björn, also one of the words for bear, and is not of one shape, but roams as a giant bear every morning and only appears in human form at night. He nevertheless sires three sons, the eldest of who are hybrids (the eldest is half elk, the second-eldest has the foot of a dog) and the third has the ability to change into whichever animal he wishes. Cf. Guðmundsdóttir, 'The werewolf', 283f. Gunter Müller, 'Zum Namen Wolfhetan und seinen Verwandten', *Fruhmittelalterliche Studien* 1 (1967), 200–212, on their names (1966) p. 310. For the fairytale of the bear-son, cf. Donald Ward, 'Bärensohn', *Enzyklopädie des Märchens* 1 (1977) col. 1232–1235. A film on the subject is entitled *Bear's Kiss*, Sergey Bodrov (dir.) (2002).
69. McGlynn, 'Bears, Boars', 152, 158f, with good cause demands a perception on the 'reality' of this international motif of the double or OBE-experience. See my

dissertation on the motif of the external soul: Christa Tuczay, *Der Unhold ohne Seele: ein motivgeschichtliche Untersuchung* (Wien, 1982), 161.

70. *The saga of King Hrolf Kraki* (London, 1998), 74–75; Apart from the *Hrolf saga Kraki* Bodvar Bjarki appears in three other sources. See MyGlynn, 'Bears, Boars', 155f.
71. For a detailed discussion see: Michael P. McGlynn, 'Bears, Noars and Other Socially Constructed Bodies in Hrófs Daga Kraka', *Magic, Ritual and Witchcraft* 4 (2009), 152–175, 163f.
72. The Celtic cauldron of rebirth also produced revenants who became 'eternal' warriors.
73. Konrad Maurer, *De Bekehrung des norwegischen Stammes zum Christenthume*, I (München, 1855), 101–119.
74. Maurer, *Die Bekehrung*, 102f.
75. Blaney, *The 'Berserkr'*, 87.
76. *The book of Settlements*, 198 In the thirteenth-century *Sturlaugs Saga Starfsama* a youth and a Lapp sorcerer fight so fast that the eye can hardly follow their movements. Suddenly they both disappear and two dogs fight in their place. These two disappear and the onlookers watch two eagles in aerial combat, until one of them falls to the ground dead. The other one flies away, with the result that no one knows which one of them has died. Even in death, he does not return to his human form. That is interesting insofar as the fictional witnesses usually know to whom the form belongs. The saga is only available in Swedish translation. Cf. H.E. Davidson, 'Hostile Magic in the Icelandic Sagas', in: Venetia Newall (ed.), *The Witch Figure* (London 1973), 20–41, 30.
77. Blaney, *The 'Berserkr'*.
78. Reier, *Heilkunde im mittelalterlichen Skandinavien*, 20.
79. Snorri Sturluson, *Heimskringla, Kongasøgur* (Tórshavn, 1961), 11.
80. *Edda*, I *Heldendichtung*, Felix Genzmer (transl.) (Jena, 1912) (=*Thule* I).
81. *The Saga of the Volsungs*, Jesse Byock (transl.) (Berkeley, 1990), 44–47.
82. *The Saga of the Volsungs*, 43.
83. Cf. Guðmundsdóttir, 'The werewolf', 284. Sinfjötli's name was considered by Martin Ninck to have been a wolf's name, 'Yellow-ankle', *Wodan und germanischer Schicksalsglaube* (Darmstadt, 1967, orig. 1935), 59. Moreover, his mother tests him when he is still a boy by sowing his coat to the skin of his arms. When he tears it off, he suppresses the terrible pain, whereupon she lets him live. She had killed her older boys for showing pain.
84. Otto Höfler, *Kultische Geheimbünde der Germanen* (Frankfurt, 1934), 200f., argues that here the motif may be pointless for the epic, but it is quite justifed in terms of the subject matter Cf. Doht, *Der Rauschtrank*, 142.
85. The throat bite of the werewolves was successfully adapted in the movie versions. The 'contagious' character of the bite was erstwhile demonstrated in the *Werewolf of London* (1935).
86. The ring-motif as means of metamorphosis appears in the Böðvar Bjarki narrative as well as the enchanted princes in the Sinfjötli tale who wear golden rings. Although both sagas do not allocate the rings a direct function as metamorphosis trigger, both accounts prove that rings originally had a central role. Cf. Rübekeil, *Diachrone Studien*, 139.
87. Cf. Guðmundsdóttir, 'The Werewolf', 285; see also Breen, 'Personal Names', 11–12.
88. Recently, particularly: Kris Kershaw, *Odin. Der einäugige Gott und die indogermanischen Männerbünde* (Uhlstädt-Kirchhasel, 2003), 143–182 (English: *The One-Eyed*

God Odin and the (Indo-) Männerbünde, Washington 2000); and Michael P. Speidel, *Ancient Germanic Warriors: Warrior Styles from Trajan's Column to Icelandic Sagas* (London, 2004), 11–51.
89. Rübekeil, *Diachrone Studien*, 121.
90. Mc Cone, 'Hund, Wolf und Krieger', 103; Speidel, 'Berserks'.
91. Åke Hultkrantz, *Conceptions of the Soul among North American Indians. A Study in Religious Ethnology* (Stockholm, 1953); Bente Alver, 'Conceptions of the Living Human Soul in the Norwegian Tradition', *Temenos* 7 (1971), 7–33; an overview of the German literature is offered by Heinrich Beck, *Das frühmittelalterliche Königtum: ideelle und religiöse Grundlagen*, Franz-Reiner Erkens (ed.) (Berlin, 2005), 38f.; see also Hans-Peter Hasenfratz, 'Seelenvorstellungen bei den Germanen und ihre Übernahme und Umformung durch die christliche Mission', *Zeitschrift für Religions- und Geistesgeschichte* 38 (1986), 19–31.
92. In the *Thidreksaga* the magical wife of Hertnit flies over the battlefield in the form of a dragon to support her husband. She is fatally wounded and when Hertnit returns home he finds his wife dying, *The Saga of Thidrek of Bern*, Edward R. Haymes (transl.) (New York, 1988).

4

Before the Werewolf Trials: Contextualising Shape-Changers and Animal Identities in Medieval North-Western Europe

Aleksander Pluskowski

In 1605, Henry Gardinn confessed to transforming into a wolf and then kidnapping, killing and eating a child with two accomplices. He was sentenced to death by burning at the stake by a Dutch court in Limburg; one of his co-conspirators 'Jan le Loup' was caught and executed a few years later.[1] Such incidents were comparatively rare and restricted to certain regions of Europe; although for contemporaries the horror of werewolfery laid not so much in the process of shape-shifting, but in the act of cannibalism. A few decades earlier, in January 1574, Gilles Garnier, a beggar from Lyons, had been burned to death at the stake after confessing to attacking, killing and eating several children as a wolf, although he admitted to strangling one boy in human form.[2] Testimonies of executed werewolves in sixteenth-century France and the Holy Roman Empire repeatedly referred to cannibalistic tendencies, particularly targeting children.[3] But an appetite for human flesh was not peculiar to early modern werewolves. Several centuries earlier, Gervase of Tilbury had compiled a miscellany of popular beliefs and folklore in his *Otia Imperialia* (completed by 1215), where he outlined a story of one Raimbaud de Pouget, a knight in the Auvergne, who became a fugitive when disinherited, wandered like a wild beast through the woods and, 'deranged by extreme fear', lost his mind and transformed into a wolf. In this form he attacked people, devouring the young and mangling the old. Eventually his paw was severed during a confrontation with a woodcutter and he was able to resume human form.[4] Even where werewolves were cast as tragic heroes in medieval literature, the horror of their cannibalistic behaviour was retained. For example, Marie de France, writing for Anglo-Norman courts in the mid-twelfth century, treats her werewolf as a sympathetic hero in her lay *Bisclavret*, but begins by reminding her audience that a werewolf is a ferocious beast, which 'devours men'.[5] The construction of the 'ferocious stereotype' of the later werewolf trials stretches back into the twelfth century, if not earlier.[6]

In medieval Christian society, cannibalism was considered a sign of absolute depravation, closely associated – in contemporary literature and art – with foreign cultures at the fringes of Europe.[7] It was a monstrous activity which cast into sharp relief correct and normative behaviour;[8] not only did Gilles Garnier kill and eat children, on one occasion he contemplated doing so 'despite the fact that it was Friday'; compounding his crime in the eyes of his accusers by desiring forbidden meat on a day when the consumption of normal meat was prohibited. Within Christendom itself, whilst incidences of cannibalism are linked to episodes of extreme starvation, such behaviour was also the hallmark of a demonic pact, typically ascribed to witches, heretics and shape-changers.[9] The persistence of the terror of being cooked and eaten was exemplified by the literary figures of the cannibalistic witch and ogre developing from the fifteenth century.[10] Cannibalistic infanticide had been associated with numerous heretical and minority sects in previous centuries, but these were a product of the literary tradition since the accusation never features in transcripts of inquisitorial trials.[11] In Middle English literature, for example, associations between wolves and Jews, typically through the medium of cannibalism, conceptually linked them to werewolves.[12] A related trope in early modern 'fairy-tales', revolving around fears of being eaten, was children's encounters with wolves in the wilderness.[13] The wolf, already perceived in various regions as a devourer of human flesh, represented the most terrifying form a shape-changer could assume.[14] From the late-eleventh century, decorated stone churches across Europe were emblazoned with representations of monstrous animals and hybrids devouring (and perhaps simultaneously regurgitating) human sinners, a motif first developed in Byzantine apocalyptic art. The idea of being consumed by animals, in this world and the next, remained a popular anxiety throughout the Middle Ages; Hell could be conceptualised as the digestive tract of a monstrous beast, or as an infernal kitchen where sinners were roasted on spits and cooked in pots, or consumed alive.[15]

However, it was only from the fifteenth century that shape-changing into wolves – and associated cannibalistic behaviour – began to feature as an occasional component of witchcraft accusations in Switzerland, France, the Low Countries, parts of the Holy Roman Empire and the eastern Baltic. They may have provided the foundation for the development of a rich corpus of werewolf folk narratives, many of which were only written down much later.[16] Werewolves in the sixteenth century testified that sometimes they attacked and consumed people in wolf form, at other times in human form. Demonologists generally concluded the process of metamorphosis was illusionary, contrary to popular belief, which they regarded as superstition. When the Dominican inquisitor Henricus Institoris considered wolf attacks on children in his *Malleus Maleficarum* (c. 1486–1487), he concluded these were all caused by real wolves possessed by the Devil, and that people could be duped into thinking they had transformed into wolves through witchcraft.[17] The *Malleus* was primarily based on earlier literary precedents with little reference to contemporary popular culture;[18] a conceptual gap between religious elites and rural communities that was also evident with the re-emergence of late medieval animal disguise. Although some churchmen complained about this, as they had done about the uses of bestial masks in the latter centuries of the first millennium, animalistic mumming would become established as a colourful fixture in the ritual calendar of primarily rural communities, whilst civic leaders promoted carnivals in urban centres, firstly in Southern Europe and then increasingly in the north. Whereas the popularity of animal forms in carnival facilitated concealment and subversive behaviour, the use of animal disguise in mumming was comparable to shape-changing, in that both sought to draw on the natural abilities of the animals whose forms they assumed. The main difference was that the former was socially legitimate, albeit potentially unsettling, whereas the latter was regarded as socially and spiritually reprehensible. At least this was the situation in parts of Western and Central Europe by the early-fifteenth century; the start of a sustained era of *die Werwolfprozesse* – the werewolf trials.[19]

How had this situation developed? During the protracted Conversion Period which saw the gradual acceptance of Christianity across Northern Europe, missionaries and churchmen encountered indigenous spiritual traditions expressing very different relations between humans and other species. Yet a model which casts the Conversion Period as the movement from a zoocentric to an anthropocentric cosmology is complicated by the persistence and re-development of animal identities within Christian society. Furthermore, the recurring use of the wolf as the subject of shape-changing on either sides of the Conversion invites a re-consideration of the chronology and contexts of shape-changing. Whilst extensive literature exists on this cultural phenomenon in different regions of Europe,[20] this paper aims to propose a chronology for the diverging traditions of medieval shape-changing, with a particular focus on the wolf in north-Western Europe. Lupine personas are situated within the broader context of animal identities in order to demonstrate their diachronic specificity. The phenomenon is both continuous and discontinuous, but it is argued that there is little connection between traditions of shape-changing in pre-Christian north-Western Europe and the early modern werewolf trials.

Animal identities in pre-Christian north-Western Europe

The immediate context of the werewolf trials is the European witch culture of the sixteenth and seventeenth centuries, whilst the first tangible evidence of beliefs in metamorphosis in these regions dates back over a thousand years and can be situated within a broadly shared north-Western European cultural horizon associated with Germanic and Celtic language groups. This horizon is complicated by the uneven adoption of Christianity across Northern Europe; ecclesiastical institutions proliferated earliest in Frankish regions, and the subsequent conquest of territories which today lie in the Netherlands and north Germany saw the imposition of the new religion by the eighth century. England had been (re-) Christianised over the course of the seventh century and the Scandinavian countries between the ninth and twelfth centuries. Whilst the chronology and the nuances of religious conversion, including the persistence of earlier practices, vary widely, many have argued the pre-Christian cultures of this region of Europe shared an ideology of transformation between human and animal forms, supported by a system of ritual magic and expressed in material practices.[21]

The most abundant body of data derives from Scandinavia, where personal names – as recorded on the earliest runic inscriptions – incorporated animals drawn predominantly from wild species, many of them predatory terrestrial and avian fauna, particularly the wolf, bear, snake/dragon, hawk, eagle and raven.[22] These names were not exclusive to or compulsory for warriors, prompting interpretations of 'totemic' affiliations, reflecting close spiritual associations between entire communities and particular species; the anonymous sixth-century *Opus imperfectum in Matthaeum* refers to barbarians naming their sons 'in accord with the devastation of wild beasts and birds of prey'.[23] These names may have emerged alongside, or inspired, the popularisation of animals in decorated personal ornament. Animals begin to be used to decorate jewellery, weaponry and other artefacts in the latter half of the second century AD in Zealand (Denmark), and a distinctive 'animal style' emerged in the fifth century.[24] Artists readily combined human and animal shapes within highly stylised and abstract forms, perhaps intending to represent or emblematise metamorphosis.[25] Animal motifs dominate Scandinavian art into the eleventh century,[26] and whilst many of the represented species are unidentifiable, wild animals – particularly predators and scavengers – feature widely,[27] particularly on artefacts and *as* artefacts, in high-status male and female graves. Alongside a possible magical or apotropaic function, the wearing of jewellery or the display of artefacts with applied zoomorphic decoration would have been a striking expression of personal and group identity;[28] indeed Nancy Wicker has argued for the widespread adoption of zoomorphic art as an expression of ethnicity, of a Pan-Scandinavian identity.[29] The specific role of these motifs continues to be debated, but is it possible to move beyond this emblematic use of animals to an 'ideology of transformation'?

Some motifs combined human and animal figures, perhaps not only reflecting a lack of clear separation between people and animals,[30] but also close metaphysical relationships with certain species.[31] This fluidity may have also been expressed through the incorporation of specific animal remains into human burials. A survey of animals deposited in cemeteries in early Anglo-Saxon England,[32] Frankia,[33] southern Scandinavia[34] and Iceland,[35] indicates a shared inter-regional preference for horses and dogs,[36] alongside joints of meat derived from all three major domesticates; teeth, particularly wild boar tusks and beaver incisors, and pelts, most notably bearskins. There is regional and diachronic variation in these material practices,[37] although the ritualistic deposition of wild mammals is more readily associated with the Sámi, particularly internments of whole bears from the Roman Iron Age through to the nineteenth century, which are described in later ethnographic accounts of ritualistic bear hunts, feasts and burials.[38] Interpretations of the role played by these species remain highly speculative; in an otherworldy context the horse's mobility may have been invoked,[39] the dog's widely documented association with death,[40] and more broadly the perception of animal skeletons in some circumpolar cultures as containers of spirits.[41] But the fact that not everyone was buried with these animals remains to be satisfactorily explained. Beyond these relatively generic archetypes, ritually killed and deposited animals most likely played important roles in the dramas enacted at the graveside.[42] More unusual combinations of human and animal identities can be found in striking depositional vignettes, such as the placement of a complete elk antler by the head of a man buried next to the door of a house at Birka, or a woman whose lower jaw had been replaced by that of a pig.[43] Funerary (or sacrificial) incinerations involving people and animals resulted in the mixture of their remains, subsequently buried as a single deposit, which some have understood as reflecting beliefs in spiritual metamorphosis.[44]

Whether the mingling of human remains with other animals can be seen as expressions of metamorphosis will remain a moot point. Later Icelandic literature, which is understood to include shared pre-Christian Norse (and more broadly Scandinavian) traditions, refers to the concept of the *fylgja*: a personal 'fetch' or spirit guardian which could assume animal form. Steadfast leaders projected the ox, goat and boar whilst the untamed projected the fox, wolf, deer and bear. Rather than a magical process, this exemplified a blurring of boundaries between species.[45] This is not peculiar to Iceland and is documented in a number of north-Western European regions including early Ireland,[46] Anglo-Saxon England,[47] and amongst the later Norwegian Sámi.[48] But the *fylgja* and other projections of the human soul were generally distinct from physical metamorphosis. The latter was related to the concept of *eigi einhamir* – 'not of one skin (or form)', with the second (animal) shape being the *hamr*, and a full individual transforming (or transferring) into a wolf referred to as *ulfhamr*.[49] The literary motif of the second skin finds its parallel in the use of physical disguises: masks, pelts and gloves. These disguises, and the link between physical appearance and socio-metaphysical status, are consistently associated with later shape-changing narratives. Although there are spatial and temporal discrepancies between the different sources of evidence – textual, iconographic and archaeological – there is convincing evidence for continuity until the period of Christianisation, followed by discontinuity and a re-contextualisation of animal identities.

Animal disguise, militarism and ritual violence

Military re-organisation in southern Scandinavia over the course of the second century AD was partly influenced by Roman culture.[50] The placement of forts with ritualistic deposits and unusual layouts in the outland (the wild landscapes between settlements and cultivated areas) from this time, as well as the mass deposition of weapons in shallow lakes, suggest a strong cultic dimension to the transforming martial culture.[51] The organisation of warfare appears to become more specialised in the latter half of the first millennium in Scandinavia,[52] as well as in England;[53] both regions share a common literary motif recorded in later narratives, where the behaviour of warriors on the battlefield was likened to that of wild carnivores, or associated with providing for carnivores, particularly wolves and raptors.[54] From the fifth century zoomorphic designs applied on weapons appear to represent wolves and raptors devouring human figures or heads.[55] By the time this motif becomes incorporated into literature its function differs between the two regional groups, with Old Norse narratives glorifying martial deeds, and Old English renditions emphasising the horror of battle.[56] In the light of the on-going debate on the religious dimensions of pre-Christian 'Germanic' warfare, perhaps this can be linked to the divergence of world views in England and Scandinavia following the relatively later acceptance of Christianity in the latter. The conceptualisation of warriors as carnivores feeding on the battle-fallen may have been extended to perceived 'ritual consumption' on the battlefield;[57] more likely a metaphorical connection with the god of battle in the absence of contemporary evidence for cannibalism in these regions.[58] Explicit identification with wild animals approaching a more tangible notion of metamorphosis is only represented on a handful of artefacts from a broadly shared cultural horizon.

One of a set of four bronze plates found at Björnhovda in Torslunda, on the island of Öland, depicts a humanoid figure with a wolf head, or wearing a wolf mask, grasping a spear and drawing a sword in the direction of a second warrior, who has a single eye, grasps two spears and wears a horned helmet terminating in bird heads (Figure 3.1). The matrix, used for stamping helmet plates, is dated to c. AD 600, and contains the clearest representation of animal disguise from Iron Age Scandinavia. Discovered and removed from their context in 1874, the plates most likely derived from a set of dwellings in Torslunda and may have been used as dies for producing decorated sheet-metal. Similar 'wolf-headed' figures have been found on contemporary weapon fittings in Alemannic Gutenstein (Figure 4.1) and

Figure 4.1 Representations of 'wolf warriors' from Gutenstein, Germany. Photo of a replica in the Römisch-Germanisches Zentral Museum, Mainz (Christian Bickel, 2007)

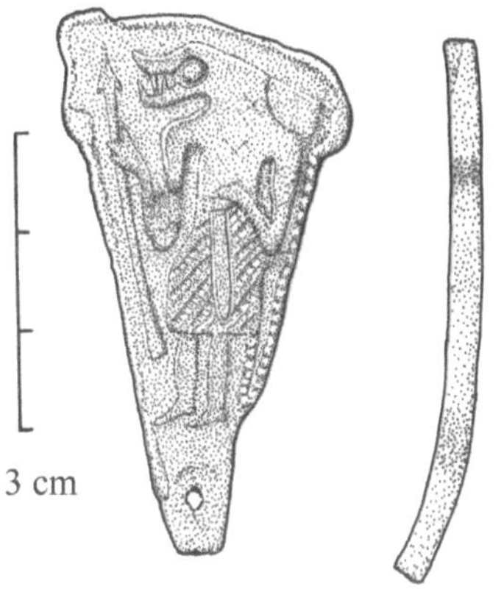

Figure 4.2 The 'wolf warrior' from Fen Drayton, Cambridgeshire, England. Cast copper-alloy die. (Reproduced with permission from the Portable Antiquities Scheme (NLM-468D41))

Obrigheim,[59] and on a copper-alloy die recovered from Fen Drayton in Cambridgeshire, eastern England, which has been broadly dated to c. AD 600–700 (Figure 4.2).[60] Seventh-century southern Scandinavian, eastern English and north-German societies were notable for their lavish material culture associated with regional centres combining political, military and religious authority. On Öland, perhaps as many as nineteen enclosed fortified camps were constructed from quarried limestone from the fourth century AD and expanded in the Germanic Iron Age (c. AD 400–700). They have been interpreted as reflecting an organised military society, which utilised some of these sites for temporary, ceremonial purposes.[61] There were additional cult sites on Öland consisting of wetlands where sacrificed animals and later weapons and precious artefacts were deposited, the most significant of which was Lake Skedemosse situated in the central part of the island. In addition to the Torslunda matrices, a number of human masks most likely manufactured on Öland and inspired by Roman and Celtic prototypes, some designed to be attached to vessels, form a striking element of the island's material culture.[62] The use of masks in Iron Age Scandinavian is increasingly understood as an important component of pre-Christian cultic dramas,[63] and the synchronicity of these finds suggests the emergence of a masking culture by the end of the sixth/seventh century, where forms of disguise were used to effect transformations in personal identity. But the first piece of textual evidence, documenting beliefs in a transformative lupine identity in this broad cultural zone, indicates these were already detached from a military context in territories incorporated into the expanding Frankish empire. In the early decades of the eighth century, the Frankish Church in Germania unsuccessfully sought to end beliefs in werewolves or the *fictos lupos,*[64] which came under criticism from German clerics into the tenth century (see below).

If the seventh century is characterised by wolf warriors on martial equipment, more diverse examples of animal disguise are dated to the ninth and tenth centuries and are found only in southern Scandinavia. A tiny bronze figure from the cremation cemetery at Kungsängen, Stockholm county, wearing a wolf mask and grasping a spear, has been dated to early Viking Age.[65] A bronze figure carrying a spear from a Viking Age cremation grave at Ekhammar, Uppland, appears to be wearing a mask with pointed teeth, variously interpreted as a canid or boar.[66] Two rolled up felt and twill masks (resembling some form of animal) were found within a sunken tenth-century ship in Hedeby harbour, Denmark (Figure 4.3).[67] Fragments of the Norwegian 'Oseberg Tapestry', included in the early-ninth century ship burial, depict a processional scene with a number of bird- and boar-headed humanoid figures.[68] The youngest example is represented on a runestone from Källby in Västergötland dated to c. 1080–1130, which partially depicts an animal-headed dancer, without any obvious military trappings. The stone was put up in memory of the sponsor's father, who may be represented by the figure.[69] Whilst more consolidated proto-states emerged in southern Scandinavia in the ninth century, there is, broadly speaking, continuity in cult practices and their underpinning cosmologies, with increasing exposure to Christianity resulting in further permutations. The ninth-century is also when the first reference to 'wolf-coats' or *ulfheðnar* is found in the poem *Haraldskvæði;*[70] warriors who carry shields and spears, and which scholars have regularly linked back to the dancing figures on the Torslunda plate.[71]

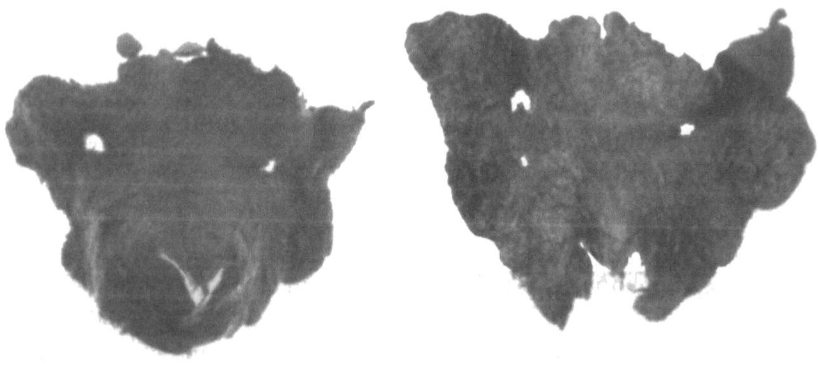

Figure 4.3 Felt mask recovered from the sunken ship in Hedeby (Haithabu) harbour (flat, top and moulded, bottom). Inga Hägg, *Die Textilfunde aus dem Hafen von Haithabu* (Neumünster, 1984), 71. (Reproduced with permission from the Archäologisches Landesmuseum, Schloss Gottorf)

Many accept some form of continuity between the seventh and ninth century examples, within a cultural milieu of identifying with animals that is gradually abandoned following Christianisation. This is not only continuity in practice, but also in the mythological underpinning of ritual animal disguise. The one-eyed horned figure on the Torslunda plate is almost certainly a representation of the god Óðinn, perhaps with his two ravens referred to in the Eddic poem *Grímnismál*. Óðinn was a complicated god; a predatory deity with strong animalistic aspects and a shape-shifter himself, he is consistently associated in Eddic poetry with the transition from the physical world to the otherworld, through ecstasy or death in battle.[72] The origins of the cult of Óðinn remain obscure, but his popularity as a motif on Migration Period bracteate art has led to interpretations of his cult as a widespread and normative 'Germanic theogony', one that created a sense of community and represented a significant political force.[73] The pairing of the horned and lupine figures on the Torslunda plate is an interesting expression of a consistent link between Óðinn and wolves. The fate of the 'army father' and 'nourisher [of carrion-eaters]' at the end of the world is, appropriately, to be consumed by the wolf Fenrir; a recurring association with 'the wolf' first explicitly mentioned on a runic inscription from Ribe, south-west Denmark, dating to the early decades of the eighth century, with older associations visible in the seventh-century wolf warriors and more ambiguous pairings of human figures and quadrupeds.[74] Assuming the various expressions of Óðinn/Woden/Wotan shared similar characteristics,[75] then the cosmological role assigned to wolves and their inspiration for ritual practices associated with this deity may have also been comparable.

Yet evidence for wolf disguises remains elusive, even in Scandinavia. Aside from the textile masks, claws from bears, and sometimes lynx, have been recovered from both cremation and inhumation graves: almost certainly the only surviving elements of outspread hides with intact paws. These have been found in fluctuating numbers from the Pre-Roman Iron Age to the Viking Age across Scandinavia,[76] and would have been placed on, under, or shrouding the deceased suggesting a special significance attached to (at least) the artefact. But rather than simply functioning as decorative items, the identity of the animal – particularly the choice of bear – has also been regarded as significant, in this case with a cultic dimension linked to Óðinn.[77] If this interpretation is accepted, it accommodated both male and female participants; at the Iron Age-Viking Age cremation cemetery in Spånga parish, just north of Stockholm, bear and lynx claws were predominantly but not exclusively associated with females, whilst the remains of cats and horses were more commonly associated with male graves.[78] Of course, the deposition of pelts in graves is difficult to link to uses of animal disguise and the commercial value of bear fur within Scandinavia may have been significant, or more plausibly the prestige associated with bear hunting,[79] but if there is a devotional aspect it would synchronise with the predominantly female representation of practitioners of *Seiður*, where worship of Óðinn was linked with specialist magical knowledge.[80] Wolf remains are noticeably absent from these contexts, and the identification of the species in many of the examples of animal disguise or hybridity discussed above can be debated. The prevalence of animal disguise in pre-Christian north-Western Europe is virtually impossible to quantify, however within societies where animal forms were used in visual display and where battlefield participants were identified with wild animals,[81] active participation in Óðinn's cult, characterised in part by the use of predatory animal disguise and an affinity with predators and scavengers, was most likely exclusive.[82]

The use of hybrids or mummers outside an obvious martial context points to their broader social role, perhaps as ritual specialists. The composition on the Torslunda matrix appears to represent a 'dance' rather than actual combat. The diverse ethnographic literature of circumpolar societies regularly links animal disguise with ritual dance, which is typically mimetic where dancers try to capture the animal's attributes; different species are widely used to affect metamorphosis in dance.[83] Whilst costumes are not essential, they help to bring dancers closer to their animal models. Many practical and magical advantages of imitating an animal in dance have been documented: from the punishment and education of children to renewing natural resources and banishing spirits.[84] The account of the *gothikon*, referring to a group of 'goths' (widely interpreted as Scandinavians) engaged in a militaristic dance wearing masks and animal skins for the Emperor in Constantinople in the mid-tenth century, points to more widespread participation in such dramatic contexts.[85] Drama in pre-Christian Scandinavian societies seems to have been a feature of initiation ceremonies and larger religious festivals,[86] as well an element of funerals and battles.[87] The recurring interpretations linking animal disguise with ecstatic states do not simply reflect mumming, but can be tied more closely to the 'shamanic' aspects of Óðinn. An implicit reference to this conceptualisation of lupine identity is found in Wulfstan's early-eleventh century *Sermo di Lupi*, where the term *werewolf* is associated with *woedfraca* meaning 'frenzied'; a reference to a familiar Scandinavian mental world where metamorphosis and ecstasy were easily linked.[88] However, the passage itself refers to the wolfish metaphor in the New Testament (Matthew 7:15), and exemplifies the shift in world-views that ultimately separate the pre-Christian and later shape-changer.

Forging new animal identities in Christian societies

Following Christianisation, at least as far as the consolidation of an ecclesiastical infrastructure, the documented response from the new religious hierarchy towards the use of animal disguise was openly hostile. In fact it had been for centuries. At the same time as the zoomorphic styles were developing in north-Western Europe in the fifth century, the festival of the Kalends – celebrating the New Year – was incorporating the use of masks, some of which were clearly zoomorphic. This prompted early Christian bishops to try to dissuade their congregations from joining in such celebrations, where people masked as 'farm animals, or as wild animals or monsters'.[89] Such a negative attitude to the wearing of animal disguise had typified the early Church's response in Southern Europe. The widespread acceptance of Christianity in the north brought with it a new worldview for younger generations, who grew up seeing themselves as a distinct and spiritually dominant species. Nonetheless, animal identities continued to play an important role in social organisation and those identified with the wolf remained marginal, shadowy figures.

The emergence of royal authorities in north-Western Europe and their associated law codes provides the first documented evidence for criminalised lupine identities. The origins of the link between Germanic *warg*, a term for a type of criminal, and wolves, remains murky.[90] In England, there are nine discrete instances of *wearg* in thirteen Anglo-Saxon bounds mostly relating to places where criminals were executed, with the earliest dating to 891 and the latest to 1046.[91] Old English *wearg* became codified as 'wolf's head' by the eleventh century and continued to be used, with its variant *warg*, as late as the mid-fourteenth century, surviving in later literature.[92] Outlaws continued to be associated with wild environments in late medieval poetry and romance literature. Scandinavian *vargr* had widespread legal resonance, with fugitives and criminals exiled into the wilderness, although by the thirteenth century it was largely used as a word for wolf, subordinating *úlfr* and ultimately superseding Swedish *ulv*. Again, the nature of the transition from humans to wolves remains unclear.[93] Whilst it is tempting to pool the literary and documentary evidence in search of a common Germanic wolfish archetype, these legal definitions of social exclusion are all examples of later Christian renderings.[94] The seventh-century wolf warriors, the criminalised *vargr* and the later Old Norse lupine personas may well be connected, but there is a striking difference between the use of wolf disguise in pre-Christian ritual drama and warfare – figures which feature as decoration on prestigious artefacts – and the overwhelmingly negative and marginal role assigned to these in Christian sources, even if these are tapping into earlier traditions or sharing a common semiotic vocabulary. The most detailed survey of the phenomenon, published by Michael Jacoby, highlighted the negative role of the wolf in Christian symbolism as the explanation for its doubling up with *vargr*.[95] The link between wolves and criminals was vividly emphasised in north Germany and Denmark. Saxo Grammaticus, writing in the early-thirteenth century, refers to the practice of hanging the condemned alongside wolves and clarifies the metaphorical connection.[96] Engelhard of Langheim refers, in passing, to a similar practice in Pomerania in c. 1200. Documented episodes then cluster in the late fourteenth and fifteenth centuries, continuing into the eighteenth century. However, with exceptional cases such as the hanging of a clothed and masked wolf as a werewolf in Ansbach in 1685, the majority of incidents involved Jews and dogs.[97]

As new metaphorical connections emerged, older ones, were abandoned. The cessation of furnished burials in north-Western Europe makes it difficult to identify any continuity in the use of animal skins as personal adornments. The wearing of luxury furs in Islamic courts, supplied from Scandinavia and Rus' lands, started a trend which spread to the Byzantine Empire and from there to Europe.[98] By the twelfth century, ermine was a staple lining of royal gowns, the apex of a hierarchy of fur valued according to colour, texture and availability.[99] At this time, the wearing of furs was so closely linked with social stratification that it came under attack from church reformers, seeking to visibly segregate clerical and secular elites. Regulation of the latter began to appear from the thirteenth century. However, there is no evidence this pan-European practice – also adopted in Scandinavian courtly circles – conferred the identity of the animals onto its wearers.[100] At the same time, over the course of the twelfth century a new set of animal identities, specifically carnivores, becomes associated with the elite class in north-Western Europe. Anglo-Norman courtly literature proliferated with animal metaphors where lions, foxes and wolves were used to represent various aspects of nobility.[101] The origins of heraldry remain somewhat obscure, but what is clear is that in the latter decades of the twelfth century there is evidence for the deployment of a limited set of emblematic animals to represent the positive qualities of nobility and martial prowess, closely aligned with contemporary religious symbolism.[102] Animals were particularly important in the formative decades of heraldic devices; between 1180–90, they made up 60% of heraldic charges, although this proportion was gradually reduced throughout the Middle Ages and varied geographically.[103] The Scandinavian aristocracy adopted codified heraldry relatively later, and whilst animals such as the wolf endured as popular personal names, their heraldic role remained extremely limited.[104] Ulf Fasi, Jarl of Sweden, who marked his coins with a wolf – a well-established icon of power in the region,[105] as well as a rebus on his name (1230s–1240s[106]) – exemplifies how, in exceptional cases, earlier uses of animal identifiers merged with the codified system. Punning was often linked to the infrequent choice to use wolves in later medieval heraldry, whilst lions and eagles became the animals of choice in the north, even in the Scandinavian countries.[107]

Whether the various lupine identities promoted in medieval Christian societies in north-Western Europe can be connected with notions of metamorphosis remains unclear. An altogether separate category of shape-changer existed in twelfth and thirteenth century literature; the broadest diversity can be found in the narratives largely produced in Iceland, encompassing transformations into bears, wolves, walrus, pigs, as well as cattle, goats, dogs, fish and even whales.[108] Echoing the prevalence of wild animals in earlier zoomorphic art, wild rather than domestic species feature as the most popular forms assumed by shape-changers.[109] Werewolves, in particular, are found in fourteen Icelandic narratives, and in two Norwegian texts known in Iceland.[110] Men (typically) were described as changing into wolves by assuming a wolf's shape or *vargshamr*, and tended to do so either physically or psychologically. Einar Sveinsson's original study of the motif has been refined most recently by Aðalheiður Guðmundsdóttir,[111] sub-dividing werewolves into two categories: an older variant characterised by an innate ability to shape-shift, associated with warfare (the best examples being the *úlfhéðnar* and individuals such as Kveldulfr in *Egils Saga Skallagrímsonar*), and the later variant which came to Iceland with Continental romances in the thirteenth century, where the transformation is caused by a spell. Magical transformation is typically associated with metamorphosis in British and Continental narratives, all of which share common motifs.[112] In *Guillaume de Palerne*, produced in the Picard dialect at the end of the twelfth century, a magical ointment is used; this requisite for transformation appears occasionally in the later *Werwolfprozesse*.[113] A ring, rather than clothing, causes the transformation in *Le Lai de Melion* composed between 1190–1204, whilst in *Arthur and Gorlagon*, a fourteenth-century Welsh narrative, a plant that is cut and used to strike the head transforms the target into a wolf.[114]

Both innate Nordic and Continental magical traditions are brought together in the thirteenth-century *Völsunga saga*, where the wolf is closely associated with the kinship and descent of the Völsungs; Sigi becomes a wolf in the first chapter, the mother of King Siggeirr transforms into a wolf and eats the nine sons of King Völsungr, whilst the most frequently cited episode is where Sigmund and Sinfjötli find enchanted wolf pelts in the forest which transform them into wolves for ten days. Whilst pictorial renditions of the Völsungs are evident from the early-mid eleventh century, as suggested by artistic renderings of Sigurd's encounter with the dragon Fafnir on public Christian monuments in southern Scandinavia,[115] the use of the wolf has been interpreted in relation to the ancestral role of Óðinn, as well as a symbol of the untamed, wild power that brings victory. At the same time, the transformation of Sigmund and Sinfjötli is also one of uncontrolled savagery which the heroes gladly relinquish.[116] The Völsung narrative, a product of thirteenth-century Christian courtly culture, is a composite of traditions, where lupine transformations are linked with family mythology, personal abilities and curses.

The results of the extensive scholarly analysis of all the high medieval werewolf texts suggest that the lupine shape-changer was a complex fusion of various sources and agendas, including contemporary theological stances on metamorphosis. The early Church Fathers – Tertullian, Ambrose – had flatly rejected pagan notions of metamorphosis.[117] However, in his early-fifth century *De Civitate Dei*, Augustine had commented on one of several

Classical werewolf stories, in this case the transformation of Lucius as described by Apuleius, to state that such metamorphosis was an illusion produced by the Devil, and that only God could transform one thing into another.[118] His position was pivotal, and influenced subsequent scholastic perspectives (e.g., Aquinas)[119] as well as Canon law (e.g., Gratian's twelfth-century *Decretum*);[120] both Augustine and Aquinas were the authorities cited in the *Malleus* to explain allegations of metamorphosis.[121] St. Boniface, Burchard of Worms and Pope Gregory VII had all reiterated Augustine's interpretation.[122] Writing at the end of the first millennium, Burchard had famously dismissed those who claimed to transform people into wolves, arguing that only God could transform one thing into another.[123] This was echoed in the penitentials of Bartholomew Iscanus, Bishop of Exeter (1161–1184) and Alain of Lille (c. 1175–1200),[124] and was even codified in law; the eleventh-century western Norwegian Gulaþing laws prescribed outlawry for those who circulated impossible tales such as the existence of werewolves.[125] These dismissals were subsequently repeated, for example, in late-fifteenth century German penitentials.[126]

The most detailed *exemplum* criticising beliefs in lycanthropy was included by Guillaume d'Auvergne within his *De universo* written in the 1230s. It outlines a story of a man possessed by a demon, which would then assume the form of a wolf, or enter a wolf and attack people and animals, convincing the possessed man he was responsible until a successful exorcism was performed by a *saint homme*. A more ambivalent werewolf story can be found in the *Vita S. Ronani*, produced in later thirteenth-century Brittany. This describes an episode where the saint was accused of transforming into a wolf by Keban, the wife of a local peasant. As a wolf, he was alleged to have devoured local sheep and Keban's daughter.[127] The popularity of the wolf as the *exemplum* of choice in these tales may well be the influence of Augustine's use of Classical lycanthropy, in accordance with the primacy of scholastic authority. Indeed, the werewolf stories associated with the Arcadians could have inspired an episode in the life of St. Patrick recorded in the thirteenth-century Norwegian *Konungs skuggsjá*, where an Irish tribe was transformed into wolves as a punishment from God.[128] But whilst these stories reinforced in varying degrees Augustine's view of metamorphosis, they suggest (as do Gervase of Tilbury's contrasting werewolf stories) that clerics were grappling with indigenous superstitions or oral traditions.[129] In the case of Old Norse literature there is little direct evidence for Augustine's theology; metamorphosis (and magical abilities in general) was perceived as pre-Christian in origin, rather than an illusion or compact with the Devil;[130] there was a general inclination to view these practices as archaisms, elements of a bygone pagan world rather than the result of supernatural intervention.[131] Despite the absence of moral judgements attached to shape-changing *per se*,[132] to a twelfth- and thirteenth-century clerical audience promoting the conversion of the former Scandinavian deities into demons,[133] there would have been an implicit diabolical undertone to stories of metamorphosis.

But untangling the relationship between literary shape-changers and popular beliefs in werewolves is highly problematic. The werewolf was certainly employed as a striking literary metaphor; Marie de France may have alluded to the social threat posed by knights in twelfth-century French society,[134] whilst Gerald of Wales' narrative promotes the idea of Irish barbarism within the context of Marcher colonisation.[135] But even in Old Norse literature it is difficult to distinguish categorically between metaphorical transformations and assumed abilities of metamorphosis.[136] Documented examples of werewolves are relatively rare in Northern Europe before the sixteenth century.[137] Bishop Patrick, writing in the late-eleventh century refers to men of the Scottish race who transform into wolves and attack sheep by leaving 'their true bodies',[138] subsequently repeated in a number of Irish sources.[139] Gervase of Tilbury refers to English werewolves twice in his compendium, stating that 'in England we have often seen men change into wolves according to the phases of the moon', and briefly mentions the etymology of Gaulish *gerulfi* and English *were wolf*, before devoting a section on the transformation where the 'daily occurrence' is re-iterated,[140] alongside his more detailed example of Raimbaud de Pouget in the Auvergne and a final description of Chaucevaire in Luc who hid his clothes before transforming. The Norwegian translator of *Bisclavret* in the thirteenth century wrote in the margin that as a child he had seen a wealthy farmer who transformed into a wolf (*vargs ham*).[141] Alongside Marie de France's preamble, these extra-literary statements hint at popular beliefs in werewolves. However, they are not easily untangled from their fictional counterparts and their origins are extremely difficult to trace. Vestiges of pre-Christian beliefs are most striking in Scandinavian regions, where the relationship between the pre-Christian past and the development of ideas concerning witchcraft, magic and shape-changing was relatively fluid – the major shift in perceptions only occurred when witchcraft was aligned with diabolism.[142]

Is there, for example, any link between shape-changing and the use of disguise? In the medieval werewolf narratives this is alluded to, whether involving changing skins or clothes, but explicit references to animal disguise, as in the case of the seventh-century representations and a recurring feature of some seventeenth-century werewolf confessions, are unusual in the medieval literature. An interesting exception is described by Saxo Grammaticus in his early-thirteenth century *Gesta Danorum* where the sons of Harald were hidden by their guardians for their own protection by assuming an animal identity appropriate for fugitives; they were concealed within a hollow oak, given the names of dogs, fed like dogs and wore wolf-claws as shoes.[143] More commonly associated with stories of metamorphosis is the use of a second skin. The motif of transformation involving the removal of clothes is central to Marie's *Bisclavret* and is also mentioned by

Gervase of Tilbury in his story of the werewolf of Luc.[144] In Gerald of Wales's account of the priest and werewolves of Meath within his *Topographia Hibernica* (c. 1187), the female werewolf demonstrates her humanity when her wolf skin is pulled back to reveal the form of an old woman,[145] and the concept was familiar enough to be incorporated into the thirteenth-century *Ancrene Wisse*, a Middle English guide for anchoresses. Here, the metaphor of lycanthropic transformation includes a 'rough pelt', and is employed in a cautionary passage concerning anger, conceptualised as an enchanter invoking a cursed and maddening state. In Icelandic literature, the most elaborate reference to enchanted wolf skins can be found in *Völsunga Saga*, mentioned above. This is different to the second skin or *hamr* of other Old Norse werewolves, but clearly both innate and magical shape-changing shared this common feature. This endured in the later folklore of Swedish werewolves, although it featured more prominently in later German werewolf trials, where pelts were often cited as part of the shape-changer's 'tool kit'.[146] But by the late medieval period the resurgence in animal disguise resulted in the eventual widespread legitimisation of specific types of metamorphosis at the heart of communal festivities.

Reviving animal disguise in the thirteenth century

Clerics had denounced the use of animal skins, antlers and horns during the New Year festivities into the tenth century,[147] although the prevalence of animal disguise is difficult to gauge since Northern European penitentials copied statements from Continental texts, which in turn had been based on much earlier Southern European sources, largely Caesarius and Martin of Braga.[148] The decline in clerical interest in animal disguise after the tenth century most likely reflected the abandonment of such mumming practices, at least in communal, public festivities,[149] and certainly in north-Western Europe at least, animal-masked mummers only began to capture the attention of commentators and artisans from the thirteenth century. Earlier customs lingered in a diminished form in some peripheral Scandinavian regions, where figures from pre-Christian cultic dramas may have manifested as bestial characters such as Grýla the ogress, described in the thirteenth-century *Íslendinga saga* and preserved in the modern mumming traditions of the Faroe Islands, Shetlands and perhaps Orkney.[150] In Icelandic literature there are only a few instances of actual animal disguise (including bears, wolves and goats), in contrast to the higher number of examples of metamorphosis.[151] The direct liturgical replacement of the Kalends – the Feast of Fools, which is documented in France at the end of the twelfth century – saw the symbolic reversal of roles within the Church hierarchy and involved condemnation of, amongst other things, the wearing of monstrous and terrifying masks.[152] But this was an ecclesiastical institution, a product of its time rather than a continuation of pre-Christian masking customs. In the following decades more innovative forms of animal disguise, disconnected from earlier practices, indicate an initial preference for horse forms. Their origins may in fact be Middle Eastern; so-called hobbyhorses are documented in Iraq from at least the ninth century, subsequently in North Africa and hinted at in Andalusian sources,[153] whilst zoomorphic dance is documented in Iran and other parts of the Arab world, as well as represented on high-status objects such as mid-thirteenth century Syrian silver-inlaid vessels.[154]

European examples are first documented in the thirteenth century. The Dominican inquisitor Étienne de Bourbon, writing in the early decades of the thirteenth century, described how at Elne in Roussillon (south-eastern French/Spanish borderland), it was the custom in the parish for young men to ride a wooden horse, and to dance in the church and churchyard on the eve of the patronal festival. This was stopped after a sermon and when one youth persisted he allegedly burst into flames after setting foot in the church.[155] In 1269, *cavallets salvatges* feature in festive games in Valencia, and in 1327, people disguised as wild horses are mentioned in an account of the coronation of Alfonso IV of Aragon.[156] In the British Isles, the earliest reference to a 'hobby horse' is in a Welsh poem by Gruffudd Gryg where it appears as something unusual and new,[157] after which the next reference is found in the Wardrobe Accounts of 1334–1335 for the court of Edward III, which mention *hobihorses pro ludo Regis*.[158] Court entertainers had been around since at least the eleventh century, but from the mid-thirteenth century began to be represented with standardised costumes which included an ass-eared hood.[159] With later additions of a foxtail and a hobby horse, the ass-eared jester was not so much disguised as a specific animal, as an eclectic fusion of zoological elements symbolising folly.[160] Horse disguises may have also been restricted to the court at this time, for equid costumes in thirteenth and fourteenth-century marginal art are virtually all donkeys,[161] with only individual extant representations of other domesticates: a dog, goat, ram and bull.[162] There is interestingly more pictorial evidence for wild species in French and Flemish manuscript marginalia, particularly deer. The earliest example is a 'staff hobby-stag', depicted in the margins of a manuscript of *Historie du Graal* (c. 1280),[163] after which the *Roman de Fauvel* (c. 1316),[164] contains the first representation of the *charivari*: a masquerade first recorded in France in the early-fourteenth century (one which did not reach Britain before the sixteenth century). Here young, unmarried men engaged in transgressive performances largely directed against second marriages; the figures in the *Roman* wearing a variety of animal masks represent folk maskers rather than professional entertainers.[165] Stag costumes, as well as a doe, feature in the margins of a late-thirteenth century Flemish psalter,[166] the Rothschild Canticles (c. 1300),[167] a north French missal (c. 1323), in the Hours of Jeanne d'Evreaux (before 1328),[168] and another early fourteenth-century Franco-Flemish book of hours,[169] The hobby-stag features again in a Flemish copy of *Roman d'Alexandre* (c. 1339–1340),[170] and another version of the same romance, illuminated in Bruges in 1344, shows two scenes of mumming: three disguised men dancing with women and wearing the masks of a stag, hare and boar, and a group in costume as an ass, bird, bull, goat and wyvern.[171] Many of these masked figures, particularly those disguised as stags, are accompanied by human musicians, suggesting that dancing was central to the use of later medieval animal disguise (Figure 4.4).[172]

In Scandinavia, these customs appear relatively late, and a recent comprehensive survey of Nordic mumming has demonstrated a break in the evidence for animal disguise between the Viking Age and the fourteenth century in Norway, and much later in Sweden and Denmark.[173] A *Jolahest* or 'Christmas horse' is first mentioned in Bergen in 1307, and subsequent references to such figures appear only from the seventeenth century.[174] Western masking traditions may have reached the eastern Baltic in the thirteenth-century. Excavations in Novgorod in northwest Russia uncovered at least a dozen, relatively simple, leather masks from waterlogged contexts, some with traces of pigment, and dated on stratigraphic grounds to the thirteenth century;[175] a further example was dated from the late-twelfth to early-thirteenth century, and unlike the others had clearly formed ears cut in profile at the top of the mask.[176] In neighbouring Livonia, a zoomorphic leather mask has been recovered from Riga and also dated to the thirteenth century. The population

Figure 4.4 Masked dancers in the margin of the Romance of Alexander. French verse – Flemish illumination, 1338–44. Oxford, Ms Bodley 264. Fol 21v. (Reproduced with permission from the Bodleian)

of Riga at the time consisted of a mixture of indigenous Livs, Orthodox Rus', Catholic German and Scandinavian colonists, and Christianisation was only being gradually adopted after the cessation of the Livonian crusades. The mask has therefore been interpreted as possible evidence for an indigenous 'wolf or dog cult', although together with the Novgorod masks this phenomenon may equally reflect a contemporary Hanseatic milieu,[177] with the earliest representations of mummers wearing animal disguise as depicted in Flemish and north French manuscripts also dating to the late thirteenth century. There is no further evidence for masking in neighbouring Estonia, Livonia or Prussia until sporadic references in town records from the sixteenth century, firmly within a Christian context.[178] Such animalistic disguises at the heart of communal festivities were paralleled in German-speaking regions by characters such as the *schoduvel* which manifested in a range of guises, some more 'bestial' than others.[179] Rather than reflecting the persistence of an earlier tradition, the Riga and Novgorod masks can be more readily linked to the revival in animal disguise which emerges alongside a general trend of elaborating communal rites, promoted within religious contexts.[180] This trend was not restricted to the north; Boccaccio described a 'wild man' hunt in his *Decameron* (finished c. 1353), which could involve someone 'dressed up as a bear, or as a wild man'; these figures were perhaps representations of malevolent wood spirits which were believed to haunt the countryside, or more broadly expressions of the deepest wilderness, 'the antithesis of the community and its culture'.[181]

The juxtaposition of representations of animal masking and agricultural activities in the mid-fourteenth century *Luttrel Psalter* suggests a potential link with harvest celebrations, and subsequent documentation consistently links mummers with seasonal activities.[182] Perhaps the reappearance of animal disguise in this particular apotropaic context during the fourteenth century was, in part, associated with anxieties concerned with food shortages; the horse and ox representing meat, dairy products and traction; deer and wild boar representing managed wild resources that were increasingly in short supply in late-medieval Western Europe. The function of these disguises could have been to invoke and – presumably – regenerate natural resources; even the images of animal masks themselves may have fulfilled an apotropaic function.[183] However, the most frequently represented animal disguises are wild mammals including predators such as foxes, which in north-Western Europe were the dwindling preserve of the nobility.[184] Why then, would peasant or town communities invoke these restricted species, unless these representations of animalistic dances were intended for elite audiences? It is also likely that the proverbial significance of these animals was more relevant for such *dramatis personae*, particularly in the case of the four recorder players dressed as wolves at the wedding of Margaret of York and the Duke of Burgundy in 1458.[185]

The consistent use of certain animals in late-medieval folk customs, contrasts with the adoption of animal identities through masking during carnival. The carnival was created in Southern Europe in the fifteenth century and quickly spread to the north; an urban phenomenon promoted by the local elite who took the lead in providing entertainments.[186] The randomness of carnival masks documented in late-medieval sources, some of which were animals, undermines any idea of mimicry. These masks simply allowed the suspension of ordinary identities,[187] and provided license for mock aggression and sexual intimacy. Clerics reacted negatively and sermonised against wearing masks or dressing up as beasts and demons, as well as cross-dressing, perceiving such practices as the survival of pagan rites and deeply sinful.[188] Ironically, their pronouncements echoed those of early Churchmen, but the violence and disruption promoted during carnivals prompted a number of German and English towns to prohibit the wearing of disguises in the fifteenth century.[189]

The use of animal disguise – legitimate, albeit controversial, metamorphosis at the heart of rural and later urban communal life – was distinctly separate to processes of magical transformation. For example, despite the prevalence of the horse in early medieval (and modern) Scandinavian ritual contexts,[190] there are no stories of mortal shape-changers transforming into horses 'echoed' in later literature,[191] whilst incidents of wolf disguises are comparatively rare; an account from Zürich in 1568 referring to sword dancers wearing wolf attire is unusual.[192] The continuous persistence of the lupine shape-changer in Scandinavian countries does not, ultimately, synchronise with post-medieval Nordic witch culture. Although shape-changing and magical practices are linked in medieval literature, there are no documented werewolf prosecutions in this region. Of course, magic was neither wholly condemned nor subject to punishment. Across much of Europe religious authorities only targeted *maleficium* and its equivalents which aimed to cause harm; from the fourteenth century inquisitors were urged to pursue only those practitioners of magic who had summoned demons and made sacrifices to them, and were therefore classed as heretics.[193] This perspective was also the norm in Scandinavian countries.[194] As a result, shape-changing was not inherently harmful; a reputed sorcerer named 'Scavius', residing in the Upper Simme Valley (Bernese Oberland, Switzerland) in the late-fourteenth century boasted about his ability to transform into a mouse, in order to escape his enemies.[195] From the practitioner's perspective there was little to be gained from lycanthropy. An unusual exception is a magical formula for transforming oneself into a wolf to gain power, which appears in the contents of a Hebrew magical text and found its way into a Latin translation in thirteenth-century Spain.[196] However in virtually all other cases, werewolves were singled out as particularly harmful and violent, whilst the Church maintained they were a diabolical illusion.

Fear and depredation in the Valais: contextualising the *Werwolfprozesse*

The werewolf in medieval north-Western Europe is typically a masculine figure whose cursed condition is linked to the actions of malevolent women. In Old Norse literature where the shape-changer features most often, there are only four instances of women transforming into wolves, three of which are voluntary transformations that can be linked to a pre-Christian motif. But in all of these cases the werewolf is a cruel and terrifying figure, whose recurring hunger for human flesh separates it from other shape-changers.[197] The first documented prosecutions for witchcraft which included associations with wolves, taking place in the Swiss Alpine regions of Basel and Valais in the early fifteenth century, were of women.[198] Hans Fründ's account of the witch hunts in Valais describes how the Devil taught witches to transform into wolves and devour cattle, whilst the accusations consistently included the use of wolves as steeds. Prosecutions of men followed shortly with the execution of Jacques de Panissière in the Vaud in 1448.[199] The particular use of the wolf in these prosecutions has been regarded as incorporating elements of local Alpine folklore, with some scholars invoking the possibility of ancient belief systems.[200] Indeed, an early-fifteenth century Swiss spell called *The Stammering Woman* refers specifically to nine wolves sent by the caster to provoke an amorous desire in the target.[201] But in the absence of any earlier evidence of shape-changing traditions, the figure of the werewolf could just have easily been invoked from a pool of motifs disseminated by ecclesiastics or those with knowledge of the relevant texts – the action of individuals and the 'personal factor' could play a significant role in the oral transference of demonological ideas.[202] At the same time, sporadic prosecutions of werewolves are attested in the sixteenth century across north-western Continental regions, albeit unevenly, whilst they are absent in Britain and Scandinavian countries – although in the case of the latter, shape-changing was encountered in confessions of accused witches.[203] What then, can be concluded about the relationship between the pre-Christian werewolf and the post-medieval phenomenon?

The early scholars of European folk customs sought to demonstrate their roots in a pagan ritual past – a thesis which, for north-Western Europe at least, is difficult to sustain largely because of the absence of evidence for continued pre-Christian cult praxis.[204] Moreover, it is extremely difficult to remove any folklore, which is, by definition, continuously undergoing metamorphosis itself, from its specific context. Whilst some beliefs concerning witchcraft documented in the sixteenth and seventeenth centuries were also found in the oldest Nordic provincial laws, the history of folk beliefs, customs and the uses of animal disguise is *discontinuous*, featuring much temporal and spatial diversity as well as the abandonment and revival of customs, and the emergence of new beliefs. Some features of the werewolf phenomenon are often imagined as a timeless sub-stratum of rural societies, emerging in times of crisis. This includes the recurring role of hallucinogenic substances to provoke metamorphosis, which regularly features in arguments about the perceived reality of shamanic experiences in pre-Christian Europe,[205] and the recurring importance of animal disguise linked to the idea of changing or removing skins.

The latter has been highlighted throughout this paper as consistently linked with mutable notions of shape-changing and invoking supernatural agency through animalised identities. Both pre-Christian and late medieval treatments of animal disguise seem to be forms of mumming, which, in contrast to the communal masking of carnival, was a performance orchestrating a confrontation between the masked and unmasked. Disturbing and potentially dangerous non-human identities were invoked, but it is possible to differentiate between these two traditions and their relationship with lupine shape-changing.[206] Acknowledging the extremely fragmentary nature of the evidence, the pre-Christian Germanic – most prominently Scandinavian – mumming traditions appear to have been linked to powerful wild animals, in turn associated with beliefs concerning souls and the roles of deities. In this instance it is plausible to link the disparate material culture with a persistent tradition of animal disguise emphasising ceremonial and particularly military functions.[207] Late medieval mumming, on the other hand, was disconnected from the theology promoted by religious authorities, and included both wild and domestic animal identities with the aim of promoting communal welfare. The werewolf is noticeably absent from these late medieval festivities, and instead appears as a physically and spiritually dangerous antagonist. It can be argued that the criminalisation of lupine identities, represented by the *warg* and its variants, begins with the introduction of Christian judicial cultures in north-Western Europe.

Old Norse literature, which provides the most abundant narratives on shape-changing traditions before the sixteenth century, is difficult to project back into the pre-Christian Viking Age, although it is widely accepted that both Eddic and Skaldic poetry contains traces of earlier practices and beliefs.[208] Whilst it may be tempting to link the incidents of shape-changing in Old Norse literature with these earlier traditions, the influence of Christianity and the dissemination of Anglo-Norman narratives are also evident. Later medieval traders encountering the Sámi in northern Scandinavia probably came into contact with shape-changing beliefs, and whilst these may have previously fed into the pool of traditions circulating more widely during the Viking Age as a result of greater cultural exchange,[209] by the later twelfth century when the *Historia Norwegie* was written, Christian institutions had distanced themselves from this worldview and its associated religious practices.[210] The *topos* of the wicked stepmother, featuring in several of the werewolf stories referred to above, was often rendered as a Sámi witch in Old Norse literature.[211]

But, despite evidence for shared ideas across Germanic and Finno-Ugric cultures, there were in fact differences between pre-Christian Sámi and Norse cultures regarding the roles attached to particular wild animals. The circumpolar societies which had contact with Norse groups in Northern Scandinavia and Greenland, attached particular significance to the bear rather than the wolf. [212] Wolves were clearly not as empowering as other species, although ethnographic data suggests they were an integral part of Sámi cosmology.[213] Alongside the disparity in the ritualistic treatment of wild and domestic animals, it is possible to suggest a distinctive southern Scandinavian role attached to the wolf, transmitted by colonists to Iceland, and which would, as a result, become prevalent in Old Norse literature. There was a recurring connection between wolves and Norse sorcery; they feature as mounts, as spirits that could be summoned for aid, or more malevolent bringers of disease.[214] In regions settled by Scandinavians the dramatic festivals involving costumed participants not only lost their cultic significance with the advent of Christianity, but some transformed into smaller-scale folk games. At the same time Christian commentators sought to clarify the boundaries between species, and realigned stories of shape-changing with the new theology. Beliefs in shape-changers were either branded as superstitious, or theologically justified as the end result of divine intervention. This perspective did not replace alternative views of metamorphosis and contrasts with the werewolves of romance and saga literature, which did not emphasise the power of God behind transformations, or even concur with this notion.[215] However, the Old French and Welsh werewolf narratives emphasised the shape-changer's humanity, concurring with Augustine's theory.[216] In summary, both clerics and poets contributed to disseminating information about shape-changing into wolves, which circulated in various permutations throughout north-Western Europe.

If the popularity of the werewolf in Old Norse literature can be related on some levels to folk beliefs that had contemporary cultural relevance, then its limited use in a discrete set of north-Western European narratives suggests that werewolves featured as enduring elements of what churchmen would label as *superstitiones* in, at least, Brittany, north-east and central France, England and Wales. What is more elusive is the antiquity of these regional traditions. Are they remnants of pre-Christian worldviews, or products of the twelfth and thirteenth centuries? Scandinavia stands out as the only European region (or rather, set of regions) where persistent beliefs in shape-changing were gradually replaced by more diffuse understandings of metamorphosis. Here the deity so closely tied to the pre-Christian shape-shifter – Óðinn and his variants – was reinvented as a demon and folk monster, whilst continuing to function in post-medieval Sweden as a promoter of a successful life.[217] Following the Reformation, the Nordic countries saw periods of intensive prosecutions of witchcraft, particularly in the seventeenth century, when a conspiratorial fear of earthly diabolical agents had replaced the medieval view of Scandinavian witchcraft as lingering paganism and a cause of occasional harm.[218] Yet, whilst witches were accused of riding wolves and shape-changing into various animals, particularly birds, they are rarely associated with transforming into wolves.[219] In Iceland, the werewolf motif barely appears in literature by the fifteenth/sixteenth centuries, and here it seems that the impact of the romance shape-changer reduced the original metamorphosing warrior to the status of a fairy-tale.[220] The lack of indigenous wolves may also have prevented any maintenance or regeneration of beliefs in werewolves active in the landscape. However, a lingering, popular understanding of earlier 'shamanistic notions' may explain why Jóns Jónsson the younger was burnt at the stake in 1656, for charges which included an assault on Jón Magnússon whilst assuming a diabolical shape.[221] Stories of werewolves are recorded in later Danish,[222] Swedish,[223] and to a lesser extent, Norwegian folklore,[224] although any direct associations with pre-Christian traditions are extremely tenuous, despite evidence for the *longue durée* of the Nordic shape-changer.[225] Ella Odstedt's comprehensive study of Swedish werewolf lore identified methods of transformation, which represent a composite of pre-Christian and Christian traditions: changes could be voluntary, hereditary or the result of a spell or curse. In this respect, Swedish folklore concerning shape-changing echoes the composite nature of medieval narratives, such as the *Völsunga Saga*, and represents a cumulative and re-contextualised set of traditions. Stories of voluntary changes were confined to inner Norrland, invariably linked to a Sámi cultural context and representing perhaps the only tangible thread of continuity with the pre-Christian past.[226]

Elsewhere, the phenomenon appears to be a truly original product of the early modern age. Sebastian Münster's account of eastern Baltic werewolves dates to c. 1550 (elaborated a few years later in Olaus Magnus' better-known account) and werewolf prosecutions in Estonia are documented only in the seventeenth century, with no evidence of earlier shape-changing traditions in this region.[227] Indeed, the era of werewolf trials marks a break with the shape-changers of the late medieval Christian past and is even further removed from the traditions of the latter half of the first millennium. For religious authorities and the judiciary, the emphasis shifted from the divine to the demonic as the explanation for shape-changing,[228] and a variety of intersecting social, political and ecological factors generated a regionally and chronologically uneven *Werwolfprozesse*.[229] In the Swiss Alps, the likelihood of wolves increasingly targeting more readily available prey – livestock and perhaps people, especially children – is a plausible outcome of the cumulative depletion of wild ungulates. Hunting was not restricted to the aristocracy, and peasants in the Valais took ibex and chamois for their meat, pelts and horns, and for paying dues to local lords.[230] In the later fourteenth century, the boom in livestock husbandry in the Valais region was paralleled by intensive hunting of bears; wolves had already been noted as attacking livestock, and the use of dogs to protect flocks from predators is documented in the fourteenth and fifteenth centuries.[231] Such episodes would have been requisite flashpoints – localised crises – in a rapidly emerging climate of anxiety and religious policing, and have been suggested as the context for the incidents of werewolf accusations in Lorraine.[232] They may go some way in explaining the uneven occurrence of the *Werwolfprozesse*. In France, the backdrop for the werewolf trials is provided by the religious wars of the sixteenth century, documented episodes of cannibalism, a renewed emphasis on the Eucharist,[233] but also the changing health and behaviour of the wolf population, which resulted in increasing numbers of documented fatal attacks on people by healthy and rabid animals, peaking at 120 in 1596–1600.[234] The depletion of wildlife in north-Western Europe corresponds with the emergence of ritual animal disguise beyond the sphere of the court.[235] Eventually these new animal identities would become standardised, with wolves typically avoided by mummers; in the minds of inquisitors, witch hunters and theologians, lupine identities, sorcery and cannibalism went hand-in-hand.[236] Perhaps because of their enduring role in Western culture, werewolves may appear, at first glance, to have been virtually timeless, pan-European figures. In fact, their uneven spatial and temporal distribution challenges us to re-consider the complexity of such phenomena in the past.

Notes

I am extremely indebted to my colleagues Willem de Blécourt, Neil Price, Sophie Page, Roberta Gilchrist and Anders Andrén for their advice and discussions on this topic, and this paper has also benefited from my participation in the extended seminar 'Nocturnal Histories: Witchcraft and the Shamanic Legacy of pre-Christian Europe' at the Radcliffe Institute in Harvard, organised by Stephen Mitchell and Neil Price in 2009.

1. J.M.H. Eversen, 'Vonnis en executie wegens weerwolverij in 1607 door de justitie te Maastricht', *De Maasgouw* 24 (1903), 95–96.
2. *Arrest memorable de la Cour de parlement de Dole, donné à l'encontre de Gilles Garnier, lyonnois, pour auoir en forme de loup garou deuoré plusieurs enfans & commis autre crimes*. Par Benoist Rigau iouxte la copie imprimée à Sens et depuis à Paris pour Pierre Des-Hayes (1574); abbreviated in: Adam Douglas, *The Beast Within; Man, Myths and Werewolves* (London, 1992), 131–132.
3. Caroline Oates, 'Metamorphosis and Lycanthropy in Franche-Comté, 1521–1643', in: Michel Feher (ed.), *Fragments for a History of the Human Body* (New York, 1988), 305–363.
4. S.E. Banks, and J.W. Binns (eds and trans.), *Gervase of Tilbury, Otia Imperialia, Recreation for an Emperor* (Oxford, 2002), 812–815.
5. Glyn Sheridan Burgess and Keith Busby (trans.), *The Lais of Marie de France* (London, 1999), 68. Sconduto argues that in most of the werewolf lays, the protagonists themselves do not consume people, see: Leslie A. Scunduto, *Metamorphoses of the Werewolf: A Literary Study from Antiquity Through the Renaissance* (Jefferson NC, 2008), 187.

6. Carlo Ginzburg, *Ecstasies: Deciphering the Witches' Sabbath* (New York, 1991), 154.
7. Jill Tattersall, 'Anthropophagi and Eaters of Raw Flesh in French Literature of the Crusade Period: Myth, Tradition and Reality', *Medium Ævum*, 57/2 (1988),240–253; Gregory G. Guzman, 'Reports of Mongol Cannibalism in the Thirteenth-century Latin Sources: Oriental Fact or Western Fiction?' in: Scott D. Westrem (ed.), *Discovering New Worlds* (New York, 1991), 31–68; Merrall Llewelyn Price, *Consuming Passions: the Uses of Cannibalism in Late Medieval and Early Modern Europe* (New York, 2003); Debra Higgs Strickland, *Saracens, Demons and Jews: Making Monsters in Medieval Art* (Princeton, 2003).
8. David Williams, *Deformed Discourse: the Function of the Monster in Mediaeval Thought and Literature* (Exeter, 1996), 145–148; John Block Friedman, *The Monstrous Races in Medieval Art and Thought* (Cambridge [Mass.], 1981), 26–36.
9. Herman Pleij, *Dreaming of Cockaigne: Medieval Fantasies of the Perfect Life* (New York, 2001), 116.
10. Charles Zika, *Exorcising Our Demons: Magic, Witchcraft and Visual Culture in Early Modern Europe* (Leiden, 2003), 445; see Elmar Lorey, 'Vom Wolfssegner zum Werwolf: Hexereiprozesse im Nassauer Land', in: Rita Voltmer and Günter Gehl (eds), *Alltagsleben und Magie in Hexenprozessen* (Weimar, 2003), 65–79, for links between wolves, cannibalism and witchcraft in early modern central Europe.
11. Norman Cohn, *Europe's Inner Demons: The Demonization of Christians in Medieval Christendom* (London, 1993), 73.
12. David I. Shyovitz, 'Christians and Jews in the Twelfth-century Werewolf Renaissance', *Journal of the History of Ideas* 75 (2014), 521–543.
13. Aleksander G. Pluskowski, 'The Tyranny of the Gingerbread House: Contextualising the Fear of Wolves in Medieval Northern Europe Through Material Culture, Ecology and Folklore', *Current Swedish Archaeology*, 13 (2005), 141–160.
14. Although documented forms assumed by shape-changers ranged from mice to tables, trees, lakes, buildings and even entire landscapes.
15. Aleksander G. Pluskowski, 'Apocalyptic Monsters: Animal Inspirations for the Iconography of Medieval North European Devourers', in: Robert Mills and Bettina Bildhauer (eds), *The Monstrous Middle Ages* (Cardiff, 2003), 155–176.
16. Willem de Blécourt, '"I Would Have Eaten You Too": Werewolf legends in the Flemish, Dutch and German area', *Folklore*, 118 (2007), 23–43; Tiina Vähi, 'Libahundiuskumused arhailise õiguse ja moraali kontekstis', *Ajalooline Ajakiri*, 1/2 (123/124) (2008): 101–154.
17. Christopher S. Mackay (ed. and trans.) *Malleus Maleficarum*, II (Cambridge, 2006), 161–162; Mackay and others have included Jacob Sprenger as the co-author of the *Malleus*, but this has been disputed, e.g. Wolfgang Behringer, Günther Jerouschek and Werner Tschacher (eds), *Der Hexenhammer. Malleus maleficarum* (München, 2000).
18. Mackay, *Malleus*, 39.
19. In Elmar Lorey's list of werewolf trials, the first is documented in 1429/30 (see http://www.elmar-lorey.de/Prozesse.htm).
20. The scholarly literature on the topic is too vast to include here. For the most recent work adopting a diachronic perspective of the medieval werewolf see: Sconduto, *Metamorphoses of the Werewolf* (as in note 5).
21. The adoption of animal identity to facilitate an ecstatic state, another way of showing that 'the shaman can forsake his human condition...to "die"', whilst reinforcing a mythological solidarity between humans and animals through the

blurring of species boundaries (Mircea Eliade, *Shamanism: Archaic Techniques of Ecstasy* (London, 1989), 93–94), is a recurring feature of circumpolar 'shamanism' and has been considered by some scholars to be a useful framework for re-contextualising pre-Christian shape-changing in the Germanic worlds; Stephen Glosecki, *Shamanism and Old English Poetry* (New York, 1989); Neil Price, *The Viking Way: Religion and Ware in Late Iron Age Scandinavia* (Uppsala, 2002), 294–295, 308; Howard Williams, *Death and Memory in Early Medieval Britain* (Cambridge, 2006); Tine Jeanette Biering, 'The Concept of Shamanism in Old Norse Religion From a Sociological Point of View', in: Anders Andrén, Kristina Jennbert and Catharina Raudvere (eds), *Old Norse Religion in Long-Term Perspectives: Origins, Changes and Interactions* (Lund, 2006), 171–176.

22. Kristina Jennbert, 'Ambiguous Truths? People and Animals in Pre-Christian Scandinavia', in: Jostein Bergstøl (ed.), *Scandinavian Archaeological Practice – in Theory, Proceedings from the 6th Nordic TAG, Oslo 2001* (Oslo, 2003) 212–30, at 219–223; Asger Jorn, Gérard Franceschi and Bente Magnus, *Bird, Beast and Man – in Nordic Iron Age Art* (volume 2) (Cologne, 2007), 11; Price, *The Viking Way*, 373.
23. Judith Jesch, 'Eagles, Ravens and Wolves: Beasts of Battle, Symbols of Victory and Death', in: Judith Jesch (ed.), *The Scandinavians from the Vendel Period to the Tenth Century: An Ethnographic Perspective,* (Woodbridge, 2002), 251–280, 275.
24. For a historical survey see Karen Hoilund Nielsen and Siv Kristoffersen, 'Germansk dyrestil (Salins stil I–III): Et historisk perspektiv', *Hikuin* 29 (2002), 15–74, and Nancy L. Wicker, 'The Scandinavian Animal Styles in Response to Mediterranean and Christian Narrative Art', in: Martin Carver (ed.), *The Cross Goes North: Processes of Conversion in Northern Europe, AD 300–1300* (Woodbridge, 2003), 531–550.
25. S. Kristoffersen, 'Transformation in Migration Period Animal Art', *Norwegian Archaeological Review,* 28 (1995), 1–18.
26. See George Zarnecki, 'Germanic Animal Motifs in Romanesque Sculpture', *Artibus et Historiae* 11, /22 (1990), 189–203; Lennart Karlsson, *Romansk träornamentik i Sverige* (Stockholm, 1976); Erla Bergendahl Hohler, *Norwegian Stave Church Sculpture* (Oslo, 1999).
27. Peter Vang Petersen, 'Warrior Art, Religion and Symbolism', in: Lars Jørgensen, Birger Storgaard and Lone Gebauer Thomsen (eds), *The Spoils of Victory: the North in the Shadow of the Roman Empire* (Copenhagen, 2003), 286–294, especially 294; George Speake, *Anglo-Saxon Animal Art and its Germanic Background* (Oxford, 1980).
28. Lotte Hedeager, 'Cosmological Endurance: Pagan Identities in Early Christian Europe', *Journal of European Archaeology,* 3 (1998), 383–397; Michel Pastoureau, 'Pourquoi tant de lions dans l'Occident médiévale?' *Micrologus: Natura, scienze e società medievali,* 8/1 (2000),11–30.
29. Wicker, 'Scandinavian Animal Styles', 546.
30. Jennbert, 'Ambiguous Truths', 216.
31. Kristoffersen, Transformation, 13.
32. Julie M. Bond and Fay L. Worley, 'Companions in Death: The Roles of Animals in Anglo-Saxon and Viking Cremation Rituals in Britain', in: Rebecca Gowland and Christopher Knüsel (eds), *The Social Archaeology of Funerary Remains* (Oxford, 2006), 89–98.
33. Michael Müller-Wille, 'Pferdegrab und Pferdeopfer im frühen Mittelalter', *Berichten van de Rijksdienst voor het Oudheidkundig Bodemonderzoek* 20/21 (1970/1971), 119–248; Julia Gerken, 'Human-Animal Relationships Reflected

in Early Medieval Horse Burials in Germany', in: László Bartosiewicz, Erika Gál and István Kováts (eds), *Csontvázak a szekrényből (Skeletons from the Cupboard)* (Budapest 2009), 65–71.

34. Price, *The Viking Way*; Kristina Jennbert, 'The Heroized Dead: People, Animals, and Materiality in Scandinavian Death Rituals, AD 200–1000', in: Andrén, Jennbert and Raudvere (eds), *Old Norse Religion in Long-Term Perspectives*, 135–140.
35. Ulla Loumand, 'The Horse and its Role in Icelandic Burial Practices, Mythology, and Society', in: Andrén, Jennbert and Raudvere (eds), *Old Norse Religion in Long-Term Perspectives*, 130–134.
36. Anne-Sofie Gräslund, 'Dogs in Graves – a Question of Symbolism?', in: Barbro Santillo Frizell (ed.), *PECUS: Man and Animal in Antiquity* (Rome, 2004), 167–176; Bond and Worley, 'Companions in Death'.
37. Chris Fern, 'Early Anglo-Saxon Horse Burial of the Fifth to Seventh Centuries AD', *Anglo-Saxon Studies in Archaeology and History* 14 (2007), 92–108.
38. Inger Zachrisson and Elisabeth Iregren, *Lappish Bear Graves in Northern Sweden: An Archaeological and Osteological Study* (Stockholm, 1974).
39. Loumand, 'The Horse and Its Role'.
40. Gräslund, 'Dogs in Graves'.
41. Patricia D. Sutherland, 'The Iconography of Palaeo-Eskimo Art', in: Neil S. Price (ed.) *The Archaeology of Shamanism* (London, 2001), 135–145; 140–141.
42. Price pers. comm; Fern, 'Early Anglo-Saxon Horse Burial', 102.
43. Sven Kalmring, 'Of Thieves, Counterfeiters and Homicides. Crime in Hedebyand Birka', *Fornvännen* 105 (2010), 281–290, 286.
44. Howard Williams, 'Animals, Ashes & Ancestors', in: Aleksander G. Pluskowski (ed.) *Beyond Skin and Bones? New Perspectives on Human-Animal Relations in the Historical Past* (Oxford, 2005), 19–40.
45. Price, *The Viking Way*, 59–60; 364.
46. John Carey, 'Werewolves in Medieval Ireland', *Cambrian Medieval Celtic Studies* 44 (2002), 37–72, 56.
47. See Glosecki, *Shamanism*; Alexandra Sanmark, 'Living On: Ancestors and the Soul', in: Martin Carver, Alex Sanmark and Sarah Semple (eds), *Signals of Belief in Early England: Anglo-Saxon Paganism Revisited* (Oxford, 2010), 158–180.
48. In Halvard Gunnarsson's *Chronicon regum Norvegiae*, c. 1606, cited in: Clive Tolley, *Shamanism in Norse Myth and Magic*, II (Helsinki, 2009), 76.
49. Gerard Breen, *The Berserker in Old Norse and Icelandic Literature*, unpublished Ph.D. thesis (Cambridge, 1999); see also note 143.
50. Birger Storgaard, 'Cosmopolitan Aristocrats', in: Lars Jørgensen, Birger Storgaard and Lone Gebauer (eds), *Spoils of Victory: The North in the Shadow of the Roman Empire* (Copenhagen, 2003), 106–125.
51. Anders Andrén, 'A World of Stone: Warrior Culture, Hybridity and Old Norse Cosmology', in: Andrén, Jennbert and Raudvere (eds), *Old Norse Religion in Long-Term Perspectives*, 33–38, 34–35.
52. Klavs Randsborg, *Hjortspring. Warfare and Sacrifice in Early Europe* (Aarhus, 1995); Johan Engström, 'The Vendel Chieftains: A Study of Military Tactics', in: Anne Nørgård Jørgensen and Birthe L. Clausen (eds), *Military Aspects of Scandinavian Society in a European Perspective, AD 1–1300* (Copenhagen, 1997), 248–255.
53. Heinrich Härke, 'Early Anglo-Saxon Military Organisation: An Archaeological Perspective', in: Jørgensen and Clausen (eds), *Military Aspects of Scandinavian Society*, 93–101.
54. Breen, *The Berserker*.

55. Petersen, 'Warrior art', 290–293.
56. Jesch, 'Eagles, raves and wolves'.
57. Price, *The Viking Way*, 382–383.
58. Examples are known from Late Neolithic and Bronze Age Scandinavian contexts; Flemming Kaul, 'Ritualer med menneskeknogler i yngre stenalder' *Kuml* (1991–1992) (1994) 7–52.
59. Karen Høilund Nielsen, 'Ulvekrigeren. Dyresymbolik på våbenudstyr fra 6–7 århundrede', in: Ole Høiris e.a. (eds), *Menneskelivets mangfoldighed. Arkæologisk og antropologisk forskning på Moesgård* (Moesgård, 1999), 327–334.
60. Kevin Leahy 'Fen Drayton' in: Helen Geake, 'The Portable Antiquities Scheme', in: Tom Beaumont James (ed.) 'Medieval Britain and Ireland in 2005', *Medieval Archaeology*, 50 (2006), 274–294, 279–280.
61. Andrén, 'A World of Stone'.
62. Ulf Erik Hagberg, 'Öland during the Iron Age and Early Middle Ages: An Archaeological Survey', in: Ulf Näsman and Erik Wegraeus (eds), *Eketorp: The Setting* (Stockholm, 1976), 12–34, 20.
63. Ing-Marie Back Danielsson, '(Un)masking Gender – Gold Foil (Dis)embodiments in Late Iron Age Scandinavia', in: Yannis Hamilakis, Mark Pluciennik and Sarah Tarlow (eds), *Thinking Through the Body: Archaeologies of Corporeality* (New York, 2009), 179–200, at 180; Price, *Viking Way*, 171–174.
64. See John-Henry Clay, *In the Shadow of Death: Saint Boniface and the Conversion of Hessia, 721–754* (Turnhout, 2010), 259–271, 384 on the geographic definition of *Germania*; Anonymous, Sermones, PL, LXXXIX, sermon 15. In *Octavi saeculi ecclesiastici scriptores maxima ex parte recensentur* (1850, cols 843–872); created in the context of Boniface's Germanic missionary field.
65. Per-Olof Ringquist, 'Tva vikingatida uppländska människofigurer i brons, *Fornvännen* 64 (1969), 287–296.
66. Price, *The Viking Way*, 372–374; Birgit Sawyer, *The Viking-Age Rune-Stones: Custom and Commemoration in Early Medieval Scandinavia* (Oxford, 2000), 149.
67. Inga Hägg, 'Textilfunde aus dem Hafen von Haithabu – Aspekten und Interpretation' *Offa*, 41 (1985), 177–188.
68. Anne Stine Ingstad, 'Oseberg-dronningen – hvem var hun?', in: Arne Emil Christensen, Anne S. Ingstad and Bjørn Myhre (eds), *Osebergdronningens grav* (Oslo, 1992), 224–256.
69. Dated stylistically to period V; Sawyer, *Rune-Stones*, 172; Samnordisk Runtextdatabas Svensk entry Vg 56 (http://www.nordiska.uu.se/forskn/samnord.htm).
70. Tolley, *Shamanism*, 199.
71. Price, *Viking Way*, 367.
72. Stephan Scott Grundy, *The Cult of Óðinn: God of Death?* (New Haven, 2014); Price, *The Viking Way*.
73. Asger Jorn, Gérard Franceschi and Bente Magnus, *Men, Gods and Masks – in Nordic Iron Age Art*, I (Köln, 2007), 50.
74. Ottar Grønvik, 'Runeinnskriften fra Ribe', *Arkiv for nordisk filologi*, 114 (1999), 102–127.
75. These deities are poorly known outside of Scandinavia.
76. Elisabeth Iregren, 'Size of the Brown Bear (*Ursys arctos L.*) in Northern Sweden During the Last Millenium', *Archaeozoologia* II/1,2 (1988), 165–178; Bo Petré, 'Björnfällen I Begravningsritualen – Statusobjekt Speglande Regional Skinnhandel?', *Fornvännen*, 75 (1980), 5–14; Oliver Grimm, 'Bear-skins in

northern European burials and some remarks on other bear-related furnishings in the north and middle of Europe in the 1st millennium AD', in: Oliver Grimm and Ulrich Schmölcke (eds), *Hunting in Northern Europe until 1500 AD: Old Traditions and Regional Developments, Continental Sources and Continental Influences* (Neumünster, 2013), 277–296.
77. Petré, 'Björnfällen'.
78. Berit Sigvallius, *Funeral Pyres: Iron Age Cremations in North Spånga* (Stockholm, 1994), 76–77, 107.
79. Sigmund Oehrl, '*Svá beitum vér björnuna á mörkinni norðr* – Bear hunting and its ideological context (as a background for the interpretation of bear claws and other remains of bears in Germanic graves of the 1st millennium AD)', in: Grimm and Schmölcke (eds), *Hunting in Northern Europe until 1500 AD*, 297–332.
80. Price, *Viking Way*.
81. Andrén, 'A World of Stone', 35.
82. Eric Christiansen, *The Norsemen in the Viking Age* (Oxford, 2002), 295.
83. Steven Lonsdale, *Animals and the Origins of Dance* (London, 1981), 11–12.
84. Lonsdale, *Animals*, 128.
85. Terry Gunnell, *The Origins of Drama in Scandinavia* (Cambridge, 1995), 72–75.
86. Gunnell, *Origins of Drama*, 76.
87. Neil Price, 'Passing into poetry: Viking-Age mortuary drama and the origins of Norse mythology', *Medieval Archaeology*, 54 (2010), 123–156.
88. Aleksander G. Pluskowski, *Wolves and the Wilderness in the Middle Ages* (Woodbridge, 2006), 175–176.
89. Meg Twycross and Sarah Carpenter, *Masks and Masking in Medieval and Early Tudor England* (Aldershot, 2002), 26–28.
90. Michael Jacoby, *Wargus, vargr, 'Verbrecher', 'Wolf' – eine sprach- und rechtsgeschichtliche* (Uppsala, 1974); Mary R. Gerstein, 'Germanic Warg: The Outlaw as Werewolf', in: Gerald James Larson (ed.), *Myth in Indo-European Antiquity* (London, 1974), 131–156.
91. Andrew John Reynolds, *Deviant Burial Customs* (Oxford, 2014), 225–226.
92. Carole Hough, 'OE *wearg* in Warnborough and Reighburn', *Journal of the English Place-Name Society*, 27 (1994–1995), 14–20.
93. J. Lindow, 'Review of *Wargus, vargr "Verbrecher" "Wolf." Eine sprach- und rechtsgeschichtliche Untersuchung*by Michael Jacoby' *Speculum*, 52/2 (1977), 382–385, 384.
94. Gerstein, 'Germanic warg'.
95. Jacoby, *Wargus*, 111.
96. *Gesta Danorum* Book V (137), Book (VIII) 232; Hilda Ellis Davidson (ed.) and Peter Fisher (trans.) *Saxo Grammaticus: The History of the Danes, Books 1–9* (Cambridge, 1980, reprinted 1999), 152, 255; section II, 89, note 137; For archaeological examples of this practice see Sigmund Oehrl, '"do hengen die drei hunde" – Richtstätten und Tiere', in: Jost Auler (ed.), *Richtstättenarchäologie*, III (Dormagen, 2012), 450–473.
97. Sigmund Oehrl, '"wer gesach ie galgen mit wirsern balgen?" Archäologische Überlegungen zur Rolle von Tieren im hochgerichtlichen Strafvollzug des Mittelalters und der Frühen Neuzeit', *SIGNA IURIS. Beiträge zur Rechtsikonographie, Rechtsarchäologie und Rechtlichen Volkskunde* 6 (2010), 107–150, 121–128.
98. James Howard-Johnston, 'Trading in Fur, From Classical Antiquity to the Early Middle Ages', in: Esther Cameron (ed.), *Leather and Fur: Aspects of Early Medieval Trade and Technology* (London, 1998), 65–79.
99. Robert Delort, *Le commerce des fourrures en occident á la fin du Moyen âge: (vers 1300 – vers 1450)* (Rome, 1978).

100. Françoise Piponnier and Perrine Mane, *Dress in the Middle Ages* (New Haven, 1997). When disguise became an integral part of the carnival *rubunte*-runs of the fifteenth century, involved participants would turn their (furred) coats inside out. Anu Mänd, 'Devils of Baltic Towns in the Context of Late-Medieval German Tradition', *Medieval English Theatre,* 23 (2001), 17–33, 23.
101. Joyce E. Salisbury, *The Beast Within: Animals in the Middle Ages* (London, 1994), 100–104.
102. Michel Pastoureau, *Traité d'héraldique* (Paris, 1993).
103. Pastoureau, *Traité d'héraldique*, 133–154.
104. Pluskowski, *Wolves and the Wilderness.*
105. Anne-Sofie Gräslund, 'Wolves, Serpents and Birds: Their Symbolic Meaning in Old Norse Belief', in: Andrén, Jennbert and Raudvere (eds), *Old Norse Religion in Long-Term Perspectives*, 124–129.
106. Coin Group VI, Uppsala; Lars O. Lagerqvist, *Svenska mynt under vikingatid och medeltid (ca 995–1521) samt Gotländska mynt (ca 1140–1565)* (Stockholm, 1970), 55.
107. Pluskowski, *Wolves and the Wilderness*, 151.
108. Davidson, 'Shape-Changing', 127.
109. Davidson, 'Shape-Changing', 140.
110. For a comprehensive survey of werewolves in Old Norse literature, including extensive references to associated texts and folklore see: Aðalheiður Guðmundsdóttir, 'The Werewolf in Medieval Icelandic Literature', *Journal of English and German Philology* 106 (2007), 277–303, at 278.
111. Einar Óláfur Sveinsson, 'Keltnesk áhrif á íslenzkar ýkjusögur', *Skírnir*, 106 (1932), 118–19; Guðmundsdóttir, 'The Werewolf'.
112. Sconduto, *Metamorphoses of the Werewolf*, 91.
113. Henri Michelant (ed.), *Guillaume de Palerne* (Paris 1876); Kratz, *Fictus Lupus*, 76.
114. Frank A. Milne (trans.) 'Arthur and Gorlagon', *Folk-Lore,* 15 (1904), 40–60; George Lyman Kittredge (ed.), *Arthur and Gorlagon: Versions of the Werewolf's Tale* (Boston, 1966). The story is extant in a single English manuscript, but is argued by Kittredge to be a direct translation of a Welsh text. Amanda Hopkins, *Melion and Bicarel: Two Old French Werwolf Lays* (Liverpool, 2005).
115. Klaus Düwel, 'On the Sigurd Representations in Great Britain and Scandinavia' in: Mohammad Ali Jazayery and Werner Winter (eds), *Languages and Cultures: Studies in Honor of Edgar C. Polomé* (Berlin, 1988), 133–156.
116. Tolley, *Shamanism*, I, 579.
117. Salisbury, *Beast Within*, 141.
118. Saint Augustine, *The City of God Against the Pagans*, 18.18, E.M. Sanford and W. McA. Green (transl.) (Cambridge, 1965).
119. *Summa Theologica* 1, Question 114, Article 4, Objection 2, Kors (transl.) (Ottawa, 1953), 68–71.
120. Brian P. Levack, *The Witchcraft Sourcebook* (New York, 2004), 34–35.
121. Dennis M. Kratz, '*Fictus Lupus*: The Werewolf in Christian Thought', *Classical Folia*, 30/1 (1976), 57–78, 66.
122. W.M.S. Russell and Claire Russell, 'The Social Biology of Werewolves', in: J.R. Porter and W.M.S. Russell (eds), *Animals in Folklore* (Cambridge, 1978), 143–182, 175.
123. *Corrector seu medicus*; John Thomas McNeill and Helena Margaret Gamer (eds and trans.) *Medieval Handbooks of Penance* (New York, 1979), 338, Question 151.
124. McNeill and Garner, *Medieval Handbooks*, 350, ft. 8.
125. Laurence M. Larson (trans.) *The Earliest Norwegian Laws: Being the Gulathing Law and the Frostathing Law* (New York, 1935), 123, no. 138.

126. Hans P. Broedel, *The* Malleus Maleficarum *and the Construction of Witchcraft: Theology and Popular Belief* (Manchester, 2003), 74.
127. Gaël Milin, 'La "Vita Ronani" et les contes de loup-garou aux XIIe et XIIIe siècles', *Le Moyen-âge* 97 (1991), 259–273.
128. Guðmundsdóttir, 'The Werewolf', 291; Alternatively This May Draw on the Same Oral Tradition as the Werewolves of Ossory, Sconduto, *Metamorphoses*, 34–35.
129. Sconduto, *Metamorphoses*, 22–23. Some have argued the reality of metamorphosis as described in medieval courtly literature diverges from the theological interpretation of its illusionary status, cf. Oates, 'Metamorphosis', 312; Richard Kieckhefer, *Magic in the Middle Ages* (Cambridge, 2000), 184.
130. Stephen Mitchell, 'Magic as Acquired Art and the Ethnographic Value of the Sagas', in: Margaret Clunies Ross (ed.), *Old Norse Myths, Literature and Society* (Odense, 2003), 132–152, 146.
131. Stephen Mitchell, *Witchcraft and Magic in the Nordic Middle Ages* (Philadelphia, 2011), 116.
132. Catharina Raudvere, 'Now You See Her, Now You Don't: Some Notes on Shape-Shifters in Scandinavian Tradition', in: Sandra Billington and Miranda Green (eds), *The Concept of the Goddess* (New York, 1996), 41–55, at 47.
133. John Lindow, 'Norse Mythology and the Lives of the Saints', *Scandinavian Studies* 73 (2001), 437–456.
134. David B. Leshock, 'The Knight of the Werewolf Bisclavret and the Shape-Shifting Metaphor', *Romance Quarterly* 46/3 (1999), 155–165.
135. Rhonda Knight, 'Werewolves, Monsters and Miracles: Representing Colonial Fantasies in Gerald of Wales's *Topographia Hibernica*', *Studies in Iconography* 22 (2001), 55–83, 74–75.
136. Catharina Raudvere, '*Trolldómr* in Early Medieval Scandinavia', in: Karen Jolly, Catharina Raudvere and Edward Peters, *Witchcraft and Magic in Europe: The Middle Ages* (London, 2002), 73–172, 103.
137. E. William Monter, *Witchcraft in France and Switzerland* (London, 1976), 147.
138. *De mirabilibus Hibernie;* Aubrey Gwynn. *The Writings of Bishop Patrick (1074–1084)* (Dublin, 1955), 62–63.
139. Carey, 'Werewolves in Medieval Ireland'.
140. *Otia Imperialia*, I, S.E. Banks and J.W. Binns (eds and trans.), *Gervase of Tilbury, Otia Imperialia, Recreation for an Emperor* (Oxford, 2002), 85–89; 813–815.
141. Guðmundsdóttir, 'The Werewolf', 294.
142. Mitchell, *Witchcraft and Magic in the Nordic Middle Ages.*
143. *Gesta Danorum* 7 (181–182), Hilda Ellis Davidson, and Peter Fisher, *Saxo Grammaticus: The History of the Danes*, I–IX (Cambridge, 1980), 201–202; Gerard Breen, '"The Wolf is at the Door": Outlaws, Assassins, and Avengers Who Cry "Wolf!"', *Arkiv för nordisk Filologi* 114 (1999), 31–43, 33.
144. Banks and Binns, *Otia Imperialia*, 814–815.
145. Rhonda Knight, 'Werewolves, Monsters and Miracles: Representing Colonial Fantasies in Gerald of Wales's *Topographia Hibernica*', *Studies in Iconography* 22 (2001), 55–83, at 74–75.
146. See Charlotte Otten (ed.), *A Lycanthropy Reader: Werewolves in Western Culture* (New York, 1986).
147. Edmund K. Chambers, *The Mediaeval Stage* (Mineola, 1996; orig. 1903), 235–9; Violet Alford, *The Hobby Horse and Other Animal Masks* (London, 1978), 20–21.
148. Twycross and Carpenter, *Masks and Masking*, 23–24, 30.

149. Ronald Hutton, *The Rise and Fall of Merry England: The Ritual Year 1400–1700* (Oxford, 2001), 62.
150. Terry Gunnell, 'Masking and Mumming Traditions in the North Atlantic: A Survey', in: Terry Gunnell (ed.), *Masks and Mumming in the Nordic Area* (Uppsala, 2007), 275–326, 285–286.
151. Gunnell, *Origins of Drama*, 80–83.
152. Twycross and Carpenter, *Masks and Masking*, 39–41.
153. Max Harris, 'From Iraq to the English Morris: the Early History of the Skirted Hobby Horse', *Medieval English Theatre*, 25 (2003), 71–83.
154. Richard Ettinghausen, 'The Dance With Zoomorphic Masks and Other Forms of Entertainment Seen in Islamic Art', in: George Makdisi (ed.), *Arabic and Islamic Studies in Honor of Hamilton A. R. Gibb* (Leiden, 1965), 211–224, 215.
155. E.C. Cawte, *Ritual Animal Disguise: A Historical and Geographical Study of Animal Disguise in the British Isles* (Cambridge, 1968), 198.
156. Alford, *Hobby Horse*, 104.
157. Cawte, *Ritual Animal Disguise*, 11.
158. Malcolm Jones, *The Secret Middle Ages: Discovering the Real Medieval World* (Stroud, 2002), 112.
159. Jones, *Secret Middle Ages*, 101.
160. Werner Mezger, 'Steckenpferd – Hobbyhorse – Marotte. Von der Ikonographie zur Semantik', *Zeitschrift für Volkskunde*, 79 (1983), 245–250.
161. The one example of a hobby horse can be found in GKS 3384 (Det Kongelige Bibliothek, Copenhagen), f. 86v (Flemish psalter, first quarter of the fourteenth century).
162. According to Lilian M.C. Randall's survey of manuscript marginalia, *Images in the Margins of Gothic Manuscripts* (Berkeley, 1966), 165.
163. MS fonds français 95 fol. 273r (Paris, Bibliothèque Nationale).
164. MS Fr. 146 fol. 36v (Paris, Bibliothèque Nationale).
165. Twycross and Carpenter, *Masks and Masking*, 46–48.
166. Add. 30029, ff. 61v, 99, 132 (British Library).
167. Rothschild MS, f. 175 (Beinecke Library, Yale).
168. Cloisters 54.1.2. f. 193v (Metropolitan Library, New York).
169. M.754. f.42v (Pierpont Morgan Library, New York).
170. MS Bodley 264 fol. 70r (Bodleian, Oxford).
171. MS Bodley 264, fol. 21v; fol. 181v.
172. MS Fr. 95. F.261 (Paris, Bibliothèque Nationale); Bodley 264 fol.70 (Bodleian, Oxford); GKS 3384, f. 11v (Det Kongelige Bibliotek, Copenhagen); W. 88, ff. 115v, 156v (Franco-Flemish hours, early fourteenth century; Walter Art Museum, Baltimore MD).
173. On Norway: Eike, Masks, 50; on Sweden: Eva Knuts, 'Masks and Mumming Traditions in Sweden: A Survey', in: Gunnell (ed.), *Masks and Mumming in the Nordic Area* (Uppsala, 2007), 107 188, at 110; on Denmark. Carsten Bregenhøj and Hanne Pico Larsen 'Masks and Mumming in Denmark: A Survey', in: Gunnell (ed.), *Masks and Mumming*, 189–274, 196.
174. Christine Eike, 'Masks in Mumming Traditions in Norway: A Survey', in: Gunnell (ed.), *Masks and Mumming*, 47–106, 50.
175. Price, *The Viking Way*, 172.
176. E.A. Rybina, 'Recent Finds From Excavations in Novgorod', in: Mark A. Brisbane (ed.), *The Archaeology of Novgorod, Russia* (Lincoln, 1992), 160–92, 181f.
177. A. Celmiņš, 'Rigas 13. gadsimta adas maska', *Sena Riga* 5 (2005): 93–103.

178. Ülo Tedre, 'Masks and Mumming Traditions in Estonia: A Survey', in: Gunnell (ed.), *Masks and Mumming*, 367–448.
179. Twycross and Carpenter, *Masks and Masking*, 49.
180. Hutton, *The Rise and Fall of Merry England*, 261.
181. Tom Pettitt, 'The Morphology of the Parade', *European Medieval Drama* 6 (2002), 1–30, at 26.
182. Michael Camille, *Mirror in Parchment: the Luttrell Psalter and the Making of Medieval England* (London, 1998), 251–252.
183. Ruth Mellinkoff, *Averting Demons* (Los Angeles, 2004); Camille, *Mirror in Parchment*, 265.
184. Chantilly 62. F. 201.
185. Laurence Wright, 'The Recorder Consort in the Renaissance', *The Recorder and Music Magazine* 1.6 (1964), 179–180.
186. Twycross and Carpenter, *Masks and Masking*, 54.
187. Twycross and Carpenter, *Masks and Masking*, 60.
188. Alessandro Arcangeli, 'Carnival in Medieval Sermons', *European Medieval Drama*, 1 (1997), 109–117, at 113.
189. Mänd, 'Devils in Baltic Towns', 24; Tom Pettitt, '"This Man is Pyramus": A Pre-History of the English Mummers' Plays', *Medieval English Theatre*, 22 (2000), 70–99, 76.
190. Alford, *Hobby Horse*, xxi.
191. Davidson, 'Shape-Changing', 141.
192. A. Margaret Arent, 'The Heroic Pattern: Old Germanic Helmets, Beowulf, and Grettis saga', in: Edgar C. Polomé (ed.), *Old Norse Literature and Mythology* (Austin, 1969), 130–199, at 138.
193. Mackay, *Malleus*, 1/39; Nicholas Eymeric, *Directorium Inquisitorium* (1376); Levack, *Witchcraft Sourcebook*, 43.
194. Bengt Ankarloo, 'Sweden: The Mass Burnings (1668–1676)', in: Bengt Ankarloo and Gustav Henningsen (eds), *Early Modern European Witchcraft: Centres and Peripheries* (Oxford, 1990), 284–317.
195. Arno Borst, 'The Origins of the Witch-Craze in the Alps', in: Brian P. Levack (ed.), *New Perspectives on Witchcraft, Magic and Demonology, Volume 2: Witchcraft in Continental Europe* (New York, 2001), 299–320, at 109–110.
196. *Liber Razielis* (book 7) refers to the achievement of goal 98 (out of 150) by this magical operation: 98 (f.199): 'ut habeas potestatem ut convertendi te in forma lupi'; 'to have power so that to transform you into the shape of a wolf'. MS Rome, Biblioteca Vaticana, Reg. lat. 1300 (s.xiv); from a Spanish Latin translation of the Hebrew text.
197. Guðmundsdóttir, 'The Werewolf', 301.
198. 1407 and 1423 in Basel; 1428–1429 in Valais (Lorey: www.elmar-lorey.de/Prozesse.htm).
199. Monter, *Witchcraft in France and Switzerland*, 149.
200. Martine Ostorero, 'The Concept of the Witches' Sabbath in the Alpine Region (1430–1440)', in: Gábor Klaniczay and Éva Pócs (eds) *Witchcraft Mythologies and Persecutions* (Budapest, 2008), 15–34.
201. Eduard Hoffmann-Krayer, 'Zum Eingang des Weingartner Reisesegens' *Schweizerisches Archiv für Volkskunde* 8 (1912), 65.
202. Liv Helene Willumsen, 'Children Accused of Witchcraft in Seventeenth-Century Finnmark', *Scandinavian Journal of History* 38 (2013), 18–41, 35–36.
203. Shapes could include cats (effected by smeared in cat's blood and wearing a cat skin), Willumsen, 'Children Accused of Witchcraft', 27.

204. Twycross and Carpenter, *Masks and Masking*, 15; Robert Darnton, *The Great Cat Massacre and Other Episodes in French Cultural History* (London, 2001), 17.
205. Summarised for Scandinavian regions in Price, *Viking Way*, 205–206.
206. Twycross and Carpenter, *Masks and Masking*, 31.
207. Tulley, *Shamanism* I, 568.
208. Judith Jesch, 'Poetry in the Viking Age', in: Stefan Brink and Neil Price (eds), *The Viking World* (London, 2008), 291–298.
209. Inger Zachrisson, 'The Sámi and Their Interaction With the Nordic Peoples', in: Brink and Price, *The Viking World*, 32–39.
210. Tolley, *Shamanism* II, 192–194.
211. Hermann Pálsson, 'The Sami People in Old Norse Literature', *Nordlit* 3/1 (1999), 29–53, 40.
212. Tolley, *Shamanism* I, 558–580.
213. Elina Helander-Renvall, 'Animism, Personhood and the Nature of Reality: Sami Perspectives', *Polar Record* 46 (2010), 44–56.
214. Price, *Viking Way*, 226–227.
215. Laurence Harf-Lancner, 'La métamorphose illusoire: des théories chrétiennes de la métamorphose aux images médiévales du loup-garou', *Annales. Histoire, Sciences Sociales*, 1 (1985), 208–226.
216. Sconduto, *Metamorphoses*, 76.
217. Mitchell, *Witchcraft*, 50.
218. Mitchell, *Witchcraft*, 204.
219. Per-Anders Östling, 'Blåkulla Journeys in Swedish Folklore', *Arv* 62 (2006), 81–122, 96–97.
220. Guðmundsdóttir, 'Werewolf', 303.
221. Kirsten Hastrup, 'Iceland: Sorcerers and Paganism', in: Bengt Ankarloo and Gustav Henningsen (eds), *Early Modern European Witchcraft: Centres and Peripheries* (Oxford, 1990), 382–401, 400.
222. Jacob Grimm, *Teutonic Mythology* (London, 1882), 1096.
223. Per Peterson, 'Attitudes and Folk Belief About Wolves in Swedish Tradition', in: Mare Koiva and Kai Vassiljeva (eds), *Folk Belief Today* (Tartu, 1995), 359–362.
224. Britt-Mari Nasstrom, 'Healing Hands and Magic Spells', in: Geraldine Barnes and Margaret Clunies Ross (eds), *Old Norse Myths, Literature and Society. Proceedings of the 11th international saga conference, 2–7 July 2000, University of Sydney, Sydney* (Sydney, 2000), 356–362, at 360.
225. In the case of Eastern Europe, traditions concerning wolves and supernatural agency are similarly interpreted as bridges between pre-Christian and modern Christian culture. For example, the figure of 'the Master of the Wolves' documented in 51 variants in primarily Slavic folklore – a role attributed to various saints as well as folk figures such as the Leši (woodland spirits) – has been linked to a pre-Christian collective Slavic heritage (Mirjam Mencej, *Gospodar volkov v slovanski mitologiji* (Ljubljana 2001); pre-Christian collective Slavic heritage (M. Mirjam, Gospodar volkov v slovanski mitologiji (Ljubljana, 2001); 'Wolf Holidays Among Southern Slavs in the Balkans', *Acta Ethnographica Hungarica* 54 (2009): 337–358). It would be useful to examine these claims of chronological continuity with a critical investigation of the cultural and ecological contexts when these tales are recorded.
226. Ella Odstedt, *Varulven i svensk folktradition* (Uppsala, 1943).
227. Stefan Donecker, 'The Werewolves of Livonia: Lycanthropy and Shape-Changing in Scholarly Texts, 1550–1720', *Preternature* 1/2 (2012), 289–322; Maia Madar, 'Estonia 1: Werewolves and Poisoners', in: Ankarloo and Henningsen (eds), *Early*

Modern European Witchcraft: Centres and Peripheries, 257–272; Merili Metsvahi, 'Werwolfprozesse in Estland und Livland in 17. Jahrhundert. Zusammenstöße zwischen der Realität von Richtern und von Bauern', in: Jürgen Beyer, Reet Hiiemäe (eds), *Folklore als Tatsachenbericht* (Tartu 2001), 175–184; see also Chapter 9 in this volume.

228. Otten, *Lycanthropy Reader*, 102.
229. Elmar Lorey, 'Wie der Werwolf unter die Hexen kam: Zur Genese des Werwolfpozesses' (2003) (www.elmar-lorey.de/werwolf/genesetext.htm).
230. Pierre Dubuis, *Une économie alpine a la fin du Moyen Age*, I (Sion, 1990), 221.
231. Franco Morenzoni, 'Note sure la presence de l'ours en Valais et dals le Chablais Vaudois à la fin du Moyen Age', in: Michel Colardelle (ed.), *L'Homme et La Nature au Moyen Age* (Paris 1996), 153–156; Pierre Dubuis. *Dans les Alpes au Moyen Age: douze coups d'oeil sur le Valais* (Lausanne, 1997), 111.
232. Robin Briggs, 'Dangerous Spirits: Shapeshifting, Apparitions and Fantasy in Lorrain Witchcraft Trials', in: Kathryn A. Edwards (ed.) *Werewolves, Witches and Wandering Spirits: Traditional Belief and Folklore in Early Modern* Europe (Kirksville, 2002), 1–24, at 6.
233. Sconduto, *Metamorphoses*.
234. Jean-Marc Moriceau, *Histoire du méchant loup: 3000 attaques sur l'homme en France XV*[e]*–XX*[e] siècle (Paris 2007), 99; Odstedt (*Varulven*) links the decline of popular werewolf beliefs in Sweden in the mid-nineteenth century with the eradication of wolves (and bears) from the lowlands.
235. For the most recent study, see the papers in Piercarlo Grimaldi (ed.), *Bestie, Santi, Divinità: Maschere Animali dell'Europa Tradizionale* (Turin, 2003).
236. Wolf costumes are documented in some regions, e.g., in Poland in the seventeenth to nineteenth centuries, Leszek Paweł Słupecki, *Wojownicy i Wilkołaki* (Warsaw, 1994), 103–104) but the origin of these customs remains obscure. In the absence of any data connections with pre-Christian cultural traditions remain highly speculative.

Interlude: Wolf-Riding

In 1865 the German artist Ludwig Pietsch (1824–1911) drew this picture of the giantess Hyrrokkin as an illustration for *Die nordischen Göttersagen* by Rudolf Friedrich Reusch (Berlin, 1865, p. 110). Hyrrokkin is a figure from Nordic mythology, mentioned in the *Edda*. While it is unclear whether her wolf-riding is in any way related to werewolves (or Fenrir, for that matter), the question remains how this reflects on the situation in Switzerland, where the connection between wolf-riding and werewolves is made, if not by the historical actors, then at least by more recent interpreters. Why are wolf-riders only found in the poetry of the thirteenth-century Icelandic Snorri Sturluson and in fifteenth-century Swiss trial sources, and based on these in *De lamiis* of 1489 by Ulrich Molitor? The answers are still outstanding, but the question remains intriguing. The image opposite is from the German 1491 edition of Molitor's book, *Won den unholden oder hexen* (p. 25), Bayerische Staatsbibliothek.

Another issue that still needs discussing a bit more is the meaning of wolf-riding. In his 1976 book *European Witch Trials*, Richard Kieckhefer professed that he had no idea.[1] On the basis of a passage in the trial reports of Lucerne, where *bluot schand* is mentioned,[2] I have suggested that this instance of wolf-riding pointed to mother–son incest; but this was met with a lukewarm reception.[3] Molitor, however, provides some basis for this explanation, as he writes that the *hexen* ride the wolves *zu irem wollust*, for their sexual pleasure. In that case wolf-riding and werewolves can be associated through the theme of illicit sexuality. WdB

Notes

1. Richard Kieckhefer, *European Witch Trials* (Berkeley, 1976), 24–25, 71.
2. E. Hoffmann-Krayer, 'Luzerner Akten zum Hexen- und Zauberwesen', *Schweizeres Archiv für Volkskunde* 3 (1899), 81–122, esp. 83.
3. Laura Stokes, *Demons of Urban Reform* (Basingstoke, 2011), 24.

5

'What about Some Good Wether?' Witches and Werewolves in Sixteenth-Century Italy

Matteo Duni

Strange tales can turn up from the files of an inquisitor, especially from those of inquisitors as keenly committed to the prosecution of witchcraft as was Bartolomeo Spina, in early sixteenth-century Italy. The Dominican friar and noted demonologist had been leading the tribunal of Modena for only about a month at the outset of his career as a judge. In the last days of December 1518, he received a series of unusual depositions from a number of women and men.[1] These testimonies had probably been instigated by the inquisitor's proclamation of a period of grace, during which anyone reporting suspect behaviours or potentially heretical beliefs would have earned special indulgences, following which period they would have incurred excommunication.[2]

If the procedure (the period of grace) and its result (a string of depositions) were common, what some of the witnesses told Spina was not. On 27 December, Apollonia, wife of Giovanni Lorenzo Villani, appeared voluntarily before the inquisitor in the church of San Domenico. When asked 'whether she knew any witch' (*striga*), she answered that about two years prior:

> [...] she had heard from several persons, whose names at the moment she does not recall, that near a village called La Bastia live three brothers of the Garutti family, one of them named Pietro, another one Giacomo, while she cannot remember the name of the third one, [and that] the aforementioned Pietro and Giacomo are witches and go to the sabbat (*sunt strigi et vadunt ad cursum*).
>
> Interrogated on what basis those people would affirm that the said brothers did such things, she answered that she heard that the wife of one of them – and she believes it was the wife of Pietro – once found him laying half-dead (*semimortuum*) on the ground, and was so distracted with fear that she fell ill and, as they say, eventually died.[3]

The information provided by Apollonia did not seem to kindle Spina's interest, perhaps because it was entirely based on hearsay. In the rest of the short interrogation the inquisitor only asked whether the witness could name any of her informants (and she mentioned one), and when exactly she had heard what she reported. Evidently he was trying to see if sufficient evidence could be gathered to justify further investigation of the story of the two brothers, but apparently Apollonia did not know anything else.

In any case, there was little time to think about her testimony since another woman came to see Spina right after Apollonia. Girolama, widow of *magister* Ludovico Spagnini, was asked whether she knew any 'witch or sorceress et conjurer of demons and things of this kind' (*aliquam personam maleficam vel incantatricem et invocantem demones et huiusmodi*). Her answer consisted in a tale she had heard from a certain Ursolina, wife of *magister* Bernardo Porrini. About two or three years before, Ursolina had found herself in the company of two peasants (*rustici*), whose names she did not remember, when one of them told the other, 'Take this belt' (*zona*):

> [...] and he rolled it up and then tied the said belt with a string (*et revolvit ipse eam in unum et deinde ligavit dictam zonam cum una stringa*) and gave it to the other one, saying: 'I want you to eat some good wether'. And as soon as he had given him the belt tied as above, he turned into a wolf (*in forma lupi conversus est*) and thus in the shape of a wolf attacked a flock of wethers. But some boys there began crying out for help and peasants rushed with sticks, killing him as he was still in a wolf's shape, while his friend, who was holding the belt, untied it.[4]

Apparently, the strangeness of the events reported did not impress the inquisitor more than the previous testimony had, since he asked Girolama only whether someone else had been present when Ursolina had told those things. Then he moved on to ask if she knew any other witches or sorcerers, actually showing more interest in an unnamed sorceress who reportedly bewitched children in Modena, as the witness had heard – a common type of offence that amounted to pure routine for an inquisitor.

However, something in Girolama's words must have aroused Spina's curiosity, since the next day he summoned precisely the person whom the woman had named as having also heard Ursolina's tale. Francesca Ferrari readily confirmed that indeed, three days before, in the house of Andrea Villani in Modena, she had heard from Ursolina the story of the peasant-turned-wolf, as Girolama had recounted.[5] But a quick question about the unnamed sorceress mentioned in the previous testimony was all that the inquisitor cared to ask to Francesca, and her very short deposition seemed to be the last of the entire, short investigation.

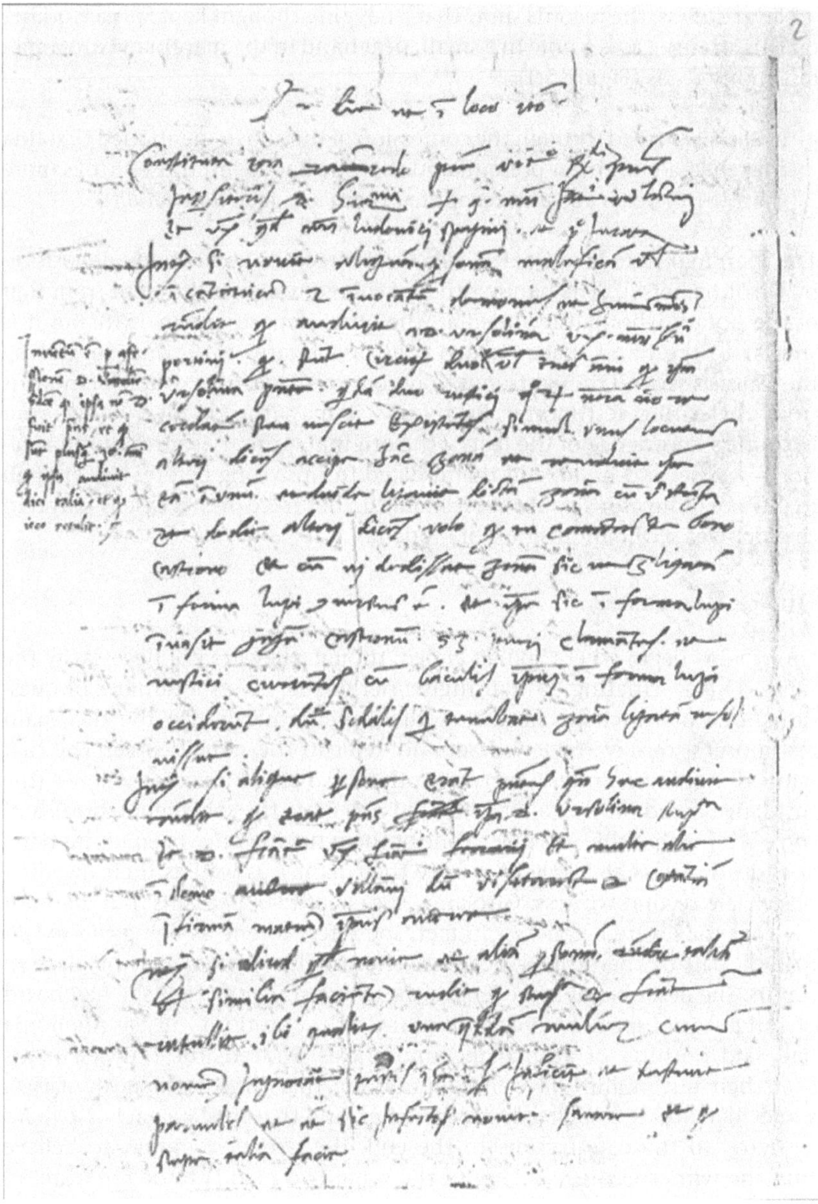

Figure 5.1 Archivio di Stato di Modena, Inquisizione, busta 2, fasc. II, *Contra Petrum et Jacobum strigos ad Bastiam*. The marginalia reads: 'Inventum est per confessionem Domine Ursoline predicte quod ipsa non fuit presens et quod sunt plusquam 30 anni quod ipsa audivit dici talia et quod ioco retulit'

Nevertheless, the records show that a nagging thought kept Spina working on this strange case. A note in a small, neat hand in the margin of Girolama's testimony reads (Figure 5.1):

> It was discovered through the confession of the above-mentioned Ursolina that she had not been present, and that she had heard those things more than thirty years ago and recounted them as a joke (*ioco retulit*).

The marginal note is not dated, nor do the records include any deposition by Ursolina herself. The handwriting of this comment is different from that of the notary who had recorded all the other depositions in 1518, but it is similar to the hand that recorded trials as of early 1519, thus suggesting that Spina's second thoughts must have occurred within a couple of months from the earlier testimony.[6] Be that as it may, the discovery that even Ursolina's knowledge of the facts – the original, primary source of information – was second-hand, and that it dated from so long before, was enough to put the inquisitor's doubts to rest: no further trace of this case is found in the archives of the Inquisition of Modena.

One or two cases?

Our doubts begin where Spina's ended, though they are not necessarily the same. This fascinating, tantalisingly brief dossier raises a number of questions, one of which we need to address preliminarily: Do the two main testimonies we have reviewed actually refer to the same people? The first witness, Apollonia, mentions two brothers as being known as witches (the third, unnamed brother apparently did not share the same reputation), and she explains that this was due to one of them having once been found lying as if dead. She is also able to specify their names as well as their dwelling place. The second witness, Girolama, speaks of two 'peasants' but does not say that they were brothers – in fact, she refers to one as being the *sodalis* (friend, pal) of the other – nor does she provide their names or place of origin. She describes at some length the shape-shifting event she had heard of, but does not speak a word about near-death conditions – just as Apollonia had said nothing of animal metamorphosis. Indeed, the only elements that their testimonies have in common are that they both speak of two, somewhat closely associated, male witches, and that they characterise them as living in the countryside. In the end, the strongest motive to believe that the witnesses are speaking of the same two men is to be found in an external fact: at the time, Spina treated these testimonies together, as if part of a single dossier (which was later given the heading 'Against the sorcerers Pietro and Giacomo from Bastia'), and thus he must have been convinced that they referred to the same case.

The inquisitor's conviction about the identity of the alleged sorcerers and the similarities between the two women's reports, though not amounting to a decisive proof, tip the scale in favour of the identification of the two unnamed peasants of Girolama's testimony with the Garutti brothers, Pietro and Giacomo, from Bastiglia. My agreement with Spina stops here, however, for the inquisitor chose to ignore Apollonia's and Girolama's words as having no value, essentially because they related things heard from someone who in turn had only heard them decades before. Friar Bartolomeo Spina could not rely (just) on hearsay since, as a judge, he was trying to ascertain whether those facts had actually happened and who had been involved. His decision unfortunately deprives us of the possibility to hear the voice of the surviving brother – if the other one was actually killed (in whatever shape) – nor can we understand whether both would have the same experiences or, if not, who would turn into a wolf and who would fall into a trance. Half a millennium later, however, the two women's accounts are important for us as an expression of collectively held and shared beliefs, even if they were not first-hand and even if we cannot ascribe all the elements described to both witches.[7]

As the testimonies present separately two clusters of themes: i.e., the lethargy in conjunction with the concept of the *cursus*, and the transformation of man into wolf, I will analyse them one by one, highlighting what their meaning might have been in the local context of Modena, and of Northern Italy in general. I will then try to see whether it is possible to read them as different aspects of the same set of beliefs.

Cursus

The first aspect of interest in this dossier is the mention of the witches' sabbat, to which Apollonia refers to with the most common of its local names, *cursus*: the ride, or course.[8] References to the sabbat are rare in the files of the Modenese Inquisition from this period. This is indeed one in only four occurrences of the word *cursus* I have encountered in the surviving records from the last years of the fifteenth to the first quarter of the sixteenth century, with all the other three dating from February 1519 (when the inquisitor was also Spina).[9] However, since *cursus* seems to have been used each time with a different meaning, I will review briefly these instances in order to define what the semantic range encompassed by the term appears to be.

The first two mentions are found in the trial against Chiara Signorini, who was accused of having bewitched Margherita Pazzani after the latter had evicted Chiara and her husband, Bartolomeo, from a farm she owned.[10] On 12 February 1519 a certain Caterina Bongardi was questioned by Spina about the reputation of the Signorini couple. She answered that she had never seen them attending mass and that she often had heard them getting up and leaving their house at ungodly hours of the night. Once, after she had jokingly reproached them for being 'like witches who go out at night' (*quod ibant ut striones de nocte*), the two became so angry that, upon hearing of Chiara's arrest, Caterina concluded that indeed they had been going at the sabbat (*iudicavit ipsos ire ad cursum*). Three days after Caterina's testimony, Spina questioned Chiara under torture 'whether she went to the *cursus* and lay with the devil and offered any animal to the devil as a sacrifice, and whether she defiled any sacrament on the devil's orders or renounced the Catholic faith or baptism'. Chiara denied all these charges and her sentence (confinement and charitable work in a hospital for a few years) condemns her contacts with the devil, but contains no mention of participation in the sabbat. It seems, therefore, that the term *cursus* was introduced by a witness who associated it generically with the going-out at night that witches were supposed to do. The inquisitor, while holding on to the local word, had been quick to define it according to his own categories, identifying it explicitly with the meeting of the devil and its minions described in stereotypical terms, although he had not insisted on having the defendant confess to participation in such a gathering.

Cursus returns, with more original and local connotations, in another testimony received by Spina a few days later. On 23 February 1519, Giovanni da Ronco reported to the inquisitor what he had heard about a certain Zilia and her mother. The two women were known to be participating in the *cursus*, a great gathering presided over by a female figure known as the *Domina cursus* – the Lady of the Ride.[11] Participants in this event would feast on the meat of an ox, which the *Domina* would eventually resuscitate by hitting the dead animal's hide and bones with her stick. We now know, thanks also to Maurizio Bertolotti, that such reports testify to an ancient and widespread complex of Eurasian folklore, typical of societies based on hunting.[12] Between the fourteenth and the sixteenth centuries a series of documents (including those from the Inquisition of Modena) show that identical beliefs were still current and meaningful for the rural populations of Northern Italy, for whom the cult of nocturnal female deities – such as the Lady of the Ride – were a guarantee of plenty for both the home and the village. Obviously unaware of (and uninterested in) such a rich texture of myths and traditions, Spina responded to Giovanni da Ronco's tale with utter disbelief, to the point that he asked the witness whether his informant, a certain Mariotto Martinelli, 'had been speaking in earnest or rather joking' (*diceret hec serio vel potius ioco*). Probably, the inquisitor doubted the authenticity of this testimony due to its striking similarity to an episode of the life of Saint Germanus, when the saint had miraculously revived a calf after having dined on it.[13] But perhaps Spina's reaction – completely different from his reactions to the testimonies of Girolama and Apollonia – was due also to his suspicion that the story of the Lady of the Ride might be like the story of the shape-shifting peasant he had heard only two months before: a second-hand tale recycled as a joke.[14] Spina decided not to follow up on Giovanni's words, but did not forget them. In his *Quaestio de strigibus* (The

Problem of the Witches, 1523, based also on his inquisitorial experience in Modena), he identified the *Domina cursus* as one of the devil's many avatars, and interpreted the resuscitation of the animal as an example of demonic delusion.[15] Through a protracted reflection he had managed to make sense of the story of the ox's bones, in the end adapting it to fit the paradigm of diabolical witchcraft.

A trance-like state

In both Chiara Signorini's trial and in the testimonies on Zilia, therefore, Spina had found the concept of *cursus* to have elements compatible, to a varying degree, with his preconceived notions of what witchcraft meant and included. However, this was not to be the case with Apollonia's mention of Pietro Garutti's apparently lifeless condition, as Spina did not try to associate it with witchcraft. One possible reason for this is that Pietro's reported state, and the description of an instance of metamorphosis in the immediately following testimony, might have reminded the inquisitor of some of the arguments put forward by the opponents of the witch-hunt. Such sceptics, Spina wrote in the *Quaestio de Strigibus*, deem that the witches are not physically transported to the sabbat, but fall into a very deep sleep after having anointed themselves with a diabolical unguent: '[...] and they dream of being transported to faraway places, to turn into cats, and to do, or to undergo, all those things which they eventually believe to have done, or to have undergone.[16] This quote makes a connection between the drug-induced catalepsy and the animal transformations, presenting them as two strong reasons to doubt the reality of many of the witches' feats. But Spina, while agreeing with the sceptics on the merely delusory nature of shape-shifting, refused to conclude on this basis that the entire sabbat was a mental illusion, and actually devoted his tract to demonstrate its reality. Perhaps he felt that any mentions of trance-like states could potentially reinforce a naturalistic interpretation of witchcraft as the fruit of artificially induced hallucinatory experiences, and thus decided not to look further into it.

While it is not so hard to think of a few reasons why Spina would have chosen to disregard a witness' peculiar account of the *cursus*, the fact that Apollonia would have so readily brought up Pietro Garutti's swoon, in an answer to the question whether she knew any witches, is very suggestive. If indeed, as one gathers from the record, she associated the practice of witchcraft and the ability to go to the *cursus* with a loss of consciousness closely resembling death, then her testimony can be read in the light of later proceedings against Domenica Burbarelli, from the village of Novi. Domenica was investigated by the Inquisition of Modena between 1532 and 1534 for having publicly expressed her determination to go to the 'ride of Diana' (*cursus Diane*).[17] Apparently, several people had kept her under a close watch in order to avert her plan, but – as one of them testified – they were unable to prevent her from falling into a sort of catalepsy. She had lain as if dead for two hours (*iacuit ut mortua per duas horas*) while they tried repeatedly to bring her back to her senses. When she finally woke up, she addressed bystanders contemptuously: 'I managed to go there, in spite of you!', and related how, at the Game (*ludus*), the 'Lady' had encouraged her to defile the cross and to dance with demons.[18]

This episode is reminiscent of a common narrative, which is found in a number of demonological treatises: the witch is convinced of having physically gone to the sabbat, but several witnesses have watched her sleeping the whole time.[19] However, this event was brought up by a witness doubtless unfamiliar with this kind of literature, and thus the deposition can be read, in conjunction with Apollonia's testimony, as evidence of the way in which early modern Modenese associated cataleptic states and participation in the *cursus*, and that someone showing such peculiar behaviour could easily be seen as a practicing witch.

Although the evidence provided by these documents is rather limited, it does not seem too far-fetched to compare the altered states of Pietro and Domenica to the belief-complex of the *benandanti*, studied by Carlo Ginzburg in his pioneering work, *The Night Battles*.[20] In the sixteenth century women and men from the Italian region of Friuli were known for falling into a swoon so deep that they seemed to be dead, but when they returned to their senses they would describe the extraordinary experiences their spirit had participated in while their body remained home unconscious. Some of them (exclusively men) related that they went during the night time to battle witches in open fields, often located far away from their homes. The outcome of the battle was crucial for their communities, since the victory of the witches would bring about scarcity and famine, while the victory of the *benandanti* would guarantee a plentiful harvest. Other *benandanti* (mostly women) could see the souls of the dead, participate in their procession and speak to them, receiving information they could refer back to the deceased's still living relatives. Both types of *benandanti* would 'go out in spirit' at particular moments of the year, usually the Ember Days: four days associated with different phases of the agricultural calendar. Both also considered their ability to stem from their having been born with the caul, and their mission to be a kind of religious duty. Inquisitors were at first puzzled by beliefs they had never heard of, but they gradually managed to recast the nocturnal processions and the battles for the crops in the light of the witches' sabbat, although capital punishment was never meted out against either type of *benandante*.

Werewolves

On such unique, out-of-body experiences the records of the Modenese Inquisition do not provide anything comparable to the wealth of information found in the Friulian documentation on which Ginzburg based his classic (but still disputed) thesis of the folkloric origin of the witches' sabbat. Yet even a brief, interrupted investigation, such as that of the Garutti brothers, opens up fascinating possibilities for unexpected connections with the myths and traditions analysed both in *The Night Battles* and in Ginzburg's more recent, ambitious and controversial *Ecstasies*.[21]

This is indeed the case with the episode of Pietro's transformation into a wolf as related by two witnesses to Ursolina's narration. It is a remarkable story, first of all as a very rare attestation of the belief in lycanthropy in this part of the Italian peninsula; to the best of my knowledge, it is the only mention in the archives of the Modenese Inquisition. References to werewolves are few and far between in contemporary Italian sources. While several demonologists speak of the witches' ability to turn into cats (usually in order to harm children), the transformation of man into wolf is not discussed.[22] Authors that do mention lycanthropy do not seem to connect it to any real, present-day case, but rather to literary examples from classical antiquity.[23] The influential jurist Andrea Alciato, writing on witchcraft in the 1520s, mentions in passing *lycaones*: i.e., men transformed into wolves (a reference to Ovid). But, it is simply an item in a list of pathologies of the imagination, which should be treated medically, just as the mental delusions of the 'poor ordinary women' who believed that they went out at night to dance with the demons.[24]

One has to turn to Johann Wier (or Weyer), the greatest of the sixteenth-century 'witchcraft sceptics', for a full-blown description of an Italian werewolf. In his *De praestigiis daemonum* (On the Tricks of Demons, 1563), Wier considers lycanthropy to be the fruit of a mental disease, which is enhanced and exploited by the demons in order to deceive men, and supports his thesis with the discussion of several contemporary cases, one of which occurred in Padua in 1541.[25] There a peasant was captured after having attacked and killed many persons in the countryside area, and had declared with strong conviction to be a real wolf. The only difference with his fully bestial companions, he had said, was that his wolf's skin had the fur on the inside instead of the outside. At this point the peasant's capturers cut off his limbs with a sword in order to ascertain whether this claim was true, predictably discovering that it was not and at the same time causing his death: those men were indeed 'true, wild and ravenous wolves', wrote an indignant Wier.[26] In his analysis of another case, he went on to argue that the supposed werewolves' confessions should never be taken at face value, since these people were sick and under the influence of the devil; rather, physicians should be called to evaluate such instances of what he diagnosed as *melancholia lupina* (or *lycaona*).

The wolfskin of the Paduan peasant has a significant parallel in the belt used by one of the Garutti. A recurrent component of the beliefs in werewolves was that donning a wolf skin or a belt – sometimes a belt made of wolf's leather – was a necessary step for the transformation of a man into a wolf. This was especially the case in Northern Europe, where such details figure prominently both in legends and records of trials.[27] In 1589, in what was certainly the most sensational episode of lycanthropy in Europe, the German werewolf Peter Stump confessed to having killed 16 people after the devil had given him 'a girdle which, being put around him, he was transformed into the likeness of a greedy, devouring wolf, strong and mighty, with eyes great and large, which in the night sparkled like brands of fire'.[28] As soon as he took off the girdle, the record continues, he would recover his human shape.

Garutti's transformation differs from Stump's (and from the other cases I know of) for two main reasons. First, apparently it was not the wearing of the belt which caused the metamorphosis, but the rolling up and tying of it.[29] Secondly – and logically – it seems that the werewolf was brought back to his human shape by the untying of the belt, not by its removal. In the transcript we read that Garutti was killed by the peasants, who were running to the rescue of the flock, 'as he was still in a wolf's shape while his friend, who was holding the belt, untied it'. Although the record is not explicit on this point, what is implied is that the irate peasants arrived before the other Garutti managed to fully untie the belt, and thus before his brother could be returned to his human condition, which would have saved him from his sad fate. Folklore had it that shedding a werewolf's blood would also provoke a counter-metamorphosis, but unfortunately the narrative of our episode is somewhat truncated and does not say anything about what happened after the killing.[30]

However, the case of the Garutti brothers has another, significant feature in common with trials against werewolves from Northern Europe. In what can be considered the 'heartland of the persecution', between eastern France, the Rhineland and the Netherlands, the slaughter of livestock (as well as of humans) was a standard allegation against werewolves. Significantly, shepherds seem to have amounted to a sizable number among those tried, to the point that accusations of being a werewolf can be considered a sort of professional liability – especially when attempts at driving out wolves from the vicinity of pastures had failed and sheep had been attacked.[31] While in the trial the Garuttis are simply defined as *rustici* ('peasants'), the setting of the brief experience of Pietro as a werewolf is clearly that of a pasture, with boys guarding a flock of sheep.

The Modenese tale nonetheless has a peculiar quality, which seems absent from the (usually) rather gruesome transalpine narratives on werewolves: a somewhat humorous tone. Garutti's words to his brother, 'I want you to eat some good wether', uttered just before turning into a wolf, sound more like an invitation to enjoy some gastronomic delicacy than the prelude to a horrific display of bestial violence. One reason for this is certainly to be found in the context in which the story had been recounted by Ursolina: according to Girolama, she told her tale at a gathering of 'many women' who were paying a visit to an ailing friend on Christmas day.[32] An anecdote of a werewolf who spoke like a gastronome was perfect to lift spirits, and indeed Ursolina declared to Spina that she had 'recounted [it] as a joke'. Such a light-hearted approach to this kind of subject does not seem to have been an isolated phenomenon. In a 1565 letter from Mantua a certain Scalzone, probably a member of the local elite, writes to an unnamed 'castellan' about the stories he has heard in the shop of an apothecary. One of them concerned the wolves that had reportedly attacked and killed many people in an area south of the city, so as to require the intervention of Cesare Gonzaga, count of Guastalla and cousin of the Duke of Mantua Guglielmo. The rest of the letter is worth quoting at length:

> Some old women (*vecchiette*) here say that these are evil wolves (*lupi rei*): i.e., men turned into wolves by magic, who greatly enjoy feeding on human blood, as do witches. I have heard that an old woman – I am not kidding – went to the Friars of Saint Dominic [i.e. the Inquisition] and reported two gardeners from the Palazzo Te (*ortolani dal Thé*). She claims that she saw them anointing themselves with an unguent, and they were immediately transformed into wolves. These rascals, according to what she says, went to take two little lambs out of her flock, and then took them to their homes and ate them in the company of their women. I hear that, because of such nonsense (*baia*), those poor men are in trouble, having been summoned by the inquisitor. Now, Your Excellency, isn't this a great joke (*bella burla*)?[33]

The Mantuan Inquisition's archive is no longer extant, and thus we cannot know whether the two gardeners were indeed questioned, or learn more about their alleged metamorphosis. The letter certainly describes their feats with a tone of mocking disbelief. Not unlike their Modenese colleague, these Mantuan werewolves are portrayed like *bon vivants*, who magically acquire an animal shape in order to supply their table with choice meat and then feast on it with their women, presumably bragging about such exploits. Scalzone's narrative is strongly coloured by social and gender allusions. An underlying misogyny makes him sympathise with the supposed lupine thieves, while the old woman is implicitly blamed for having caused their troubles with her silly accusations. As far as social status goes, Scalzone is clearly voicing upper-class condescension for the foolishness of the populace, as well as incredulity toward the belief in werewolves, which seems to him nothing more than an object of derision.

Werewolves and trance

Renaissance Italians, then, could and would laugh about werewolves, perhaps more than was usual north of the Alps.[34] However, this does not mean that we should discount their tales. Unlike inquisitor Spina, we can see how the dossier on the Garutti brothers includes an element which cannot be explained simply by the joking intentions of one witness: the connection between trance-like states and animal metamorphosis.

This aspect was first brought to the attention of scholars by Carlo Ginzburg's research on the *benandanti*, which highlights the striking similarities between the beliefs of the Friulian ecstatic defenders of the crops and those of Livonian (i.e., Latvian) werewolves. The members of both groups would go out at specific moments of the year, in a trance or in animal form, to an otherworldly dimension, where they would fight the witches and defend the crops, or anyway guarantee abundance of food to the community. Ginzburg considered the lethargy of the *benandanti* and the animal metamorphosis of the werewolves to be functionally equivalent, and to represent the soul's exit from the body and its journey to another dimension. Therefore, according to this hypothesis (developed in his later book *Ecstasies*), a same, very ancient shamanistic core would be at the root of both these belief-systems. Through the analysis of a wide range of Eurasian folkloric traditions and mythologies, the Italian scholar identified a number of other figures who, like the *benandanti* and the werewolves, were believed to be able to reach another world while their body was in a sort of trance, and to derive from this journey the ability to protect their communities from evil forces.[35]

Ginzburg's theory has been critiqued because of its questionable methodology, and it has sparked a broad and long debate on, among many other things, the definition of the category of shamanism and its applicability to werewolves.[36] In recent years, Willem de Blécourt has proposed a reconsideration of the case of one Livonian werewolf, whose supposed similarity to the *benandanti* has been crucial to Ginzburg's research. A closer reading of the 1692 trial of Thies, as the man was called, shows that his experiences do not have any clear shamanistic connotations. Whereas Ginzburg had claimed that Thies' wanderings in animal form took place in a 'mythical' dimension, while in a ecstatic state, the transcripts themselves do not mention any elements relative to a state of altered conscience, but actually make precise references to areas in which Thies was said to have roamed, all in the vicinity of the man's town. De Blécourt has, therefore, argued that 'the equation of trance and shape-shifting' proposed by Ginzburg does not have any real basis in this case: even if the souls of shamans would sometimes take the shape of an animal during catalepsis, 'it does not follow that every metamorphosis was a trance experience'.[37]

While de Blécourt's approach is more soundly anchored in the sources and has evidenced weak points and non sequiturs in Ginzburg's reconstruction, it has neglected substantial evidence in favour of some aspects of Ginzburg's thesis: the episode reported by German humanist physician Kaspar Peucer (1525–1602) in the second edition of his *Commentarius de praecipuis generibus divinationum* (An analysis of the main types of divination, 1560). In this vast and learned work, Peucer aims to illustrate the different kinds of divination and to ascertain which ones are of divine origin, which ones are based on purely natural causes, and which are superstitious or outright diabolical. In his analysis of 'theomanteia' – a false type of prophecy in which the devil causes the diviner to lose his or her senses and deludes him/her with false visions – Peucer describes the *lykaones* (or *lykanthropoi*) as men who believe themselves to turn into wolves, to wander in the countryside in animal form and to slaughter cattle.[38] Although these stories may sound like ridiculous tall tales, writes Peucer, they should not be dismissed as totally false, since (according to several trustworthy witnesses) such things do happen regularly in Livonia during the 12 days between Christmas and Epiphany. At that time, 'those who are slaves of the devil' – thousands of them – are summoned and led in wolf shape by a lame child toward the banks of an enormous river, while a tall man armed with an iron whip strikes at those who try to escape or are too slow. They attack and devour sheep, but cannot do any harm to men; one of them (a 'rustic Lykaon') told of his battles against witches who had changed into butterflies. But how did the transformation of men into wolves take place?

> Those who are transformed fall abruptly to the ground as if struck by a sudden disease, and they lie deprived of their senses, as if lifeless. They do not move from the place where they fell, nor do they actually assume the shape of a wolf, rather they become like corpses and do not give any visible signs of being alive. This has given origin to the conviction that their souls are drawn out of their bodies and sent into the fictitious images of wolves and that eventually, after having done what the devil decided, they are returned to their bodies, which are then restored to life. The *lykaones* confirm this idea, affirming that their bodies are not deprived of their human shape, nor are they given a lupine form, but only their souls are forced to fly into the bodies of the wolves, and by these latter they are carried around for a predetermined time. [...] However, the soul does not really get out of the human body and migrate into the lupine one [...] nor do bodies lose their form or shape-shift; rather, the souls, overcome by a most profound sleep, precisely as in an ecstasy (*prorsus ut in ektase*), not distracted by the task of governing the body, are kept busy by the deceitful phantasms created and put forth by the devil.[39]

While Peucer argues that the metamorphosis is just a diabolical deception, his rich and detailed description of Livonian traditions, based on first-hand local testimony – only partially quoted and utilised by Ginzburg – shows that the existence of a link between ecstatic trance and animal transformations cannot be ruled out completely, and it also permits a fuller appreciation of our Modenese story in both its distinctiveness and problematic nature.[40] Indeed, no other court document I know of provides such clear evidence of an individual who would fall into a lethargy and also turn into a wolf. While the Livonian Thies, as we have seen, did not fall into a trance, the *benandanti* did not transform into wolves.[41] Furthermore, trials of French, Dutch, and German werewolves (such as those studied by Caroline F. Oates and de Blécourt) do not contain any significant references to ecstatic states.[42]

Can we thus conclude that the case of the Garutti brothers, further substantiated by Peucer's account, actually supports Ginzburg's thesis? I would not be so sure. A careful reading of the Modenese records indicates that trance and metamorphosis, even if they might have been experienced by one or both brothers, do not seem to be connected, at least not in the way they were for the Livonian 'lykaones' discussed by Peucer. On the one hand, Garutti's transformation into a wolf did not take place while he was in a trance-like state but when he was fully awake, right after a conversation with a real person, and through a physical gesture (the tying of the belt). On the other hand, his swoon on another occasion does not seem to be associated in any way with shape-shifting; rather, it was interpreted by a witness as an indication that Garutti was a witch and went to the sabbat. The problem is, the transcripts do not provide any information, either direct or indirect, about the types of vision Pietro Garutti (and his brother Giacomo?) might have had during his (their) trance. This constitutes a serious limitation for any attempt at interpreting the nature of such events (a difficult task in itself), and it makes any theory about their connection with animal metamorphosis hypothetical at best.

While the information we can glean from the Garutti dossier is only fragmentary, it is nonetheless worth analysing in the context of other Inquisition documents from Modena, with the goal of eventually discerning a common background of myths and traditions. The natural starting point is the lethargic state of Pietro Garutti, which is described with the same words that will recur, some 15 years later, in the proceedings against Domenica Burbarelli. But, unlike Pietro, Domenica was questioned by the inquisitor and recounted her participation in the 'game', where the 'Lady' (whom other witnesses referred to as 'Diana') presided over the defilement of things sacred and demonic dances: i.e., over a stereotypical witches' sabbat. While the figure mentioned by Domenica, the Lady of the Game, is obviously reminiscent of that which recurs in the deposition against Zilia, the Lady of the Ride, in this latter document the Lady still had original, benevolent characteristics as a deity capable of resuscitating animals; but, in Domenica's description she seemed to have acquired a predominantly diabolical identity. Perhaps the increase of the Inquisition's activity in the years between the two documents was beginning to graft some features of the sabbat onto ancient mythologies, as the inquisitors would do with the *benandanti*'s tales later in the century.[43]

The investigation of Domenica shows a couple of elements at least partly foreign to inquisitorial theories, over and above the presence of the Lady instead of the Devil. One of the witnesses who accused her of being a witch, Antonia Magnani, reported that one night, while Domenica was a guest in her house and actually in her bedroom, she had heard 'a loud noise in the attic, with different, horrible voices'. Antonia then had lighted the lamp and saw a ram (*montonus*), which immediately put out the lamp. When she lighted the lamp again, the ram, 'which had horns', had vanished, and at that point Domenica told her that the animal actually 'was her grandmother' (*quod ipse montonus erat eius ava*).[44]

This very brief reference to a human being appearing in animal form is obviously another element which echoes the Garuttis' shape-shifting abilities. But establishing a direct parallel between the two is not as easy as with Domenica's catalepsy and Pietro's trance. First of all, as far as I know, the ram was not associated in any way with a group, as were the *benandanti*, the Livonian werewolves, the Hungarian *táltos*, etc., whose members were considered to be protectors of their communities and mediators with the otherworld. None of these were believed to turn into a ram, or to be transported by a ram to a night battle or any special mission during their swoons.[45] However, in the sudden materialisation of the ram in Antonia's bedroom some connections with Pietro's transformation might be detected if we consider Domenica's participation in the 'ride' of the Lady, insofar as she is a deity often identified with Diana/Artemis. By the beginning of the early modern era this nocturnal goddess was still the protagonist of the visions of mostly (but not exclusively) female followers, who would apparently attend her celebrations during a trance. In the otherworldly dimension reached when in a state of ecstasy scholars have recognised the world of the dead, adding thus a strong funereal component to the cult of the goddess.[46]

The Modenese documents considered could thus be read as a description of the different types of experiences undergone by those partaking in such beliefs: some would feast on the meat of animals (an ox, Zilia), others would see the dead in animal form (a ram, Domenica), or would turn into animals (a wolf, Pietro).[47] An association between the three cases is possible, positing as their common centre the surviving myth of the ancient goddess, who had retained some features of her primeval form: Mistress of Animals, expression of a deeper communion between human beings and animals, bearer and protectress of abundance and fertility, also through a close connection to the otherworldly dimension. To assert such a common denominator would require some cases sharing one or more elements found in the other cases: such as, Garutti having visions of the Lady's game when in a trance, or seeing the souls of the dead. Admittedly, this kind of reconstruction would have a mainly conjectural value, and it would pose an additional, major problem: by absorbing the stories about Zilia, Pietro and Domenica into the same, hypothetically reconstructed mythology, the specific traits found in each of them would be diluted.

Garutti's metamorphosis in particular presents features which seem largely absent from Italian documentary evidence on visions of the goddess or ritual battles for fertility. The belt used for transformation and the attack on sheep, invite comparison with northern European 'wolfsmenschen' rather than with the Friulian *benandanti*. Indeed, while the latter believed themselves (and were believed by fellow villagers) to be beneficial figures with a markedly protective function, Garutti's episode as a wolf was described as having only harmful effects. Perhaps the witnesses in our document associated a werewolf with witchcraft precisely because in the Modenese region (and possibly in a wider area in Northern Italy) such a creature always had negative connotations. Being a witch, in turn, was seen as a condition which included behaviour that was not necessarily seen as evil *per se*, such as the phenomenon of the trance. However, although both shape-shifting and deep swoons could be interpreted, in other cultural contexts, as special gifts thanks to which benevolent figures could defend their communities, in this part of Italy such abilities were mostly seen in a negative light, at least in the early sixteenth century.

If there is a lesson to be drawn from the study of these few documents, it is that the great *variety* of local traits found in witchcraft trials should never be overlooked for the sake of more convenient generalisations.[48] In this brief dossier, the contours of what Pietro and Giacomo Garutti did and were believed to be doing remain fuzzy: any interpretation cannot avoid being tentative. While the possibility of connections between the different strands of beliefs here explored should not be ruled out, only a wider, more comprehensive survey of lycanthropy in early modern Italy (or a series of in-depth local studies) would enable a less impressionistic assessment of their hypothetical common background. Until then, the link between the man who caused the death of his wife with his own apparent death, and the man who was killed after having turned into a wolf, will remain unavoidably elusive.

Notes

1. On Spina see Gabriella Zarri, 'Spina, Bartolomeo della', in: Richard M. Golden (ed.), *Encyclopedia of Witchcraft. The Western Tradition*, (henceforth *EoW*), IV (Santa Barbara, CA, 2006), 1081; Matteo Duni, 'Spina, Bartolomeo', in: Adriano Prosperi, Vincenzo Lavenia and John A. Tedeschi (eds), *Dizionario storico dell'Inquisizione*, III (Pisa, 2010), 1471–1472
2. It was common for inquisitorial edicts to be issued during the cycle of sermons held in December during Advent. The length of the period of grace was normally between six and 12 days for denunciations, and thirty for self-denunciations. See Francisco Bethencourt, *L'Inquisition à l'époque moderne. Espagne, Portugal, Italie XVe–XIXe siècle* (Paris, 1995), 164–171; Christopher F. Black, *The Italian Inquisition* (New Haven, London, 2009), 57.

3. Archivio di Stato di Modena, Inquisizione, busta 2, Processi 1489–1549, fascicolo II (henceforth: Inquisizione), *Contra Petrum et Jacobum strigos ad Bastiam,* 27 December 1518/19. Documents produced in Modena in this period were still dated according to the 'stile della Natività', in which December 25 was considered the first day of the year. Thus dates such as 'Dec. 27, 1519' actually refer to the previous year. The village of Bastia (*Villa Bastia*), at present called Bastiglia, is located in a rural area about 12 km. north of Modena. This and all following translations from the court records, as well as from other texts originally in Latin or Italian, are mine.
4. Inquisizione, *Contra Petrum et Jacobum*, 27 December 1518/19.
5. Ibid., 28 December 1518/19. According to what Francesca remembered of Ursolina's tale, the first peasant had asked the other one: 'Do you want me to make you eat some good wether?' (*'Vo' tu che te facia manzare d'un bono castrone?'*, in the vernacular, while Latin is used in the rest of the document), and the answer had been: 'I do' (*'Voio'*).
6. The rest of the proceedings against Pietro e Giacomo were written by Giovanni Niccolò Morani, the notary most active with the Inquisition of Modena in the period 1518–1520. The marginal note looks as if it could have been written by a friar Giovanni, who seems to have acted as a notary in the deposition of Giovanni da Ronco about a woman named Zilia on February 23, 1519 (see *infra*, p. 126)
7. This of course does not mean that the quality of the depositions (direct testimony or hearsay) or the intentions of the witnesses (joke or earnest tale) are irrelevant for our understanding of the document, and I will try to keep both into account in my analysis.
8. In Northern Italy other common names of the sabbat were *ludus/zogo*, the game, and *striacium/striazzo/strigozzo*, the meeting of the witches (*stric*). On this variable terminology see most recently Marina Montesano, *Caccia alle streghe* (Roma, 2012), 80–85.
9. One trial from 1499, however, includes a reference to *striacium*: see Matteo Duni, *Tra religione e magia. Storia del prete modenese Guglielmo Campana (1460?–1541)* (Firenze, 1999), 35. In any case, significant gaps in the archive of the Inquisition's early years give to my remark a mostly impressionistic value. The files contain about forty trials or proceedings from the period 1495–1523, 1495 being the year of the earliest Modenese trial in the archive, but there are no documents in the years 1504 through 1516. Then there is another gap during 1524–1530. Starting from 1531 the documentation is complete; however, mentions of the sabbat are few and far between even in this later period, as only two trials from the 1530s and one from the 1560s include a description of the witches' meeting. I will refer to one of those from the 1530s later in this essay. On witchcraft and its prosecution in Modena see Matteo Duni, *Under the Devil's Spell. Witches, Sorcerers, and the Inquisition in Renaissance Italy* (Florence, 2007), 42–45 and *passim*. On the archive of the Modenese Inquisition see Giuseppe Trenti (ed.), *I processi del tribunale dell'Inquisizione di Modena: inventario generale analitico, 1489–1874 [i.e. 1784]* (Modena, 2003).
10. Inquisizione, 12 February 1519. On Chiara's trial see Carlo Ginzburg, 'Witchcraft and Popular Piety: Notes on a Modenese Trial from 1519' in: Idem, *Myths, Emblems, Clues* (London, 1990), 1–16.
11. Inquisizione, *Contra Ziliam*, testimony of Giovanni da Ronco, 23 February 1519.

12. Maurizio Bertolotti, 'The Ox's Bones and the Ox's Hide: A Popular Myth, Part Hagiography and Part Witchcraft', in: Edward Muir and Guido Ruggiero (eds), *Microhistory and the Lost Peoples of Europe. Selections from 'Quaderni Storici'* (Baltimore, 1991), 42–70.
13. Bertolotti, 'The Ox's Bones', 50.
14. Perhaps Spina had discovered that Orsolina had meant his tale to be a joke in the same days of Giovanni da Ronco's deposition, as both records appear to have been written by the same notary (see *supra*, note 6).
15. Bartolomeo Spina, *Quaestio de strigibus, una cum tractatu de praeeminentia sacrae theologiae, et quadruplici apologia de lamiis contra Ponzinibium* (Romae, 1576), 3.
16. Spina, *Quaestio de strigibus*, 5–6.
17. Archivio di Stato di Modena, Inquisizione, busta 2, Processi 1489–1549, fascicolo III, *Contra Dominicam filiam Joannis Burbarelli de Novo*, collective deposition of several witnesses, 24 November 1534. Domenica's case was also mentioned in: Carlo Ginzburg, *The Night Battles. Witchcraft and Agrarian Cults in the Sixteenth and Seventeenth Century*, (Baltimore, 1983), 16 (first published in Italian as *I benandanti*, 1966). Novi is about 35 km. north of Modena – and only 25 from Pietro Garutti's Bastiglia.
18. Ibid., depositions of Domenica, 13 October 1532, and of Antonia Magnani, 24 November 1534. Domenica did not speak of Diana, but of the 'Lady of the game' (*Domina ludi*). While some witnesses reported her intention of going to the *cursus Diane*, others referred to the mysterious gathering with the expression *ludus strigarum* (the game of the witches).
19. One of the earliest examples of this narrative is found in Johannes Nider's *Formicarius* (c. 1436), ch. II, 4, in: Martine Ostorero, Agostino Paravicini Bagliani, Kathrin Utz-Tremp and Catherine Chène (eds), *L'imaginaire du sabbat. Édition critique des textes les plus anciens (1430 c.–1440 c.)* (Lausanne, 1999), 134–7. Spina, *Quaestio de strigibus*, 4, analyzes a similar episode, but then (80–82) argues that it does not prove the unreality of the witches' flight or of the sabbat.
20. See *supra* note 17.
21. First published in Italian as *Storia notturna. Una decifrazione del sabba* (Torino, 1989). I will be referring to the English edition: *Ecstasies. Deciphering the Witches' Sabbath* (Chicago, 2004). For a comprehensive, wide-ranging critique of Ginzburg's findings and method, see: Willem de Blécourt, 'The Return of the Sabbat: Mental Archaeologies, Conjectural Histories or Political Mythologies?', in: Jonathan Barry and Owen Davies (eds), *Witchcraft Historiography* (Basingstoke, 2007), 125–145.
22. While Spina mentions the witches' supposed ability to shift into cats, Giovanni Francesco Pico, in *Strix, sive de ludificatione daemonum* (The witch, or on the deceits of demons, 1523), discusses transformations of humans into birds and donkeys on the basis of classical texts such as Lucian of Samosata's *Lucius, or the Ass* and Apuleius' *Metamorphoses* (also known as *The Golden Ass*), denouncing all metamorphoses as diabolical illusions: cfr. Jean-François Pic de la Mirandole, *La sorcière. Dialogue en trois livres sur la tromperie des démons*, Alfredo Perifano (ed. and transl.) (Turnhout, 2007), 140f. The impact of literary/mythological figures from classical antiquity on the formation of the image of the witch, and particularly on the connection between witchcraft and metamorphosis in Renaissance art, has been highlighted by Charles Zika, 'Transformation, Death and Sexuality in the Classical World', in: Idem, *The Appearance of Witchcraft.*

Print and Visual Culture in Sixteenth-century Europe (London–New York, 2007), 125–155.

23. This is the case with Jacopo Sannazzaro's celebrated poem, *Arcadia* (1504), in which the learned, venerable shepherd Enareto is able to turn into a wolf thanks to some 'very powerful poisons' he knows, but makes use of this ability only to spy on real wolves and prevent their raids, thus protecting humans and cattle (this was likely a variation on a story told by Virgil, *Eclogues*, VIII: the shepherd Moeris transforms into a wolf via some magical herbs). Cfr. Gianfranca Ranisio, *Il lupo mannaro. L'uomo, il lupo, il racconto* (Roma, 1984), 69–70. Ranisio also reviews (99–101) other Italian authors' tales of shape-shifting (Franco Sacchetti in the thirteenth century, Straparola in the sixteenth, among others), and argues for the mainly literary character of their works.
24. Andreae Alciato, *Parergon iuris* (The Accessory of Jurisprudence), bk. VIII, chap. XXII: *De Lamiis seu Strigibus scitum non indigna* (Things worth knowing about witches, c. 1525), in: Idem, *Opera Omnia* (Basileae, 1558), vol. II, ff. 406–408. The myth of Lykaon, king of Arcadia transformed into a wolf by Jupiter, was narrated by Ovid in the *Metamorphoses*: see Charlotte F. Otten, *A Lycanthropy Reader. Werewolves in Western Culture* (Syracuse, NY, 1986), 227–230, as well as chapter 2 of this book. On Alciato and witchcraft see Matteo Duni, 'Alciati, Andrea', in: *EoW*, vol. I, 29–30.
25. *De praestigiis daemonum*, lib. IV, cap. XXIII, English edition in: George Mora, Benjamin Kohl (eds), *Witches, Devils, and Doctors in the Renaissance. Johann Weyer, De Praestigiis Daemonum*, John Shea (transl.) (Binghamton, NY, 1991), 342–344. See also a fine overview of early modern literature on lycanthropy in Paolo Lombardi, *Streghe, spettri e lupi mannari. L''arte maledetta' in Europa tra Cinquecento e Seicento* (Torino, 2008), 125–150.
26. The cutting off of limbs, however, might have been a form of punishment similar to the dismemberment inflicted to Peter Stump, discussed *infra*: on the rationale for such gruesome treatment see: Willem de Blécourt, 'The Werewolf, the Witch and the Warlock: Aspects of Gender in the Early Modern Period', in: Alison Rowlands (ed.), *Witchcraft and Masculinities in Early Modern Europe* (Basingstoke-New York, 2009), 191–213, esp. 197.
27. Already in the eight century Boniface of Mainz mentioned a Germanic tradition according to which the metamorphosis would be provoked by donning a wolf's skin or a belt of wolf's leather: the text is quoted in A.M. di Nola, *Animale, trasformazioni*, in *Enciclopedia delle religioni* (Firenze, 1970), vol. I, col. 403. Robert Eisler, *Man Into Wolf. An Anthropological Interpretation of Sadism, Masochism and Lycanthropy* (New York, 1951), 149, speaks of Norse traditions in which the *ulf-har* (=wolf's hair) is a 'belt of wolf's leather, evidently representing the wolf's pelt or *ufhamrs* or 'wolf's skin', which needs to be donned to effect the transformation. Cfr. De Blécourt, 'The Werewolf, the Witch and the Warlock', which discusses several examples of this belief.
28. *A Lycanthropy Reader*, 69 (the text is quoted from the English translation of the trials' transcript, dated 1590, which gives the werewolf's name as Stubbe Peeter).
29. Willem de Blécourt, 'I Would Have Eaten You Too: Werewolf Legends in the Flemish, Dutch and German Area', *Folklore*, 118 (2007), 23–43, mentions a version of a German folktale in which the removal of the belt, not its donning, is described as the cause of the metamorphosis.

30. Willem de Blécourt, 'Wolfsmenschen', *Enzyklopädie des Märchens* 14 (2013), col. 975–986.
31. Cfr. Gábor Klaniczay, 'Hungary: the Accusations and the Universe of Popular Magic', in: Bengt Ankarloo and Gustav Henningsen (eds), *Early Modern European Witchcraft. Centres and Peripheries* (Oxford, 1990), 219–255, 254; de Blécourt, 'Wolfsmenschen'; Peter Dinzelbacher, 'Lycanthropy', in: *EoW*, vol. III, 680–682.
32. Interrogated by Spina, Girolama said to have heard Ursolina's story in the company of Francesca Ferrari 'and many other women in the house of Andrea Villani, while they were paying a visit to the ailing Donna Costanza, mother of the said Andrea'. Francesca Ferrari, in her December 28 testimony, had declared that the gathering had taken place three days before: that is, on December 25.
33. Roberto Navarrini, 'Vita religiosa nella diocesi di Mantova tra Cinquecento e Seicento', in: *San Maurizio in Mantova. Due secoli di vita religiosa e di cultura artistica* (Brescia, 1982), 11–48, 34 (the full text of the document, in the vernacular, is on p. 157). The letter is dated 1 July 1565; the Palazzo Te was the suburban residence of the ruling Gonzaga family (many thanks to Molly Bourne for help on Mantuan topography and genealogy).
34. However cf. Willem de Blécourt, 'A Journey to Hell: Reconsidering the Livonian "Werewolf"', *Magic, Ritual, and Witchcraft*, 2 (2007), 49–67, which discusses a 1691 Latvian trial in which judges apparently laughed at a werewolf's detailed confession (but eventually they decided to prosecute him).
35. Ginzburg, *Ecstasies*, 153–173.
36. Cf. Caroline F. Oates, *Trials of Werewolves in the Franche-Comté in the Early Modern Period*, doctoral thesis (University of London, 1993), 72–75.
37. De Blécourt, 'A Journey to Hell', 53–55 and *passim*.
38. Kaspar Peucer, *Commentarius de praecipuis generibus divinationum* (Witebergae, 1560), 141v–142v.
39. Peucer, *Commentarius*, ff. 141v–142v, 144v.
40. Ginzburg, *Ecstasies*, 156–157, discusses Peucer's text and persuasively identifies its source as having been Hermann Witekind, a colleague of Peucer at the University of Heidelberg, who was Livonian by birth and described his encounter with a werewolf in a book he published in 1585. However, Ginzburg strangely does not make any reference to Peucer's lengthy analysis of the trance the werewolves would fall into before experiencing their metamorphosis.
41. Only in very rare cases *benandanti* would mention the soul re-entering their body in the shape of a mouse: cf. Ginzburg, *The Night Battles*, 19–20.
42. De Blécourt, 'Reconsidering the Livonian Werewolf', 64–67. A few, 'very faint' mentions of trance-like conditions is all Oates has found in her extensive research: cf. Caroline F. Oates, 'Metamorphosis and Lycanthropy in Franche-Comté, 1521–1643', in: Michel Fehrer (ed.), *Fragments for a History of the Human Body*, vol. I (New York, 1989), 304–363, 309–310.
43. Cfr. Duni, *Tra religione e magia*, 38–39.
44. *Contra Dominicam filiam Joannis Burbarelli de Novo*, deposition of Antonia Magnani, 24 November 1534.
45. Ginzburg, *Ecstasies*, 165–173. It is true, however, that in the Judeo-Christian tradition the ram did not have the strongly negative connotations of the similarly horned goat, one of the animals most commonly associated with Satan (see e.g., Zika, *The Appearance of Witchcraft*, passim). On the contrary, in the Old

Testament rams are the animals chosen for sacrifices to God (most famously, Abraham sacrificed one in the stead of his son Isaac).

46. Ginzburg, *Ecstasies*, 100–103 ff.
47. But the records do not explicitly state that Domenica's grandmother was dead at the time of the apparition.
48. Cf. De Blécourt, 'A Journey to Hell', esp. 64–67, rightly inviting researchers not to lose sight of the strong regional variants. See also the first chapter of this book.

6

'Species', 'Phantasia', 'Raison': Werewolves and Shape-Shifters in Demonological Literature

Johannes Dillinger

This chapter discusses various concepts of animal transformation in the demonological literature of the late Middle Ages and the early modern period. It focuses on five most prolific advocates of witch-hunting: Heinrich Kramer, inquisitor and author of the *Malleus maleficarum*; Jean Bodin, celebrated jurist, intellectual founding father of the modern state and ardent propagandist of witch hunting; Nicholas Rémy, a career administrator de facto in charge of law enforcement in the duchy of Lorraine for several years; Martin Delrio, a theologian and jurist who wrote an almost encyclopaedic treatise on magic; and Pierre de Lancre, a judge with extensive experience with concrete witch trials and a marked interest in werewolfery.[1] Other authors will only briefly be dealt with. As this chapter concentrates on authors who contributed to the witchcraft doctrine the medicinal debate about lycanthropy as a mental condition is largely left out.[2] Firstly, the text will briefly sketch the conditions of the debate on the werewolf: It will review fundamental theological statements concerning animal transformation and early demonological writings provoked by the nascent witch-hunts of early fifteenth-century Switzerland. After that, the chapter will deal with each demonologist in turn in chronological order, giving the main arguments of each and exploring the interrelations between authors. The chapter focuses narrowly on animal transformation, especially werewolves, but tries to integrate the discussion of shape-shifting in the wider context of concepts of reality and plausibility in the witchcraft doctrine.

Early demonologists

Demonologists always had to face the fact that the canonical law and the confessional literature of the Middle Ages were comparatively critical about the belief in magic. The *Canon Episcopi*, which condemned the belief in magical flight, is well known in witchcraft research. A rather neglected passage of the *Canon* explicitly anathematised belief in transformation: 'Whoever believes that it might happen that any agency other than that of the Creator who created everything transforms any creature to its advantage or disadvantage or changes it into another form (species) or into another image is certainly an infidel and worse than a heathen'.[3] This would clearly suggest that demons, let alone magicians, could not transform human beings into animals. The essential form ('species') of all creatures was determined by the will of the Creator and thus unchangeable. It would seem that all adherents of the witchcraft doctrine who defended the existence of werewolves would have to fight an uphill battle against this piece of canon law. But matters were even worse for them. Burchard of Worms, also well-known for his rejection of the nightly flight of magicians, addressed shape-shifting in an equally uncompromising way: It is forbidden 'to believe that women whom the common folk call the Fates (Parcae) exist or possess the powers that are attributed to them: At the birth of a man, they destine him to whatever end they please, even in such a way that this man is capable of transforming into a wolf at will – this the common stupidity calls the werewolf – or into any other form'.[4] Lecouteux was probably right to suggest that Burchard's text was based on the *interpretatio Christiana* of a Germanic concept of the soul. According to this concept, the soul could temporarily leave the body in the form of some animal.[5] At any rate, Burchard's aim was to fight two elements of folk belief: The belief in god-like beings that decided the fate of human beings – a belief that blatantly contradicted the Christian idea of free will – and the belief in transformation, especially the belief in werewolves. It seems that Burchard addressed the werewolf as an obvious and outrageous example of common superstition that would help to show to what ridiculous extremes the belief in the Parcae had been taken by the uneducated. This means that in popular culture – or at least in Burchard's representation of popular culture – the ability to transform was a kind of talent: People were evidently born with it. They did not choose to become werewolves, they simply were werewolves and they did not seem to be able to get rid of that quality, even if they wanted to.

For the *Canon* and Burchard alike, the bottom line would be: Magical shape-shifting is impossible; true Christians are not supposed to belief in (animal) transformation. For practical purposes of pastoral care and church discipline, this was probably all that mattered. Of course, in the more sophisticated theological literature, things were not that simple. Augustine explained that Circe did not really change Ulysses' companions into pigs, she just caused the hallucination that the men had turned into animals.[6] In a few meagre sentences, Thomas Aquinas explained that demonic transformations were based on deception: The demons could not manipulate the free will but they could form images in the mind or rather activate the memories of objects stored in the mind in order to tempt humans. They might, Thomas argued, manipulate the mind in such a way as to present memories as actual perceptions, thus abusing human 'phantasia' in order to cause hallucinations. The demons could also simply corrupt the senses and thereby cause delusions. In a way, Thomas too fought the belief in magic. When he explained shape-shifting as a delusion he did not yet think – as later demonologists would – of demons in league with human beings and the criminal machinations of witches that could use just that power to deceive.[7] Bernardino of Siena confirmed the view that all animal transformation, as well as magical flight, were just diabolical delusions.[8]

During the early Swiss witch-hunts of the 1420s some suspects admitted that they had changed into wolves.[9] One of the earliest authors, who combined demonological thought with news about actual witch trials, was Hans Fründ. For Fründ, matters were somewhat equivocal: 'Some were amongst them [i.e., the culprits of the witch trials of Wallis] whom the devil had taught how to turn into wolves, in such a way that they themselves thought that they were wolves; indeed they did not know any better. And whoever saw them at that time did not know any better either ... And they also caught sheep, lambs and goats, and ate them, too, raw, in the shape of a wolf. And when they wanted to, they turned back into the human beings they had been'.[10] For Fründ, animal transformation was clearly another element of demonic magic, evidently a rather complicated one: Not all the witches were capable of assuming animal form, but only those that had been taught the technique by the devil. In a similar way, the devil taught some witches how to become invisible. This would imply that the witches did not really turn into wolves: Witchcraft only altered their appearance. The invisibility magic was of course not supposed to make the witches disappear utterly and completely; it did not annihilate their existence. It just kept other people from seeing them. Thus, we might think that werewolf magic did not really transform human beings into animals. They just seemed to be wolves. Even the witches themselves were deceived by that trickery: They thought themselves that they had become animals. Even though Fründ's text invited this interpretation he did not say in so many words that the werewolf transformation was just some delusion caused by the demons. Apparently, Fründ wanted to keep things open or did not dare to suggest an unequivocal interpretation. Maybe he knew that even though people had suffered capital punishment for witchcraft including werewolfery animal transformation was – according to the theological tradition – just not possible, but did not care to challenge either the judges responsible for the trials nor any theological authorities.

Nider mentioned that the archmagician Scavius transformed into a mouse whenever his enemies tried to apprehend him. Nider did not criticise the strange tales about Scavius explicitly, even though he presented the story as hearsay and couched it in somewhat cautious terms. Nider wrote that he learned about Scavius' shape-shifting from the judge Peter who was responsible for the witch trials of Berne. It might be that Nider did not want to contradict his informant.[11] Nider did not mention werewolves but he compared the witches to wolves: Witches supposedly ate their own children, a practice that was so base it set them apart from all other creatures aside from wolves which, according to Nider, devoured their own offspring, too.[12] The *Malleus maleficarum* would later repeat this bit of misinformation.[13]

Martin le Franc, an influential courtier-theologian who became bishop of Lausanne after writing his anti-demonology *Champion des Dames*, showed no respect whatsoever for the 'facts' supposedly unearthed during the early witch-hunts. For Martin the werewolf was just another bit of crass 'superstition': He listed the 'varous' (i.e., the werewolves) in the second sentence of his work together with the 'luitons' (i.e., the fairies) as bits of folk belief he ridiculed. Later on, he mentioned them together with Melusine.[14] For Martin, fear of such beings was simply below the dignity of educated Christians.

The preacher Johann Geiler von Kaysersberg, who like Kramer was one of the incredibly productive Alsatian school of theologians of the late fifteenth century, addressed the problem that some wolves attack humans and seem to be more cunning than others. Geiler suggested seven explanations: These wolves were starving and thus made desperate attacks, they suffered for some reason and therefore became more aggressive, they might be old and incapable of catching their usual prey or they might have the rabies. Geiler speculated whether some wolves preferred to eat human flesh because they like the taste. The preacher stressed that God sent aggressive wolves as punishment for the sins of mankind. So far, Geiler had avoided explanations that implied diabolism or magic. However, he did write that the demons could take the form of wolves and in this guise ravage the countryside. Geiler even mentioned briefly an episode by Guillaume d'Auvergne about a man whom the devil had tricked into believing that he was a werewolf (see below). Nevertheless, Geiler's short text on uncommonly aggressive wolves focused on natural explanations, not on demonological ones. Werewolves in witch-trials were obviously not his topic. Geiler clearly suggested that a man-eater wolf was most likely to be just an animal.[15]

Malleus maleficarum

It was Heinrich Kramer who managed to make some sense out of the contradictory ideas about werewolves. Kramer brought the message of the trial records and the theological tradition together. The price Kramer was willing to pay for this was reality itself.

Kramer presented the wolf as the harmful animal per excellence. The very nature of sheep supposedly enabled them to distinguish between (friendly) dogs and (hostile) wolves. For the inquisitor, the animals God used to punish mankind for their sins according to Lev. 16: 21–22 and Deut. 32: 24 were clearly wolves. When Kramer explained how human imagination worked he referred to people sighting a wolf: Persons who spotted a wolf ran away not because of the animal's 'ugliness of colour and figure' that corresponded to a memory of a wolf's appearance stored in the imagination. They ran away simply because human nature itself dictated that the wolf was an enemy that was to be avoided. Kramer had no doubt that wolves came into houses to prey on children and adults alike. Together with Nider Kramer credited wolves with the same quasi auto-destructive trait witches were supposed to have: They ate their own offspring.[16]

Explaining shape-shifting, Kramer referred to Thomas Aquinas. As all imagination rested in a bodily organ, angels, including Satan's fallen angels, had power over it, because the spiritual power of angels was greater than that of mere bodies. As the demons were able to manipulate the humours and the vital spirits freely, they could cause whatever hallucinations they pleased in human beings.[17] The idea that the demons could cause hallucinations, their virtually unlimited capability to create 'prestigia', i.e., glamours, was a passepartout apt to explain things that otherwise would have been unexplainable. All elements of magic that were – according to the church's teachings – impossible in reality could be explained as demonic hallucinations. Kramer referred directly to Augustine's interpretation of Circe's magic as a mere hallucination. He added another example from the legend of St. Macharius, about a rejected lover who had a Jewish magician bewitch the woman he had fancied. The – no doubt demonic – magic of the Jew caused the woman to believe that she had been turned into a horse. In addition to that, magic caused all other people to hallucinate the image of a horse whenever they saw the woman. Even the Jewish magician and his employer evidently believed that the woman really had become an animal. Kramer stated namely that only St. Macharius 'because of his holiness' was not fooled by the demons' glamour and realised that the woman had not been transformed at all.[18] Obviously, even though hallucinations could explain basically all strange phenomena associated with witchcraft, this explanation itself tended to become complicated. One had to assume that demons caused analogous hallucinations in the minds of all people concerned: the person supposedly transformed, all the witnesses and the magician himself.

After Kramer had elaborated on this basic concept of shape-shifting as a demonic deception, he addressed the problem of wolves that 'eat…adults and children and prowl about with such cunning that nobody can injure or catch them no matter what art or power one might use'.[19] Even though is seems clear in the context what Kramer thought about werewolves he admitted – following Albertus Magnus and Nider rather than Geiler – that uncommonly aggressive and intelligent wolves might be just that. Wolves were capable of this behaviour when driven by hunger, cold, extraordinary fierceness or the care for their young – here, Kramer did not realise that he contradicted his own statement about the supposed cannibalistic tendencies of wolves. There were, however, other explanations for wolves behaving in an uncommon way. Kramer stated that wolves could be possessed by demons. Their extraordinary abilities resulted from the fact that the intelligent mind of a demon had taken over the mindless body of the animal. Kramer backed this statement with two Biblical episodes: that of the boys who mocked the prophet Elisha and were therefore devoured by bears (4 Reg. 2: 23–24), and that of the pseudo-prophet who was killed by a lion for disobeying God (3 Reg. 13: 23–26).[20] It seems strange that Kramer picked these episodes, as in both of them the demonic agency is far from obvious. Kramer ignored the palpable example of the herd of pigs that had had to play host to the demons driven out of the demoniac of Gadara (Mt. 8: 28–34, par.). Apart from animal possession without the involvement of magicians there was yet another possible explanation for the extraordinary behaviour of some wolves. Kramer quoted an episode from Guillaume d'Auvergne about a man who from time to time hid in a cave. While he was in that cave he thought that he had changed into a wolf and attacked men and farm animals. However, in reality a demon which possessed a wolf had caused all the damage. The point of this deception was to spread the erroneous – indeed, as Kramer insisted, the pagan –belief that human beings could really change into animals. Even though Kramer presented this episode as an example of magicians causing hallucinations with the help of demons, in the story itself – in Kramer's version as well as in William's original – no magician was mentioned.[21] Even though the entire *Malleus* was carelessly written, the inquisitor seems to have worked on the werewolf passages with extraordinary haste. This might be due to the fact that Kramer had not come across accusations of werewolfery in his practical inquisitorial work.

Nevertheless, the episode by Guillaume d'Auvergne had addressed and indeed answered another question: If animal transformations were just diabolical delusions of the witnesses, the person transformed, and, if such a person was involved, the magician who performed the transformation magic, how could these hallucinatory pseudo-transformations have any material effects? In the example by William: If the man in the cave had not really changed into a wolf who or what had wounded people and cattle? How could there be any real harm if no real transformation had taken place? The answer was, of course, that the demon caused this harm. Kramer illustrated this point further: Quoting episodes from Augustine he explained that some people had thought themselves transformed into horses or other beasts of burden. As such, they had performed work vastly beyond anything the human body was capable of. As Kramer needed to uphold his explanation of all transformations as mere hallucinations he claimed that invisible demons performed everything seemingly done by transformed persons.[22] Here, Kramer even dropped the idea of animal demoniacs: He did not write that a demon possessed a horse to do the work supposedly done by the person whom everybody thought to have been turned into a horse. Instead, Kramer emphasised the idea that a demon intervened in invisible form and carried the burdens in the transformed person's stead.

All of that would have taken Kramer exactly nowhere in the theological debate had he not been able to criticise the *Canon Episcopi* that 'had anathematised the belief in animal transformation. Kramer knew exactly and said quite openly why the simple rejection of shape-shifting in the *Canon* had to be wrong: 'We say that the Canon we quoted cannot exclude these transformations because authority, reason and experience and matters concerning certain proof related by Augustine [Augustine's rendering of the story of Circe and similar episodes]...demonstrate them'.[23] In other words, there had to be animal transformations, or at least hallucinations, virtually indistinguishable from them because experience taught that they existed. On the first pages of the *Malleus* Kramer explained that the *Canon* must not be taken at face value because 'the witches of today are often turned into wolves or other fierce animals by the work of the demons'.[24] The witches 'of today' were of course the victims of recent witch trials Kramer had learned about. While Fründ and Nider had still hesitated to bring 'facts' discovered and proven by concrete witch trials and old theological teaching together, Kramer did just that. In a way, this even made some kind of sense. Because there were stories about werewolves, it was not good enough to reject shape-shifting as impossible. The best 'stories' – the best reports about werewolves, and thus the most striking proof that something that was in effect very like animal transformation did happen – were, of course, the early witch trials themselves. Kramer outwardly respected the church's traditional teaching and yet made it compatible with the new and for him indisputable 'facts' unearthed by the early witch trials. To be sure, Kramer was a very scion of this particular way of demonologist thinking. He still had comparatively few trial records at his disposal he could – or rather had to – base his arguments on. The trials themselves would furnish later demonologists with more and more explicit materials concerning all kinds of magic, including shape-shifting. The episodes of Ancient History Kramer still had to deal with at some length lost some of their validity for demonology vis-à-vis the evidence provided by the records of recent witch trials. The *norma normans non normata* Kramer – and other demonologists – felt impossible to ignore were the 'facts' supposedly unearthed by witch trials. Kramer said so very clearly in the opening chapters of the *Malleus*: Aside from the papal bull *Summis desiderantes* it was 'the teacher experience' that proved the existence of the diabolical conspiracy by means of the juridical confessions of the witches themselves.[25]

On the background of Kramer's interpretation of shape-shifting one might as well call reality itself into question. If such a diabolical set-up – at once drastic, complex and virtually indistinguishable from reality – was possible, how could anyone still trust his senses? How could anybody know that what he perceived was not another elaborate sham of witches and demons? It might indeed be that the demonological doctrine of hallucinations gave rise to Descartes' idea of a malevolent god out to deceive anybody all the time. Against this treacherous god – is this not a thinly disguised allusion to the devil? – only systematic doubt and the refuge of one's own thoughts provided any help.[26] Kramer himself did not show any way out of the diabolic mirror maze of hallucinations. There were only passepartout arguments in the *Malleus* that could be used to fight the suspicion that all our perceptions were just demonic illusions. The first was the *permissio Dei:* One might speculate that God would not allow the demons to deceive everybody all the time. In addition to that, Kramer suggested a kind of 'Ockham's Razor' for demonologists: The demons were supposed never to do anything unnecessary. One might argue that it would be superfluous and pointless to surround everybody always with hallucinations. Therefore, the demons did not do so.[27] Neither of these arguments was very reassuring. The conclusion that Kramer's interpretation of shape-shifting questioned all perceptions and lastly reality itself is unavoidable. What Thomas Aquinas had meant as a defence against the belief in magic became in Kramer's interpretation a highly aggressive ferment apt to dissolve common sense and the perception of the material world.

Bodin and Rémy

As far as werewolves were concerned, Jean Bodin appears to have been the least critical of the major demonologists (Figure 6.1). Bodin opened his account of shape-shifting with the contradiction we are already familiar with: No element of witchcraft seems more incredible than animal transformation but it is well documented in recent witch trials and in history.[28] Bodin did not address the theoretical difficulties first, but a number of recent witch trials in which animal transformation had featured prominently. The most striking of Bodin's examples was the trial of Gilles Garnier that had taken place in 1574; i.e., just six years prior to the publication of Bodin's demonology. Bodin backed the well-known story about the werewolves of Livonia with the testimony of a prominent person he knew personally: None less than the renowned diplomat, lawyer and Protestant theologian Hubert Languet, who had been to Livonia and confirmed that everybody there unquestioningly believed in the existence of werewolves. Bodin even quoted a letter by a diplomat in the service of Henri II who claimed that the grand duke of Moscovy believed that Livonia was plagued with werewolves.[29] According to Bodin, European traders with contacts in Asia reported that werewolfery was very common in the East. In 1542, no less than 150 werewolves had allegedly terrorised Osman Constantinople.[30] Only after painting this alarming picture of a world about to be overrun by shape-shifters did Bodin recite the usual examples of animal transformation from Antiquity.

How could all this be possible? Up to this point, Bodin had not given any indication that he did not believe in the corporal reality of the

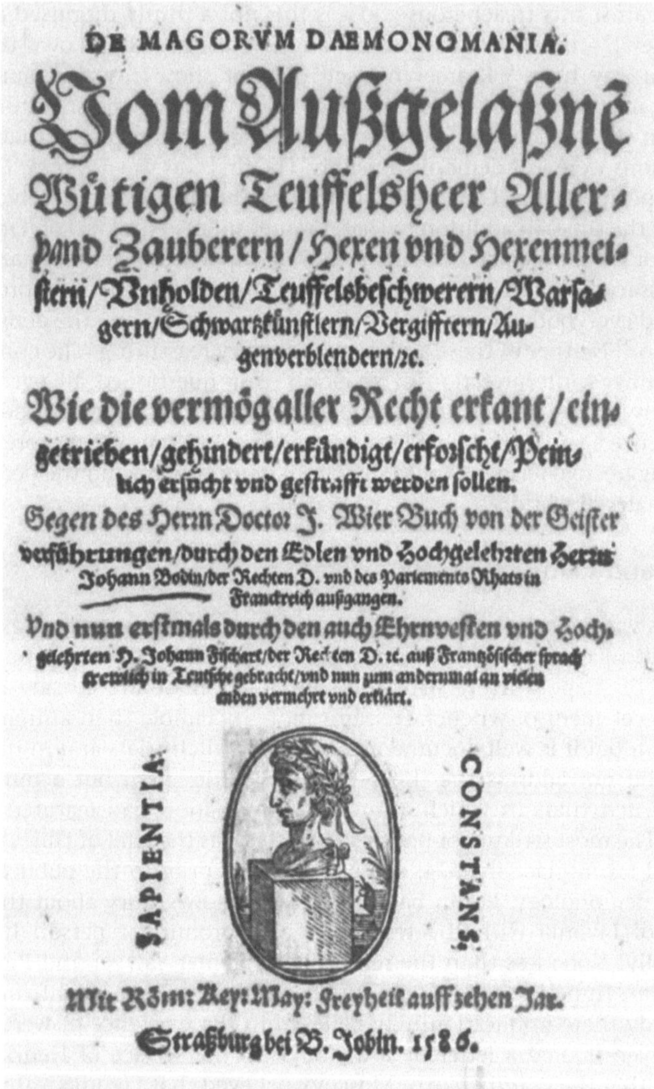
DE MAGORVM DAEMONOMANIA.
Vom Außgelaßnē
Wütigen Teuffelsheer Aller-
hand Zauberern/Hexen und Hexenmei-
stern/Unholden/Teuffelsbeschwerern/Warsa-
gern/Schwartzkünstlern/Vergifftern/Au-
genverblendern/etc.
Wie die vermög aller Recht erkant/ein-
getrieben/gehindert/erkündigt/erforscht/Pein-
lich ersucht und gestrafft werden sollen.
Gegen des Herrn Doctor J. Wier Buch von der Geister
verführungen/durch den Edlen und Hochgelehrten Herrn
Johann Bodin/der Rechten D. und des Parlements Rhats in
Franckreich außgangen.
Und nun erstmals durch den auch Ehrnvesten und Hoch-
gelehrten H. Johann Fischart/der Rechten D. etc. auß Frantzösischer sprach
trewlich in Teutsche gebracht/und nun zum andernmal an vielen
enden vermehrt und erklärt.
SAPIENTIA. CONSTANS.
Mit Röm: Key: May: Freyheit auff zehen Jar.
Straßburg bei B. Jobin. 1586.

Figure 6.1 Title page, Jean Bodin, translated and edited by Johann Fischart, *Von Außgelaßnen Wütigen Teuffelsheer* (Straßburg, 1586)

metamorphoses. Indeed, he seemed to be bent on proving the reality of the shape-shifters in which people from all over the world and from different periods of time had clearly believed. Bodin explained that the Ancient Persians, Egyptians, Chaldeans, the Pythagorean philosophers – whom he did not in any way criticise as unreliable or credulous pagans – had thought transformations possible. Indeed, the Biblical episode of Nebuchadnezzar proved that shape-shifting was real. Bodin ridiculed the thought that werewolfery could be explained as a mental illness; only as an afterthought Bodin admitted briefly that some people might hallucinate about shape-shifting. The devil used both techniques: hallucinations as well as real transformations. Bodin's main point, however, was the corporal reality of transformation. He clearly deviated from the explanation established in the *Malleus*. Thus, the jurist had to sidestep the *Canon* even more deftly than Kramer. Bodin reassured his readers that the devil was incapable of transforming human nature into the nature of an animal. That was, according to Bodin, the only point the *Canon* had wanted to make. Bodin interpreted human nature, the essence of humanity, in a peculiar way. According to him, the essential human form was reason. Reason remained unchanged and unchangeable. This was, according to Bodin, the true meaning of the *Canon*: 'The evil spirits do not have the power to change forms: Because the essential and true form of the human being which is reason ("la raison") is not transformed in any way. Only the figure and outer appearance of the human being are transfigured'. Even if witches turned into animals their reason remained intact. In a very strange way, Bodin let reason triumph over the body, the senses and imagination.[31]

Why did Bodin interpret shape-shifting in this rather questionable way? Does his insistence on reason as the true form of humanity make him a strange predecessor of Descartes' subjectivity? Of course, one might speculate about the philosophical concepts of the mind Bodin could have thought of when he wrote his *Demonomania*. However, it seems more helpful to bear in mind that Bodin was first and foremost a lawyer. His insistence on the reality of the transformation served a twofold purpose, which had some bearing on the witch trials. In effect, Bodin tried to save the reality that was about to disappear behind the demonic smokescreen of delusions described by Kramer. Bodin's solution was certainly a crude one. But its very crudeness provided relief from the speculations of the *Malleus* school. To be sure, the game of hallucinations had been designed by Kramer and others to reconcile the 'facts' discovered in witch trials with theological doctrines. But still, jurists and judges might long for something more tangible, a more simple explanation that left absolutely no room for any doubt concerning the suspected werewolf's guilt. Bodin's suggestion that reason was the unchanged form of humanity had juridical overtones. Even though he did not say so in so many words, the jurist Bodin was probably mindful of the fact that, in order to be fully responsible for all their actions, were-animals had to be capable of rational thought at all times. There could be no insanity defence for them, if the metamorphosis did not affect their rational mind.[32] Bodin's angry rejection of lycanthropy as a mental illness fitted well into that picture.

Binsfeld explicitly rejected Bodin's interpretation. He dealt with animal transformation but briefly and much in the vein of Kramer.[33]

Rémy largely followed Kramer. However, he had some new and characteristic points. He scoffed at examples of animal transformation from Greek and Roman antiquity. He said quite clearly that it was not his 'intention to bring the ass of Apuleius again on the stage, or to adduce fresh examples to support the old tales of the poets ... but only to bring forward such instances as are attested by the evidence of many witnesses and are proved by actual experience'. Indeed, Rémy quoted some witch trials from the 1580s. The old tale about the witch who was found suffering from injuries that had a short while before been inflicted on a wolf, Rémy retold on the authority of a noblewoman he knew personally. She claimed that such an occurrence had taken place a short time ago in one of her villages.[34] However, Rémy seems to have felt uncomfortable with this journalist-style demonology after all. Even though he had declared that he did not have the time for examples from Antiquity, he quickly added some of them. The reason for this might be that Rémy, even though he evidently considered recent cases of shape-shifting more important and more convincing, wanted to demonstrate that he was familiar with the discussion of animal transformation. He might have thought that he needed to do that because he was about to defend Kramer's multiple hallucination explanation of animal transformation against the dissident Bodin. After addressing new and 'classical' examples of metamorphoses he mounted a rather aggressive attack on the 'absurd' opinion that corporal transformation was real. Everybody endowed with 'natural intelligence and power of reasoning' knew that no creature could change 'its own shape and appearance ... except by its death' and that it was impossible to 'interchange bodily forms'. For anyone who had advanced to Christian knowledge it was plainly 'madness' to assume that the devil, who was just another of God's creatures, could act against the obvious will of the creator by changing the appearance of human beings, the crown of all creation. That Rémy's outpourings were directed against Bodin is obvious. While Bodin had placed great stock in the stories of the Livonian werewolves as proofs for real transformations, Rémy used them as proofs for hallucinations: The Livonian shape-shifters were said to steal beer and to jump over the walls of a ruined castle by way of exercise. Did this rather human behaviour not prove that they had not become real wolves? That Rémy fought Bodin's opinion becomes even more evident in his next argument. According to Cicero, it was wrong to believe that 'a soul endowed with reason can, even for a moment, dwell or reside in a body which is altogether unadapted to the use of reason'. Thus, Bodin's suggestion that bodily metamorphoses were possible because a human being's unchangeable form was reason was absurd. Even if one was willing to accept that the immutable form the *Canon* had spoken about was reason it was still unthinkable that reason could work, indeed survive, in a physiological environment that was just not suitable for it, for example, the body of a mindless animal. Simply put: There can be no human mind in a wolf's brain, thus reason could never survive the transformation of the body. Rémy did not hesitate to go one step further: The idea that the soul of a man 'so largely and variously blessed can be put to such ludicrous humiliation as to be transferred into the carcass and entrails of the baser animals...cannot be consistent with true religion'. It clearly contradicted the *Canon*.[35] In other words, Bodin's text on werewolves smacked of heresy.

If Bodin's interpretation of animal transformation was meant to free witch trials from complications and possible obstacles, his fellow jurist Rémy did not share his concerns. Maybe he did not see Bodin's objective, or just thought it better or more important to stick to established demonological opinions. Rémy simply defended the orthodox interpretation of metamorphoses according to Kramer. But why did Rémy attack Bodin so viciously? It seems safe to assume that the controversy had no political overtones but was really about a point of demonology. Rémy's overtly aggressive tone might be due to the fact that he, the mediocre jurist from Lorraine with the brilliant career, felt after all inferior to Bodin, the brilliant jurist from France with the mediocre career.

Delrio and de Lancre

Delrio confirmed the orthodox view that described shape-shifting as demonic trickery. Delrio told two versions of the story of the witch found with injuries inflicted on an animal, only in these variants the animals were a bird and a toad. Again, the occurrence was supposed to have happened only recently, and Delrio claimed to have it on the authority of trustworthy men he knew personally.[36] However, Delrio was most interested in witches changing into wolves. For him, there were essentially four types of deception used in the alleged werewolf transformation: A demon might possess a real wolf. The devil could create for itself a body made from air that looked like that of a wolf. The third variant was somewhat similar: The devil might do the same thing for witches. He could surround their natural human bodies bit by bit with the likeness of a wolf's body made from air. In order to deceive the witches and to make them believe that they really change into wolves the demon might even order them to smear their bodies with some ointment and to cover themselves with wolf pelts as if these per se useless acts could bring about a metamorphosis. Delrio was familiar with cases in which magicians were supposed to have done just that and had – seemingly – turned into animals. This explanation was not what Kramer had suggested: Delrio's demon did not interfere so much with human perception. He did not use hallucinations as a 'short-cut' to the brain but rather presented pseudo-bodies made by demonic power from condensed air to the – per se healthy and undeceived – senses. Kramer himself had explored that line of argument but finally rejected it: He thought it hard to accept that so volatile an element as air would always stay in the same place quasi surrounding the witch's body like a full-body costume. Kramer's argument was less than convincing – it has been pointed out repeatedly that the *Malleus* was a somewhat shoddy book – as he did not have a problem with the idea that the demons used bodies manufactured from condensed air than they approached the witches.[37] Nevertheless, Delrio was close enough to Kramer; he too was apparently willing to sacrifice the reliability of sensory perception in order to be able to explain the outlandish feats of witches discovered in witch trials. In Delrio's 'airy' explanation of shape-shifting the perception of reality became only marginally more reliable than in Kramer's hallucination concept.[38]

Finally, Delrio presented a fourth explanation of werewolfery that did not refer to magic at all. He admitted that some people could be mistaken for werewolves who simply suffered from 'an excess of black bile which has caused their mind to be taken over by a kind of wolfishness'. This malady was known as lycanthrophy or as 'insania lupina', or simply as melancholy. Persons under the influence of 'insania lupina' might attack livestock and people most ferociously but they were no magicians and not personally responsible. Only if other people too imagined that the alleged shape-shifters had turned into animals, one might suspect witchcraft.[39] Delrio adroitly drew conclusions from this state of affairs for the practical witch trials. He suggested that all confessors should ask suspected witches in jail 'to what extend [they]...consider themselves to be werewolves? To what extend are they thought to be such by other people?' Within the framework of Delrio's argument, only the answer to the second point would suggest any answer to the central question if the defendant was guilty of witchcraft.[40] By including werewolfery into the standard catalogue of questions to be asked of any suspected witch, Delrio emphasised the importance and significance of animal transformation. Thereby, he stressed the demons' ability to deceive and to manipulate human perception even more. In consequence, it became even harder to base any criticism of witch hunting, let alone any criticism of the belief in witches as such, on any material evidence. The suggestion that what seemed to be material was but demonic trickery was too strong.

De Lancre essentially echoed Delrio. Explicitly rejecting Bodin, he repeated Delrio's idea about three different kinds of deception used by the devil to bring about the werewolf effect. He also admitted the fourth possibility of lycanthropy: People who suffered from this mental condition needed a physician, not a judge.[41] When de Lancre wrote about witches he relied most heavily on the witch-hunts in the Labourd, where he had worked as a judge. What he had to say about werewolves centred on the case of one Jean Grenier, a 13-year-old boy who in 1603 had been condemned to lifelong confinement in a monastery because he allegedly had a pact with the devil and had – seemingly – transformed into a wolf. De Lancre had visited Grenier in the monastery. Thus, he was the only major demonologist who could boast personal acquaintance with a werewolf even though he never came across another werewolf in the Labourd.[42] It seems that de Lancre was the only 'classical' demonologist who referred explicitly to the werewolf's obsession with the moon, the favourite topic of today's popular culture of werewolfery: Grenier had told him that 'he runs in the moonlight'. De Lancre's comment made it clear that the moonlight had no real effect whatsoever: 'This is a devil's trick to make him more susceptible to illusions. Or, as St. Jerome and St. Chrysostom say, he does this to defame the moon and disgrace the creation of God'.[43]

De Lancre seemed to believe Grenier when he said that he regretted his former crimes. However, he presented the former werewolf not so much as a reformed criminal or repentant killer but rather like a harmless mental patient or a person of extraordinarily weak intellect. De Lancre stressed that Grenier had been examined by a number of doctors who had reported that he did not suffer from lycanthropy. Even though Grenier had been in the monastery for several years, he confessed to de Lancre that he still loved to see wolves and that he craved human flesh, especially that of children. This jarred strangely with the fact, also noted by de Lancre, that Grenier seems to have been a favourite of children, whom he entertained with his remarkable ability to walk on all fours.[44] De Lancre emphasised not only this skill but also the fact that the nails on Grenier's hand were worn down and discoloured. 'This clearly shows that he was indeed a werewolf, and that he used his hand both for running and for grabbing children and dogs'. The teeth of the former shape-shifter showed marks of unusually heavy wear. When Grenier came to the monastery he shocked the monks by wolfing down innards of fish, apparently raw.[45]

These curious details were most interesting: They indirectly contradicted de Lancre's explanation of werewolfery as demonic delusion. Why should Grenier still have an obvious and abnormal affinity to wolves – i.e., prefer strange food, wish to see wolves, walk adroitly an all fours – if his shape-shifting had just been a demonic delusion? To be sure, demonology explained that the devil made sure that all the wounds inflicted on the were-animal could later be found on the body of the witch. But would the devil see to it that years after Grenier had confessed he would still have bodily marks that suggested that he had been an animal or at least submitted his body to some extraordinary extortions? It must be borne in mind that de Lancre said very clearly that Grenier was no lycanthroph; he had never, not even under the influence of a mental illness, really behaved like a wolf. So where did Grenier's strange urges, skills and bodily marks come from? In de Lancre's account of Grenier the boundary between delusion and fact broke down completely. Hallucinations were evidently supposed to leave traces in reality. Were hallucinations more real than reality itself or the blueprint for reality? Or were the marks on Grenier's body de Lancre saw just imaginations themselves? In that case, it was no longer possible to say where the hallucinations finally ended and the material world of facts, clear perceptions and sharp senses began. It made matters worse that this was supposed to be the state of affairs years after the trial, the confession and the beginning of the 're-education' of the alleged werewolf. It seems as if Grenier's shape-shifting had muddied the water of perception and reality permanently. De Lancre's patronising tone seems to mask a strange situation: The demonologist did not visit a former witch who once suffered from delusions. Rather, Grenier granted the demonologist a quick glance at the scratches in a thin layer of factuality over diabolical undercurrents of unending delusions.

Conclusion

One might call werewolfery the cutting edge of witchcraft. The animal transformation was the most outrageous, the most incredible form witchcraft could take. It contradicted learned theological doctrines and escaped physiological explanations. Nevertheless, the experience of witch hunting demanded that animal transformation was more than superstition. Some kind of explanation was needed to justify trials against werewolves. These explanations included complexes of trickery and hallucinations that dissolved reality itself. Clear perceptions, material facts, and common sense conventions of the real disappeared irretrievably under layers of deceptions. The belief in witches, especially in werewolves, did not just erode social trust. It eroded the trust in reality itself.

Notes

1. Heinrich Kramer (Institoris), *Der Hexenhammer, Malleus Maleficarum* (Strasbourg, 1486), Wolfgang Behringer and Günter Jerouschek (eds) (Munich, 2000); Jean Bodin, *Vom außgelasnen wütigen Teuffelsheer* (Fischart translation of *De la démonomnie des sorciers*, Paris 1580) (Straßburg, 1591), reprint Graz 1973; Nicholas Rémy, *Demonolatry* (*Daemonolatria*, Lyon, 1595), E.A. Ashwin (transl.), Montague Summers (ed.) (London, 1930), reprint London 1970; Martin Delrio, *Investigations into Magic* (*Inquisitiones magicarum libri six*, Louvain 1599) Peter Maxwell-Stuart (transl., ed.), (Manchester 2000); Pierre de Lancre, *On the Inconstancy of Witches* (*Tableau sur l'inconstance des mauvais anges*, Paris 1612), Gerhild Scholz William (transl., ed.) (Tempe, 2006).
2. It was almost a characteristic of older texts that in dealing with the theory of the werewolf they combined medicinal and demonological literature, cf. Rud. Leubuscher, *Ueber die Wehrwölfe und Tierverwandlungen im Mittelalter. Ein Beitrag zur Geschichte der Psychologie* (Berlin, 1850), reprint Allmendingen 1981; Wilhelm Hertz, *Der Werwolf. Beitrag zur Sagengeschichte* (Stuttgart, 1862), reprint Wiesbaden, 1973; Sabine Baring-Gould, *The Book of Were-Wolves: Being an Account of a Terrible Superstition* (London, 1865), reprint London, 1995; Frank Hamel, *Human Animals* (London, 1915), reprint New Hyde Park, 1969; a similar approach, in spite of the title in: Nicole Jacques-Lefèvre, 'Such an Impure, Cruel and Savage Beast: Images of the Werewolf in Demonological Works', in: Kathryn A. Edwards (ed.): *Werewolves, Witches, and Wandering Spirits. Traditional Belief and Folklore in Early Modern Europe* (Kirksville, 2002), 181–198.

3. As this is a key text it might be worthwhile to give the *Canon* in the Latin original verbatim: 'Quisquis credit posse fieri aliquam creaturam aut in melius aut in deterius immutari aut transformari in aliam speciem vel in aliam similitudinem nisi ab ipso creatore qui omnia fecit et per quem omnia facta sunt procul dubio infidelis est et pagano deterior', Kramer, *Hexenhammer*, 150–152, 429.
4. Claude Lecouteux, *Witches, Werewolves, and Fairies. Shapeshifters and Astral Doublers in the Middel Ages,* Rochester 2003, 110 (transl. of: *Fées, sorcières et loups-garous au Moyen Age: histoire du doubl*e, Paris, 1992).
5. Lecouteux, *Witches*, 103–125. Of course, Lecouteux's interpretation that insists on Burchard addressing the Parcae just because this 'clarifies things', Lecouteux, *Witches*, 110, contradicts the source. The Parcae were clearly Burchard's most important topic.
6. Augustinus, *De civitate Dei*, 18,17, text available at http://www.thelatinlibrary.com/august.html; cf. Kramer, *Hexenhammer*, 278.
7. Thomas Aquinas, *Sentences*, 2,8,1,5,4 and ad 4; Thomas Aquinas, *Summa*, 1,114,4; all texts available at http://www.corpusthomisticum.org.
8. Martine Ostorero, et al. (eds): *L'Imaginaire du sabbat. Edition critique des textes les plus anciens* (Lausanne, 1999), 497.
9. Ostorero, *Imaginaire,* 87–88.
10. Ostorero, *Imaginaire,* 36.
11. Ostorero, *Imaginaire*, 168. This Peter was probably not Peter von Greyerz, Ostorero, *Imaginaire*, 18.
12. Ostorero, *Imaginaire*, 152.
13. Kramer, *Hexenhammer*, 371.
14. Ostorero, *Imaginaire*, 453, 467.
15. Baring–Gould, *Book*, 261–266.
16. Kramer, *Hexenhammer*, 250, 284, 246, 283.
17. Kramer, *Hexenhammer*, 276–278.
18. Kramer, *Hexenhammer*, 278–279.
19. Kramer, *Hexenhammer*, 283.
20. Kramer, *Hexenhammer*, 284.
21. Kramer, *Hexenhammer*, 285; for William cf. Lecouteux, *Witches*, 117.
22. Kramer, *Hexenhammer*, 430–433.
23. Kramer, *Hexenhammer*, 430.
24. Kramer, *Hexenhammer*, 152.
25. Kramer, *Hexenhammer*, 176.
26. Wolfgang Behringer, *Witches and Witch-Hunts* (Cambridge, 2004), 184.
27. Kramer, *Hexenhammer*, 178–179.
28. Bodin, *Außgelasnen*, 119.
29. For the Livonian werewolves cf. Willem de Blécourt: 'A Journey to Hell: Reconsidering the Livonian "Werewolf"', *Magic, Ritual, and Witchcraft* 2 (2007), 49–67; Stefan Donecker, 'The Werewolves of Livonia: Lycanthropy and Shape-Changing in Scholarly Texts, 1550–1720', *Preternature* 1(2012), 289–322.
30. Bodin, *Außgelasnen*, 121–123.
31. Bodin, *Außgelasnen*, 125–129.
32. H.C. Erik Midelfort, *A History of Madness in Sixteenth-Century Germany* (Stanford, 1999), 182–227.
33. Peter Binsfeld, *Tractat von Bekantnuß der Zauberer vnd Hexen* (Munich, 1592), 32–33, orig. Trier, 1590.
34. Rémy, *Demonolatry*, 108–109.

35. Rémy, *Demonolatry*, 110–113.
36. Delrio, *Investigations*, 101–102, 196.
37. Kramer, *Hexenhammer*, 275, 396–397, 432–433.
38. Delrio, *Investigations*, 98–100.
39. Delrio, *Investigations*, 99.
40. Delrio, *Investigations*, 241.
41. Delrio, *Investigations*, 336–338.
42. De Lancre, *Inconstancy*, 326–327.
43. De Lancre, *Inconstancy*, 315.
44. Admittedly, the text is rather unclear here. De Lancre might refer to another witch boy whom the devil had taught to walk on all fours. Even if the latter part of this passage was really about this other witch, the fact remains that according to de Lancre, Grenier had been astoundingly capable of walking like an animal when he came to the monastery.
45. De Lancre, *Inconstancy*, 326–334.

7

The Judge's Lore? The Politico-Religious Concept of Metamorphosis in the Peripheries of Western Europe

Rita Voltmer

Under the reign of Louis XV, the southeastern part of France, especially the Périgord and the Auvergne, was driven by several wolf panics. The mysterious case of The Beast of Gévaudan attracted the greatest public interest, mediated by numerous pamphlets and newspapers.[1] Even in the dawn of the enlightenment, some people were convinced that The Beast was a devilish man-slaughtering werewolf of immense strength. Between 1764 and 1767, the killing of more than 100 people was attributed to The Beast. Very soon, the French King took an active interest in the affair. At the beginning of the century, the region had been involved in the rebellion of the Camisards, which had been crushed by Louis XIV. The peasants still were thought to be secret Huguenots, who could start disturbances again. Thus, the Jansenist bishop of Mende interpreted the attacks of The Beast as signs of God's wrath against the heathen peasants of the Gévaudan. At the same time, the King stationed a regiment of dragoons in the area. Allegedly, the soldiers should hunt The Beast, but the hidden purpose of this *dragonard* was to control and discipline the restless populace. With the final killing of the wolves, which were thought to constitute 'The Beast', the king's Catholic authority was re-established.

Whereas the affaire of Gévaudan shows the politico-religious dimension of wolf or werewolf hunting rather clearly, modern witchcraft research only marginally recognises this dimension. In the critical edition of Jean de Nynauld's tract about lycanthropy, Nicole Jacques-Chaquin emphasises that the learned dealing with werewolves at the end of the sixteenth century should be interpreted as a kind of handling of daily experiences of war and turmoil.[2] Likewise, Gerhild Scholz-Williams is firmly convinced that Pierre de Lancre orchestrated his hunt for witches and shape-shifters to bring the notorious Labourd under the firm control of the French state.[3] In Danish Schleswig-Holstein, wolf packs terrorised the land after the Thirty Years War. The wolf and his magical counterpart, the werewolf, became the symbol of death and evil itself. Hunting wolves remained a royal privilege, which extended lordly superiority over land and peasants. Consequently, the hunting of witches (together with werewolves) strengthened princely authority.[4]

Even in the northern countries and in the Baltic area, debates about the Livonian werewolves should be seen in the context of war, political conflict and unrest. Recently and for the first time, Stefan Donecker has scrutinised the narratives of the famous Livonian werewolves in the context of a broad debate.[5] Olaus Magnus seems to be the first who had constructed a certain image of 'northernness', which included the concept of Evil, being settled in the northern countries. Bodin, de Lancre and others agreed that witches and shape-shifters dwelt in the peripheries of Europe, in the forest and mountains. The discourse about the Livonian werewolves fits neatly into this framework. Moreover, the Baltic German nobles, as well as the Christian missionaries – be it the Jesuits or Lutheran superintendents like Paul Einhorn – considered the Livonians and Estonians as non-Germans: barbarous, deceitful pagans, who were still practising their heathen rituals and magic under a thin Christian whitewash. Thus, the German noble elite in the Baltic used the charge of 'werewolvery' well: a territory, inhabited by witches, werewolves, and rebellious pagans, needed strong, rigorous control. The label 'shape-shifter' marked the boundary between Christian civilisation and pagan wilderness, where humans became beasts. Furthermore, Willem de Blécourt argued that charges of werewolvery targeted at men (and sometimes at women) who were thought guilty of immorality, sexual deviancy (such as homosexuality and bestiality) and of perverting the virtues of a Christian housefather.[6]

Demonologists defined diabolic witchcraft as the rebellious crime of Satan. In this framework, shape-shifting, lycanthropy, and werewolvery, inverted godly order in a most horrible way. It became a neat metaphor and a symbol for political crisis in a world turned upside down. Werewolves and devilish shape-shifters were seen as the most barbarous, heretical rebels against godly and, consequently, kingly, princely, religious and the housefatherly order. Therefore, ruling patriarchal authorities were obliged to cleanse their territory of witches and werewolves in order to maintain control, peace, order, welfare, and the true religion.

In ignoring the politico-religious dimension of werewolf and shape-shifter hunting, a body of folklore and anthropological studies are at hand, linking the werewolf-myth with deep-rooted fears, shamanism, soul travelling, drug abuse, or de facto illnesses like porphyria. Based on unreliable source material, mainly compounded of trial records and demonological tracts, these debates remain rather fruitless.[7] Most vital questions are yet to be answered. Firstly, did indigenous imaginations about shape-shifting linger in the minds of the unlearned populace before they were infected with the elaborated concepts of learned judges and scholars? The myth of the Livonian werewolves, as somehow shamanistic fighters against so-called witches, seems to give an insight; but, again, anthropological generalisations are over-stretching this indeed singular finding.[8]

Secondly, physicians, jurists and theologians from different confessional parties debated about the 'realities' of shape-shifting. Physicians such as de Nynauld and Wier talked of hallucinations, the effect of drug abuse, illness or melancholy. Bodin wrote of a real physical transformation from human body into an animal body, whereas the human ratio remained intact. In line with the leading authority of Saint Augustine most discussants opted for devilish delusion, which didn't change the bodily essence at all. Like discussions about transvection and copulation with Satan, sceptics and believers shared the same arguments.[9] Whatever discussions Johann Wier, Jean de Nynauld, Jean Bodin, Nicolas Rémy, Henri Boguet, Peter Binsfeld or Pierre de Lancre were involved in, did these partly contradictory debates have any impact on the respective witchcraft and werewolf trials? Certainly, Rémy, Boguet and de Lancre were judges, demonologists and witch-hunters themselves. Yet what happened in the courtrooms all over Europe, especially in the Holy Roman Empire, where local judges and their lords were determined to wipe out witches, werewolves and other were-animals? Were they acquainted with the already published concepts about werewolvery and shape-shifting? How did they apply these ideas to criminal procedure?[10]

Thirdly, whenever anthropological or historical research considers the topic 'werewolf trial', a hotchpotch of all ever-mentioned cases is blended together.[11] De Blécourt has estimated up to 300 cases of 'werewolves' (mostly men) all over Europe in 200 years.[12] At the same time, he emphasises the lack of research on the topic, especially concerning the Holy Roman Empire, where samples of witch trial records have not yet been scrutinised. Furthermore, since the demonologists debated lycanthrophy as part of metamorphosis, including cats, dogs, serpents or hares, we should not separate artificially the werewolf cases from other charges of shape-shifting. Thus, in addition we have to scrutinise witchcraft trials with the inherent charge of shape-shifting. Villages and lordships with a Saint Hubertus patrocinium seem worthy of further research, since the saint worked against rabies and the bites of werewolves.[13]

More open questions concerning the geographical spread of werewolf trials are apparent. Some European territories, like England and Scotland, never saw charges of werewolvery. Other places faced chain witchcraft trials, including charges against were-animals. So, did the fear of werewolves emerge all over Europe? Or did it only occur in certain mountain areas in certain times of crisis, for example, war or when wolf packs roamed? De Blécourt and others suggest that predominantly men were accused of being werewolves, whereas women more regularly had to confess to shape-shifting into cats or other minor animals. But in regard to the as yet unread source material we should avoid any gender generalisation. Furthermore, ideas about werewolvery and were-animals came from various narratives about shape-shifting, which could be interpreted differently. It is likely that the werewolf paradigm, with its specific narratives, travelled with the help of the media (e.g., pamphlets, sermons, tracts) and thus infected even those regions without any wolf population.[14]

Fourthly, at the beginning of the seventeenth century, the basic demonological concepts about shape-shifting were at hand.[15] The tracts of Jean Bodin, Nicolas Rémy, Henri Boguet, and Pierre de Lancre had been published in several editions. Except for Boguet's work, they had been translated into different languages.[16] These notorious witch-hunting authors shared the long-standing classical and medieval narratives about metamorphosis, but dressed them up by adding the names of current locations and the accounts of so called eyewitnesses to gain authenticity. Only a few cases of werewolf trials could be added to the examples of the well-known stories. Nicolas Rémy counts only two new cases of werewolvery.[17] Robin Briggs, who looked for werewolves and shape-shifters in the rich source material coming from the French-speaking *bailliages* in Lorraine, found about seven trials with confessing 'werewolves'.[18] On the other hand, Nicolas Rémy ascertained that most female witches confessed to transforming into cats.[19] Like him, most demonologists exemplified their debates about metamorphosis and shape-shifting with narratives about were-animals (mostly cats) in general and werewolvery in particular. Pierre de Lancre tells at length about the one prominent example of the juvenile werewolf, Jean Grenier, who was banished to a monastery by the Parliament of Bordeaux in 1603. Against that background, Henri Boguet was rather successful, since he sentenced eight 'werewolves' to death between 1598 and 1599, whilst one suspected woman was lynched by the peasants.

It seems that the aforesaid questions cannot yet be answered, simply because historical research in cases of devilish shape-shifting is lacking, especially in the Holy Roman Empire with its hotspots of witch-hunting. Needless to say, even this paper cannot focus on all topics mentioned above, but it attempts to give some food for thought by following the theses of Elmar Lorey and Willem de Blécourt, which pointed to the fact that charges of diabolic metamorphosis as part of the witches' crimes were mostly introduced by the inquiring judges. Thus, the confessed narratives about were-animals, manifested in the trial records, were not the result of deep-rooted folklore about werewolves, but, quite to the contrary, were triggered by the inquiring judges' assumptions about werewolves and were-animals, presumably inspired by demonological tracts. The men in charge, judges, notaries, witch commissioners and their commanding lords, took a special interest in introducing the metamorphosis concept into their courtrooms. Therefore, my second thesis is that, in transgressing gender and status boundaries,[20] the shape-shifting was implied to mark the suspected persons as far more dangerous than a simple witch. Being already a metaphor of crisis, conflict, and war, the hunting of werewolves and other were-animals (like were-cats) became an integral part of a politico-religious programme, which sought to purify society from heretics, from so-called superstition, from paganism, from deviant morality, sin and debauchery, from vagrant malefactors – and from any rebellious resistance against increasing lordly authority. Therefore, the juridical conquest of werewolves and were-animals granted ultimate social prestige to witch-hunters and greater legitimisation to lordly authorities.

In discussing the thesis about the 'judge's lore', three analytic steps are taken: Firstly, a short tour d'horizon into the context of one of its origins in the mountainous peripheries of France (Valais, Franche-Comté, Labourd). Secondly, the politico-religious dimension of the notorious Stump-case. And thirdly: a case study in the werewolf and were-animal trials in some Eifel territories. Thus, new source material is scrutinised.

The metamorphosis concept and political demonology – a very short tour d'horizon

Polemical debates during the French war of religion inclined to demonise the political adversary. Already in 1563, in his famous apologetic *Response,* Pierre Ronsard invented the exorcism of a Calvinist werewolf, whereas after the Saint Bartholomew massacre Agrippa d'Aubigny branded the Catholics as having demoniac powers.[21] Moreover, since the important study of Jonathan Pearl we know that, in the French kingdom during the Wars of Religion, a political (or better: politico-religious) demonology was established. Doubtlessly, the Spanish Jesuit Juan Maldonado has to be named as one of its main 'inventors'.[22] Strongly attached with the French monarchy as well as with the dukes of Lorraine, Maldonado developed a clear interpretation of heresy in general and of Protestantism, Lutheran or Calvinist, in particular. Like the Hussite movement, heretics were chaperoned by swarms of witches, sorcerers and magicians. Religious sectarianism was doomed to collapse into diabolic witchcraft, debauchery and mere sin. At the same time, the existence of superstition, witchcraft, sorcery and magic indicated deep-rooted heresies. At the advent of religious conflicts in France, a heretic Huguenot was marked as a witch, a devil's minion, a traitor and a rebel against the sacred Catholic king, renouncing godly order. Thus, neither the French monarchy nor other authorities could tolerate the heretics and witches; instead, they were obliged to root them out.

In his famous sermons about the nature of angels and demons, given at the college of Clermont in Paris 1572, Maldonado, who was generally praised and accepted as a theological authority, answered the question, *si les corps peuvent estre changez en diverses formes par les démons*? In citing well-known examples from classical literature, the Jesuit explained that demons conducted metamorphosis in three different ways: As a real mutation as when the Egyptian sorcerers had changed their staffs into serpents. However, for Maldonado it remained impossible for demons to transform a human body in such a material way because of its soul and reason. Thus, the demons achieved metamorphosis either as an apparition, which deluded both, the enchanted and the bystander, or as a delusion which deceived only the enchanted. Maldonado stated that disbelievers in these facts of lycanthropy and shape-shifting acted like Calvinists who denied the transubstantiation. Once and for all, the Jesuit labelled all sceptics of shape-shifting as heretics and blasphemers.[23] Thus, Maldonado established a rigorous new belief system. Only about 63 years previously and with a slightly shifting attitude, another radical Catholic theologian and reformer, the preacher John Geiler of Kaysersberg, had stated that believers in a substantial metamorphosis committed the deadly sin of blasphemy, since all kinds of shape-shifting were only the result of devilish delusion.[24] Later on, Peter Binsfeld used this argument against Jean Bodin.

Maldonado's lectures as well as the theatrical public exorcisms staged by him, reached a wide audience. Pierre de Lancre was probably one of his pupils and copied down his sermons. Martin Delrio – related to De Lancre and Montaigne – used other manuscripts of Maldonado's sermons in working on his *Disquisitionem magicarum*. Both de Lancre and Delrio were highly influenced by Maldonado's politico-religious demonology.[25] In taking up the argument of Maldonado, Delrio and De Lancre likewise rejected Jean Bodin's definition of lycanthropy as a real physical transformation, as described in his *Démonomanie* (published 1580).

I assume that the politico-religious demonology of Maldonado lived on among the French Jesuits. Temporarily, in 1595 the order was banned in France, and members of the Dijon college migrated just beyond the borders and found a new home in Dole, the seat of the provincial government of still Habsburgian Franche-Comté,[26] which belonged to the diocese of Besançon. Since 1586, Ferdinand de Rye held the seat as archbishop. Simultaneously, he was abbot of Saint-Claude (Jura) and its *terre*. Archbishop de Rye, as an ardent pursuer of post-tridentate Catholicism, favoured the Jesuits, whose order was established in Dole (1582), Besançon (1597), Vesoul (1610) and Gray (1622). The Jesuits in Franche-Comté, joined post-1595 by French refugees, were explicitly acting as missionaries in a territory, which had been affected by protestant heretics. The reformed Catholics thought the people of the mountainous regions in the Jura to be badly instructed. Among others, the Jesuits re-established the pilgrimage to Saint Claude. It seems natural, therefore, that the order, battling at the very frontline against superstition, so called pagan beliefs and popular religious malpractices, worked together with the grand judge of Saint-Claude, Henri Boguet. This is more than likely, since the ordinance of Philipp II, Spanish king and archduke of the Spanish Netherlands (and thus, of Franche-Comté), in 1592 had commanded to root out all kinds of magic, superstition, pagan lore, sorcery, and witchcraft.[27]

Therewith, joint actions were on top of the counter-reformational agenda, which helped to trigger the fear of witchcraft and werewolves.

Accordingly, Boguet's *Discours des sorciers* (Figure 7.1), with its vivid descriptions of female and male werewolves, was granted print permission from Father Michel de Coyssard, rector of the Jesuit college in Besançon. In his opinion, the book contained nothing against the Catholic faith and nothing against morality. Other high theologians from Dole and Besançon joined in the chorus of approval.[28] Moreover, Boguet's treatise was printed in Lyon – illustrated with the Jesuit Imprimatur – by Jean Pillehotte, the official printer of the Jesuit order. In the prefaces, which were adjusted for the several revised editions of his tract, Boguet clearly explained his politico-religious mission. In referring firstly to Ferdinand de Rye, and later on to his nephew, François de Rye, coadjutor of Besançon, Boguet explicitly declared war against witches (including werewolves) to uproot this disorderly sect of malefactors. The grand judge painted the picture of thousands of witches, swarming under the standard of Satan to win the battle against godly ordered kingship. We find this argument already in the chronicle of Hans Fründ, who probably gave the first report of witches, transforming into werewolves, which was connected with the great witch-hunt in the French and German speaking Valais (1428–1436). According to Fründ, the charged witches had confessed that if the prosecution had not stopped them, they might have elected them a king to oppose Christianity, authority and criminal court procedure.[29] Lambert Daneau and others would repeat this argument. Logically, rebellious witches and werewolves were the inborn enemies of well-functioning statehood.[30] Boguet even suggested to install the specific office of a single judge, who would deal only with witch-hunting, since the ordinary judges were much too overburdened with their daily work to prosecute witches with the necessary strictness. In a third preface to the archdukes of the Spanish-Habsburgian provinces, Albert and Isabella, thus, Boguet thanked them overwhelmingly for having issued an ordinance against witches in 1606. Therein, the archdukes had argued against malpractice in witch trials to underline that *la main souveraine* alone had to guide proper witch-hunting. To guarantee that princely prerogatives in witchcraft trials were not violated, they proposed to nominate a superintendent for witch-hunting in each of their provinces. Perhaps the very ambitious Boguet had hoped to win such a position in Franche-Comté for himself.[31]

All in all, in the small territory of the abbey St. Claude, under the chair of Boguet, about 35 persons were accused of witchcraft, 28 were executed (including eight 'werewolves'), four died in prison, two others were banished, and one she-werewolf was lynched.[32] A persecution structure, similar to other hotspots of witch-hunts, repeated itself. Children who were afflicted with demonic possession, were exorcised by Jesuits and consequently accused adults of being witches, who had tormented them with demons or had taken them to the witches' sabbat.[33] Comparable incidents occurred in

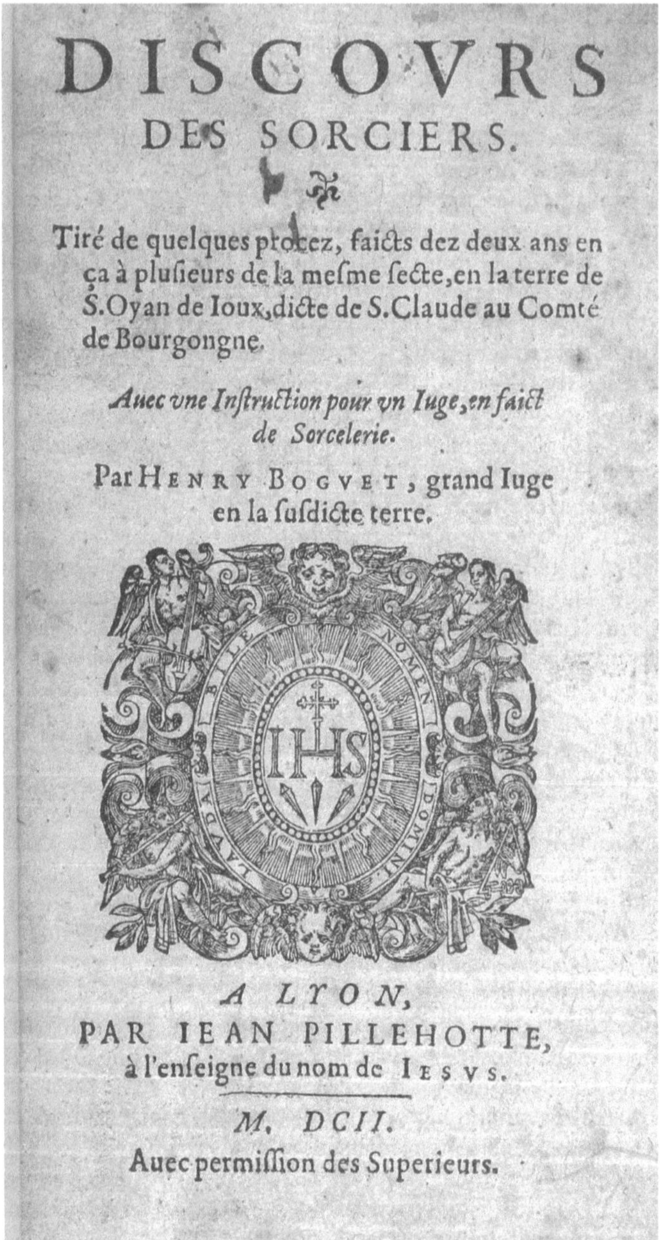

DISCOVRS
DES SORCIERS.

Tiré de quelques procez, faicts dez deux ans en ça à pluſieurs de la meſme ſecte, en la terre de S. Oyan de Ioux, dicte de S. Claude au Comté de Bourgongne.

Auec vne Inſtruction pour vn Iuge, en faict de Sorcelerie.

Par HENRY BOGVET, grand Iuge en la ſuſdicte terre.

A LYON,
PAR IEAN PILLEHOTTE,
à l'enſeigne du nom de IESVS.

M. DCII.
Auec permiſſion des Superieurs.

Figure 7.1 Title page, Henri Boguet, *Discours des sorcies* (Lyon 1602), printed by Jean Pillehotte on behalf of the Jesuit order (see the vignette in the middle of the page)

Saint Claude.[34] Yet, we do not know if Jesuits had personally taken part in the Saint Claude trials, but there are some striking indicators: Boguet had read the treatise of Peter Binsfeld, suffragan bishop in Trier, who had argued strongly for using child witches as witnesses, because only with their help the most wicked leaders of the witches' pack could be exposed. Furthermore, contrary to Bodin's argument, Binsfeld had declared the witches' alleged shifting into wolves or cats to be a mere outcome of *praestigium* and devilish delusion: either the devil created the illusion in the human mind or he cloaked and masked the human body so that it appeared to look like a cat or a wolf.[35] Boguet, for his part, followed Binsfeld's argument straightforwardly: if the human body could not be shifted materially into an animal's body, but if the human body stayed intact, veiled by devilish disguise, how did the alleged were-animals kill their victims? Boguet found a rather pragmatic answer. They attacked children and cattle with instruments, with knives and swords, they strangled them or they simple dragged them heavily through rocks and stones, until they were dead. Thus, the so-called werewolves had committed the crimes in person,[36] not the devil with their consent in the shape of a wolf or their wandering spirit, which had shifted into an animal form.

In the counter-reformatory milieu, created by the Jesuit's missionary work, and against the legal background of ducal ordinances, several factors stipulated the trials of Saint Claude. Firstly, Boguet happened to be an ardent persecutor of witches and an ambitious man, seeking social profit. Secondly, a wolf epidemic ravaged the Jura, which forced the populace to press for trials. And finally, but most important, we have to set the trials in the legal framework of this distant Spanish-Habsburgian province. Already in 1573, the duke of Alba had accused the parliament in Dole of conducting criminal trials inefficiently. These complaints had led to a very quick trial of the werewolf Gilles Garnier. With his execution the provincial government demonstrated its legal competence and its control over the enemies of order, authority and God.[37] Thus, the werewolves and witch trials in the Terre of Saint Claude satisfied the appetites of Boguet as well as the provincial government. In general, the anti-witchcraft-movement joined the forces of the Jesuit's missionary work, the secular justice authority and the supervising parliament in purifying the land of all rebels against godly and lordly authority.

Next in line, Pierre de Lancre, educated by Jesuits, followed the politico-religious demonology of Maldonado, Boguet, and Delrio. He accepted the narratives of Boguet and included the arguments in his own report about the Labourd cases. Additionally, he relied on the arguments of Peter Binsfeld. Pierre de Lancre's key case was the one against a young boy, Jean Grenier, who confessed to be a werewolf and a cannibal. Moreover, Grenier claimed to be the bastard of a priest and as such, he symbolised the prominent folk devil of post tridentate Catholicism, preached by Jesuits.[38]

In summarising the French writing demonologists Bodin, Boguet, and De Lancre, shape-shifters and werewolves could be either male or female. Mostly, they belonged to the lower class; they were vagrants, beggars, healers, religious and sexual deviants, nearly pagans. Sometimes they committed their abnormal crimes alone, but mostly they acted in a pack. At the bottom of the social ladder, they rebelled against every godly given order, against the true religion of Catholicism, against the rules of Christian morality and finally against kingship itself. Thus, the evil quality of werewolves and were-animals was considered the most wicked, most sinister and most vicious.

The shape-shifter concept as part of the political demonology gained ground in other parts of the boundary area between France and the Old Reich, mostly in the strongholds of Catholic princes. We have to remember that the House of Lorraine attached itself to the Bavarian House of Wittelsbach and with the dukes of Jülich-Kleve-Berg. For example, in 1581 the prince bishop of Lüttich, Ernst of Bavaria, was installed in the princely abbey of Stavelot-Malmedy. Educated by Jesuits in Ingolstadt, he became the elector of Cologne in 1583 and the bishop of Münster in 1584. His sister-in-law was Renata of Lorraine, the wife of Wilhem V of Bavaria. His cousin, Jakobe of Baden, married the insane duke Johann Wilhelm of Jülich-Kleve-Berg. After her being murdered, in his second marriage, Johann Wilhelm was linked with Antonie of Lorraine, the daughter of Charles III of Lorraine and a niece of Renata. Thus, via the bishoprics of Lüttich and Cologne, backed by the duchies of Lorraine, Bavaria and – finally after 1609 – Jülich-Berg, a strong network of Catholic princes, educated and advised by Jesuits, was established. It included the founding of new colleges, schools and universities, controlled and occupied by the Jesuit society.

The contacts between Maldonado and the Dukes of Lorraine has been noted already. In forming a Catholic bulwark against the Protestants in the Netherlands and in German territories, the dukes of Lorraine engaged very eagerly in counter-reformation. A Jesuit university and college were installed at Pont-à-Mousson, visitated by Maldonado in 1578.[39] In Lüttich, the Jesuits were established in 1582. Between 1591 and 1594, Martin Delrio gave lectures in moral philosophy at this college, where he began writing the *Disquisitionem magicarum*. In Lorraine, the so-called 'bewitched duchy',[40] witch-hunts emerged and crossed the border to the Old Reich via the German-speaking offices in Lorraine and entered the electorate of Trier.[41] The chief prosecutor, Nicolas Rémy, boasted in 1590 of his executions in the duchy. His treatise was well known by Boguet and by De Lancre.

The impact of Bodin's *Demonomanie* on the French witchcraft trials seems to have been of no great importance. As William Monter has pointed out recently, 'it is impossible to believe... that Bodin's arguments about the legal treatment of accused witches had even the slightest effect on French jurisprudence'.[42] Jean Bodin's arguments failed in his homeland, but probably its translations gained prominence. A Calvinist, François du Jon, had translated it into Latin; and thus, the Scottish King James VI could have relied on Bodin in writing his *Daemonologie, in Forme of a Dialogue* (1597). Nevertheless, the king considered shape-shifting to be a mere delusion, affected by melancholy.[43] Only Joseph Glanvill, an Anglican clergymen, in the second half of the seventeenth century, shared Bodin's belief that it was a substantial metamorphosis of humans into animals.[44] Far from becoming a canonical authority in France, it is still debatable whether Bodin's treatise gained any importance in German-speaking regions via the German translation, which appeared in 1581, manufactured by the Lutheran Johann Fischart. Evidentially, Jonathan Schütz has argued that Fischart's text has to be 'recognised as a text on its own, with its own history of reception that exists independently of the French version'.[45] So far, there is little evidence that Bodin's treatise had any impact on charges of shape-shifting and werewolvery in the German lands. Since 1596, the *Démonomanie* (translated in Italian in 1587) was listed on the Roman-Catholic Index of forbidden books.[46] Thus, Catholic scholars avoided making too many references to it. Peter Binsfeld declared Bodin to be a heretic, since the belief in a substantial metamorphosis was considered blasphemous. Martin Delrio advised that greatest care should be taken in dealing with Bodin's work. Nevertheless, both Binsfeld and Delrio made use of untainted narratives from Bodin.[47]

Presumably, those German-speaking judges who took an interest in hunting witches and werewolves, considered Fischart's treatise, which included some of the well-known narratives of metamorphosis.[48] Later on, Fischart became a bailiff in the Lorraine office of Forbach. There is striking evidence that in 1587 he supervised several witchcraft trials, in which female witches confessed their transformation into cats. Nicolas Rémy kept the trial records at his desk and included their narratives in his demonology.[49] If and how the metamorphosis concept was transferred into the Lorraine and German courtrooms before 1580, is still debated.[50] Likewise, it has yet to be established without doubt that the later trials against werewolves in German Protestant lands, which were aimed at shepherds and wolf banners, had been influenced by Fischart and Bodin.[51]

A key case of political werewolf hunting – Peter Stump

The most notorious werewolf trial took place in Bedburg near Cologne when Peter Stump was convicted on a charge of werewolvery and executed on 31 October 1589.[52] It was the first case to attract great popular interest in the Old Reich, even though there is no legal evidence that the case ever happened (Figure 7.2). Yet, for the year 1589, the general-vicar in Lüttich, Jean Chapeauville, annotated in his *Gesta Pontificium Leodiensium* (published in 1616) – also including the witch-hunts in the electorate of Trier and the execution of Dr Dietrich Flade – the most notorious case of a werewolf, executed for having devoured 13 people.[53] Chapeauville did not mention

Figure 7.2 Coloured broadsheet from 1589, depictinging the execution of Peter Stump (Stump Petter), the 'werewolf'. (Hessische Landes- und Hochschulbibliothek, Hs. 1971, vol. 24, fol. 5). This is one in a series of at least five broadsheets about the case. Cf. the Geneva drawing (Figure 1.1) which shares three elements with the Stump print

the name of this beast, but the execution year and the number of persons killed indicate that it had to be the story of Peter Stump. Thus, Chapeauville, besides Hermann Weinsberg and Arnoud van Buchell,[54] probably is one of the first to record this notorious case. Although there are vital arguments that the trial really did occur, there is still a very great uncertainty concerning the actual crimes of Peter Stump. All in all, the information given by the different broadsheets has to be treated with great suspicion. Nevertheless, we find strong evidence regarding the political dimension of the Stump case.[55] In 1589, Bedburg returned to the hands of Count Werner of Salm-Reifferscheid, an ardent Catholic lord, who supported Archbishop Ernst of Bavaria in the so-called Cologne War. In Bedburg, Werner immediately took action to re-Catholise his manorial district and to re-establish his seigneurial prerogatives. The trial against a werewolf, who had allegedly committed rapes, infanticide, mass murder, incest, debauchery and cannibalism, amplified the lordly authority of Werner. Besides, the count played the multifunctional instrument of witchcraft trials with a sleight of hand in the different political contexts of his fragmented lordships. He has to be seen as a key figure in the witch-hunts that were conducted in the Eifel region.

The Stump case gained great publicity all over Europe. Several German pamphlets (published in 1589) were accompanied by broadsheets from Antwerp (1589), from London (1590) and Copenhagen (1591). Most remarkably, the case attracted the attention in bi-confessional Augsburg, in Lutheran Nurnberg and in Protestant communities in the Netherlands, England and Denmark. Traditionally, the Netherlands had strong connections with Cologne, so news about the case spread easily. The politico-religious dimension of the Stump case seems apparent. In England, fears of popish conspiracies were common, even after the defeat of the Armada. In 1591, *The News from Scotland* appeared in England, with stories about diabolic witchcraft and a conspiracy against the King of Scots and his Danish Queen. Because it had happened in a village near the most Catholic city of Cologne, the Stump broadsheet fitted neatly into this milieu of fear about conspiracies, combined with both confessional conflicts and witchcraft panics. The werewolf Stump sensationally reflected ugly popish crimes. Analogous to Maldonado's political demonology, but under different religious terms, its Protestant counterpart interpreted witchcraft as a crime infected by popery that targeted the godly law of truly reformed Christians. In Protestant propaganda, witchcraft and popery were two sides of the same coin: namely, superstition and pagan lore. In England and Denmark, the case of Peter Stump seemed to be a symbol for popery mixed up with superstition and adultery. Pointedly, Ronsard's Calvinist werewolf had turned into a popish werewolf.

Obviously, the French demonologist judges remained quiet about the Stump case. Nicolas Rémy did not refer to it directly but he noted a narrative about shape-shifting, which allegedly he had heard from count Paul of Salm-Badenweiler, the descendant of another branch of the House of Salm. Count Paul was the father-in-law of Francis II, duke of Lorraine. Probably because Rémy did not refer to the Stump narrative, neither Boguet nor De Lancre took it up. Likewise, Jean Beauvois de Chauvincourt showed no interest in the Stump story, although he connected the horrors of the French wars of religion with the horrors of lycanthropy.[56] On the other hand, in the debate about lycanthropy, published by the Franciscan Claude Prieur from Louvain in 1596, we find a reference to the notorious case.[57] Likewise, Martin Delrio commented on the Stump affair in 1599.[58] I assume that the Antwerp broadsheet, as well as communication channels from Cologne, had introduced the narratives about the werewolf Stump to the northern Spanish-Habsburgian provinces as well as to the Netherlands. Thus, in 1595 the provincial court of Utrecht conducted witchcraft trials against several people from Amersfoort, and – as de Blécourt has argued with good reason – introduced the werewolf concept, following the narratives around the Stump case.[59] Notably, in France and its French-speaking peripheries the Stump case made no significant impact since there was no French or Latin translation of the narrative.

In 1591 another German broadsheet featured werewolves, with the horrid story of a pack of she-werewolves in the duchy of Jülich. Published by a sensationalist printer in Augsburg, the story claimed that 85 women were charged and executed. Most historians dispute that such a mass execution could have taken place in the duchy, whose former duke Wilhelm V was thought to be highly influenced by Johann Wier. Probably the broadsheet simulated (perhaps ironically?) the Stump case, but with a pack of women being accused and executed.[60] However, we should keep in mind that in 1591 the insane duke Johann Wilhelm ruled the duchy of Jülich. He was possessed by terror of Protestants and witches. I speculate that poor Johann was driven deeper into panic by the innumerable exorcisms that he underwent. Johann's father in law, Charles III of Lorraine, sent his own trusted exorcist, the Ambrosian monk Francesco Maria Guazzo. Guazzo could not help Johann, but was inspired to write his *Compedium maleficarum*, in which he dedicated a long chapter to the crime of diabolic shape-shifting, re-quoting amongst others Delrio and Rémy. Notably, the Ambrosian monk used the narrative of 'Stumphius Petrus' from Bedburg, not as example for werewolvery, but in the discussion about the substantial reality of Incubi and Succubae. According to his comment, Guazzo obtained the information about the case from broadsheets.[61]

The judge's lore – the metamorphosis concept as part of anti-witchcraft movements

Besides anthropological interpretation, the metamorphosis concept was part of the politico-religious demonology that emerged in the French Wars of religion and was disseminated in those peripheries of France, the Netherlands and the Old Reich that were affected by political and religious turmoil. Shape-shifters and werewolves were branded as the most horrible rebels against any kind of God-given order. Thus, the mere existence or allegations of shape-shifters produced ambivalent readings. On the one hand, they symbolised deep human and social disruption caused by paganism, superstition, heresy (likewise popery in the Protestant context), and rebellion. At the same time, these religious and political disorders had to be responded to with a hard hand by ardent lords (like Werner of Salm-Reifferscheid) and judges (such as Boguet). Moreover, the mere detection of a werewolf or other were-animals (like cats) by ambitious judges demonstrated how very necessary must be the implementation of an anti-witchcraft-policy. In entering the laboratory of the Eifel lordships, we can test our assumptions with source material, which never before has been scrutinised for shape-shifters and werewolves.[62]

The south-western areas of the politically fragmented Eifel region belonged to the Spanish-Habsburgian province of Luxembourg. The northern and eastern parts were members of the Old Reich, and like the imperial counties of Manderscheid, they sought political autonomy. These parts neighboured the electorate of Cologne and the duchy of Jülich. Between 1580 and 1650, extensive witch-hunts pervaded the whole Eifel region, with hot spots in the small lordships of the Luxembourgian province, in the counties of Manderscheid-Kail, Manderscheid-Blankenheim, and Manderscheid-Gerolstein. Likewise, minor lordships participated in hunting witches, such as Bürresheim, Neuerburg, Wildenburg and Schmidtheim. In the whole Eifel region, about 900 executions are known to have taken place, with a possible estimate of around 1500. Thus, the thinly populated Eifel Region has to be marked as a center of witch-hunting, where the counts and lords initiated and promoted the trials themselves, assisted by legally trained judges, and so-called 'witch commissioners'.

Doubtless, the witch-hunts in the Eifel region reflected certain politico-religious interests. In several lordships, an anti-witchcraft-policy was established, wherein the metamorphosis concept gained prominence. A most striking example is found in the very small lordship of Schmidtheim, which contained only one village of approximately 50 households. Between 1597 and 1637 around 60 people were executed as alleged witches, with a peak in 1630–1 of around 48 victims (Figure 7.3). Recent research has pointed to the specific political background.[63] The lords of Schmidtheim, Reinhard the Younger and his son Bertram Beyssel of Gymnich, were seeking for autonomy from their overlord, the count of Blankenheim. Anti-witchcraft policy played a prominent role, since arresting, torturing and executing alleged witches demonstrated the claimed privilege of high justice as core of sovereignty. The chance for Reinhard Beissel von Gymnich the Younger to engage in witch-hunts came in 1597 when in Blankenheim a notorious thief and robber was apprehended. At first, Clas Kramp, who was originally from Schmidtheim,[64] was charged with murder and robbery and then made to confess to further crimes such as sodomy, incest and witchcraft, which he claimed to have learnt from his mother Susanna. Subsequently, in February 1597, the Lord of Schmidtheim arrested the old widow Susanna, her second son Hans, as well as her daughter Anne, followed by four other suspected village people in the period up to April 1597. The Kramp family certainly had a dubious reputation, since the late father was a cunning man and had been followed in his trade by his son-in-law, who in 1605 likewise was executed in Schmidtheim as a witch, but not as a werewolf or as a were-cat.

Since Reinhard the Younger was determined to conduct the trials, not only exclusively in Schmidtheim, but also on a solid, unchallengeable basis, he sought legal advise in Trier, in whose surroundings a vast witch-hunt had just come to a temporary halt. The lord was concerned about what penalty a person should be sentenced with, if he has confessed to theft, robbery, murder, adultery, sodomy, incest, and finally, shape-shifting. Obviously, the lord was referring to Hans Kramp, who had allegedly admitted maleficia, committed in the shape of a cat. The legal answer from Trier has not been

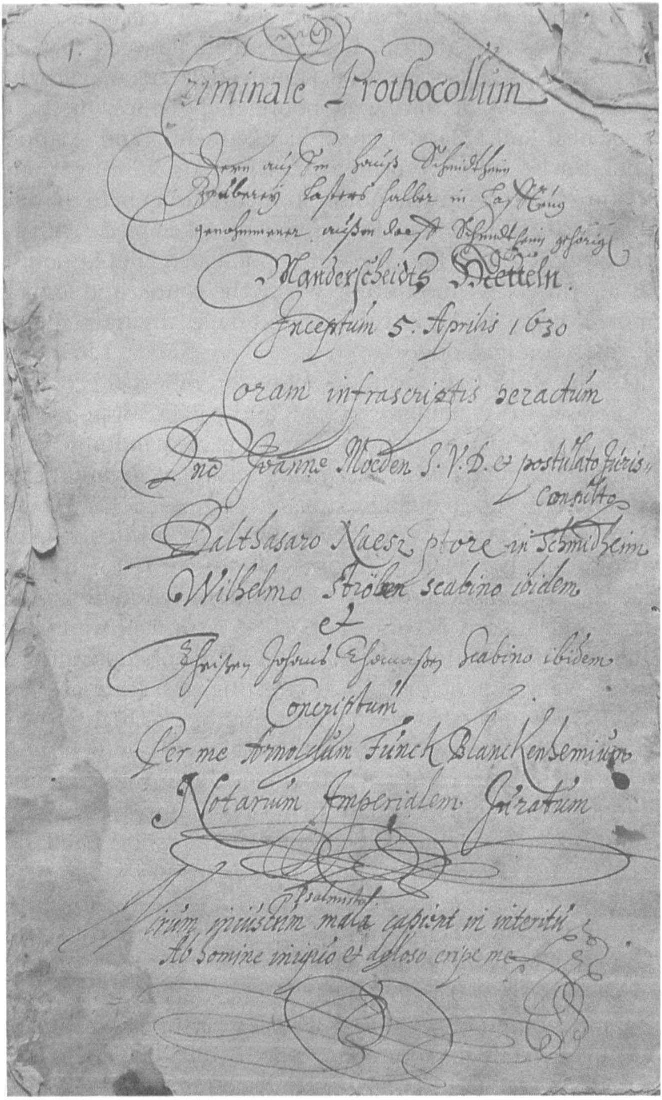

Figure 7.3 Frontpage of the trial records, concerning so-called *Manderscheids Mettel*, a female witch and werewolf, executed on 10 April 1630 in Schmidtheim (Eifel)

preserved, but I assume that it was rather inspiring, because in 1597/1598 (one year in advance of the werewolf hunting in the Terre of Saint Claude) five witches and shape-shifters were executed in Schmidtheim. Most interestingly, the gender distribution and the shape-shifting confessions do not fit into the framework indicated so far by historical research. Hans Kramp confessed to having been transformed into a cat with the help of a salve, which the devil had given to him. He re-transformed himself after simply washing the ointment away. With the help of the devil, his old mother allegedly shifted into a werewolf, whereas Maria, one of her accomplices, confessed to changing either into a cat or a werewolf by saying a special spell and returning into a human by praying. Another woman, suspected of belonging to the pack, allegedly changed into a werewolf, whereas the fourth 'witch' supposedly did her mischief as a were-cat. The forced confessions show that the images of werewolvery and shape-shifting were not fixed at all. However, the influence of the Stump broadsheet seems to be marginal, since no girdle for transforming was mentioned.

I assume that Reinhard the Younger had been well informed about the political dimensions of werewolf and shape-shifter hunting, either as bailiff in a former district of the electorate of Cologne (Nürburg), or in networking with the greater territorial lords in the Eifel: such as, count Werner of Salm-Reifferscheid, who some years later would become guardian of the young count of Blankenheim, Johann Arnold. Triggered by the denunciations of Clas Kramp, three women were sentenced to death in Blankenheim in Mai 1597 as witches, who had transformed themselves into cats and werewolves.[65] Notably, the concept of shape-shifting and werewolvery was included amongst other charges of evil witchcraft. The marginalised family Kramp fitted all too well into the image of rebellious devilish shape-shifters and most horrible villains, who had violated godly and familial order. With good reason, the lord and his judges had introduced the concept of metamorphosis into the courtroom. No one could blame the lord of Schmidtheim for rooting out these beasts. Clearly, Reinhard the Younger had two aims: firstly, to gain sovereignty from Blankenheim, and secondly, to expand lordly prerogatives in Schmidtheim. Thus, in 1612, he supplemented the witchcraft trials with a police ordinance to improve obedience, discipline and the weak morals of the peasantry. Most of all, the lord wanted to root out increasing adultery.

From about 1604 to 1614, Arnold II, Count of Blankenheim, took up witch-hunting in his county. To succeed his childless brother Hermann, he had to give up his office as dean of the Cathedral chapter in Trier, where he had been involved in the prominent witchcraft trials. The count, who in Trier had been denounced as a witch himself,[66] ordered the parish priests to make a general inquiry in order to report all damages to men and beasts, which were suspected of being caused by witchcraft.[67] Long lists were sent to the count's chancery, but only two cases can be vaguely related to shape-shifting. In one village, a cat had suddenly scratched a little girl, who had died afterwards. In the same village on one night, a wolf had broken into a sheep pen, carrying away eight sheep to the garden. The beast had bitten five sheep to death and one other animal was hardly alive. Most strangely, the wolf had buried two other sheep alive with their heads free above the ground. According to the reporter, the wolf must have been a very strange beast indeed since it had left no traces in the sodden garden soil. Apart from these two very far-fetched allegations, we cannot find any more charges coming from the populace concerning shape-shifting. Furthermore, there are no complaints about roaming wolves, which could have precipitated popular demand for hunting werewolves. This finding underlines once more the thesis that the metamorphose concept, which dominated five to eight witchcraft trials in 1597/1598, had been introduced by the lord and/or the judges.

Post 1625, the worst phase of witch-hunting began in the Eifel region, whose backgrounds cannot be discussed here in detail. However, in the counties of the Eifel and in parts of the electorates of Trier and Cologne so-called witch commissioners made a great impact on the dynamising of witch-hunts. These commissioners were trained lawyers, accredited by one of the superior courts of the sovereign. Theoretically, lords, villages and courts which wanted to pursue witches, installed a witch commissioner who should supervise the trials to avoid misconduct, but most of these witch commissioners initiated devastating witch-hunts in villages and lordships. Some of them were freelancers, advertising their service at several different lordships in order to establish an anti-witchcraft business. One of these soldiers of fortune was the most notorious Dr Johann Möden, who was personally responsible for more than 300 death sentences between 1627 and 1653 in the Eifel region and in the Rhine-Moselle region.[68] In Blankenheim, Gerolstein, and Schmidtheim, Möden and his colleague Arnold Funk established a well-functioning anti-witchcraft business, which cooperated with the lordly anti-witchcraft policy. The biography of Möden reveals interesting details. Educated by Jesuits in Coblence and being a member of a Marianian congregation, he showed himself to be an ardent pursuer of the new post-tridentate Catholicism. In 1613, he started his study of jurisprudence at the catholic University of Würzburg, founded by Julius Echter of Mespelbrun and led by Jesuits. Back in the Rhineland, Dr Möden settled in Münstereifel, where in 1625 the Jesuits founded a college and seminary.[69] Thus began a close relationship between Möden and the Jesuits, who were headed by Father Heinrich Rinkop from Cologne.

As in northern and southern France, in Lorraine and in Franche-Comté, the Jesuits in Münstereifel took a position as a bulwark against the spread of Protestant heresies; particularly, because a small community of Protestants had settled in the city. Moreover, the late count of Manderscheid-Schleiden and his late wife had shown great sympathies to Protestantism. In general, the Jesuit order worked out its plans for an internal mission against the so-called 'Indians' in the peripheries of Europe, including the Eifel. The term 'Indian' was applied to every practitioner of devil-worshiping idolatry – heretics, witches, Protestants, and even 'Old Catholics', who were entangled in religious errors. The thinly peopled mountains, valleys and forests of the Eifel region, which so far had not benefitted from post-tridentate reformed Catholicism, turned out to be an ideal playground for missionary campaigns to free the peasants from superstition. Moreover, the purifying campaigns targeted a parish clergy, whose members were ill learned, living in concubinage, and suspected of practicing magical, superstitious deeds. The Jesuits' spiritual anti-witchcraft mission called for secular support. Like Boguet in the Terre of Saint Claude, the ambitious witch commissioner Möden was more than eager to take up the battle against witches. From the start, the cooperation worked out efficiently, since the Fathers served as confessors and exorcists in the witch trials conducted by Möden or Funck in the county of Blankenheim. In Schmidtheim, anti-witchcraft mission, anti-witchcraft business, and anti-witchcraft policy joined forces. Unsurprisingly, the metamorphosis concept appeared again in the courtrooms. Between 1630 and 1635, 50 witches stood trial in Schmidtheim, among them 22 shape-shifters, seven men and 15 women. Altogether, nearly every second witch in Schmidtheim was accused of being a shape-shifter. The gender-mixed character of the concept prevailed. The male category included a were-cat, werewolves, and men who changed into both and into other animals. Likewise, the female group consisted of were-cats and she-werewolves, among several women transforming into cats and werewolves; and one who pretended to change into a cat, hare or werewolf. In general, the charged persons differed greatly in age and in social status. Some of the men even belonged to the village elite, being former members of the manorial court. Not one shepherd or wolf banner was included.

With close reading of the trial records, three background frameworks can be established in which the metamorphosis concept emerged.

The unmasking of conspiracies: In 1630, the inquiring notaries allegedly uncovered a witches' conspiracy against the noble family Beissel of Gymnich. The accused witches from Schmidtheim confessed under torture that they had plotted against the young lord Bertram, who 'wanted to become a greater lord at the back of the poor peasants'. To stop his harshness and his ambition, the witches had (allegedly) planned to kill his new-born first son. Therefore, they changed into cats, sneaked into the castle and poisoned the child. On other occasions, they changed into werewolves to kill horses and other animals. The metamorphosis happened in different ways, either with a girdle or a fur, given by the devil, or ointment, applied by the devil. The lord attended the inquisition of the witches in person, and showed himself determined to extirpate his rebellious and murderous subjects. The same conspiracy concept was utilised in Blankenheim by Möden, where in 1630 a plot of about 20 shape-shifting witches against the count of Blankenheim was 'detected'. In 1631 once more a plot of about four werewolfing witches was unmasked by the witch commissioners. This time the witches had planned to murder the last and youngest son of lord Bertram.

Disciplining the village élite: Also in 1630, three men, originating from the village elite, were accused of being witches, werewolves, and shape-shifters. I assume that they belonged to the manorial assembly, which tried to resist the young lords attempts to widen his prerogatives. Unsurprisingly, these men fitted likewise the image of rebellious shape-shifters, the enemies of all godly and princely orders.

Denouncing members of the élite: Apparently, the forced denunciations of male and female werewolves and shape-shifters were directed against two significant groups of leading men, who probably resisted the anti-witchcraft movements of Möden, the lords and the Jesuits. Firstly, members of the court jury or their relatives in Schmidtheim and Blankenheim were denounced as the witches' accomplices. Even the noble district magistrate in Manderscheid-Gerolstein, Heinrich von Mülheim, who had already been executed in 1629 as a witch, was still named as a member of the witches' pack. Secondly, alleged werewolves and shape-shifters were forced to bear witness against parish priests, who were denounced as Satan's servants, conductors of black masses and perverters of holy rituals. In the church districts of Manderscheid-Blankenheim and Manderscheid-Gerolstein at least eight parish priests were executed as supposed witches, including the parish priest of Schmidtheim. The arguments of the Jesuits missions against the unreformed clergy had bolstered this specific persecution. The Jesuit Heinrich Rinkop himself re-baptised the parishioners, who originally had received the holy sacrament from the hands of the corrupted priests.

But how was the very delicate problem of shape-shifting demonology resolved in the Eifel courtrooms? Bodin and his colleagues had debated about the materiality of the metamorphosis. Pragmatically, Boguet had launched his idea about 'werewolves', who under the veil of devilish illusion had committed their crimes as entirely human beings with the help of metal tools. In Blankenheim and Schmidtheim, the concept of metamorphosis as a devilish masquerade was elaborated even further. Already in April 1629, in the presence of the count, a man in Blankenheim confessed that he had transformed himself into a werewolf with the help of a girdle, but committed his crimes only after the devil had equipped his mouth with additional iron teeth. In Schmidtheim, the illusionary transformation was caused either with ointment, or a girdle, or a fur, or a cloth from sheep's wool, or a wolf's fur with four paws. Moreover, the devil supplied the witches with iron teeth, iron claws and iron knifes. With these instruments, smeared with deadly poison, the naked witches, masked as were-animals, scratched children, men and cattle to death. With the help of iron rakes, they tore out the throats of animals.

The most elaborate werewolf and shape-shifting narrative was given by young Angelica in 1635, whose mother and other relatives had already been executed as witches in the previous years. Clearly, she showed herself acquainted with the stories about the Schmidtheim werewolves and responded all too well to the leading questions. And she told a fantastic story. After the witches had undressed, the devil wrapped each of them in a wolf's fur with paws, so that they were able to walk on all fours. Moreover, the devil covered them in a hare's fur. In the hare's disguise, they ran into a field with cows, where they stripped the hare's fur from their heads revealing the wolf's fur underneath. Additionally, the devil provided Angelica with two iron claws, which she held in both of her hands. With these instruments she ripped open the belly of a cow so that the guts spilled out. After the killing, the witches replaced the hare's fur to run back to the place with their human clothes. On another occasion, the witches' pack burst into a stable, one of the wolves holding a candle in his ear to give light. Using knives they butchered, skinned and devoured several sheep. After their 'transformation' Angelica still had the sheep's meat between her teeth, which caused her to vomit several pieces.

Prospects

The politico-religious dimension of hunting werewolves and shape-shifters fitted well into the framework of lordly anti-witchcraft-policy. Demonological debates about metamorphosis, emerging in the French wars of religion, could easily be adapted to regions rife with religious conflicts and battles for political autonomy. A label was at hand which could be attached conveniently to alleged enemies of godly and princely authority. It needed no response in the populace or any roots in popular beliefs. But with its help judges and their lords established legitimately a persecutory context, in whose machinery little chance for the once accused was left.

The similarities between the witch-hunts in Saint Claude and in Schmidtheim are striking. Furthermore, we find parallels with the discourse about the Livonian werewolves and shape-shifters: The Jesuit's missionary work, as well as the lord's purifying policy, was in existential confrontation with a peasantry whose members were thought to be weak, nearly pagan Christians: deceitful, malevolent, lewd, hateful, and rebellious. Marked as werewolves and shape-shifters, their barbarous beastly nature revealed itself. These 'Indians' needed to be tamed with strict control in which post-tridentate reform was allied to the out-rooting of paganism, witchcraft and werewolvery. A strong anti-witchcraft movement was established, fuelled by an ambitious free lancing witch-hunter and his operation, by the missionary programme of the Jesuits, and by the lord's policy to gain autonomy and to purify his lordship. Demonological ideas about werewolvery and metamorphosis seemed to have made an impact. The Latin term *'lycanthropia'*, presumably stemming from a respected treatise, is mentioned three times in the trial records. At least, Möden was certainly acquainted with the work of Peter Binsfeld, which is cited in one of the first trial records. Probably, the Münstereifel Jesuits, coming from the Cologne branch, had not only read Delrio, but were also informed about the Stump case. Besides, between 1627 and 1630, the Imperial Free City of Cologne faced 33 witchcraft trials, in which likewise the Jesuits were involved. A final assumption may be made. Perhaps, though its not very likely, one of the witch-hunters – be it Möden, Funck, or the lord himself – had come across the tract of Boguet, because the grand judge had been the demonologist whose devil had equipped the werewolves and shape-shifters with iron tools. At the very least, the noble networks of the Eifel lords helped to transmit both ideas about witches and werewolves, including ideas about how witchcraft trials (and the metamorphosis concept) might be conducted and instrumentalised in the battle for outer sovereignty and inner authority.

Did the very small entity of Schmidtheim create the only hotspot in the Eifel region, wherein a death toll of nearly 50 percent proved the terrible efficacy of a multi-layered anti-witchcraft-movement? Recent and ongoing research suggests that Schmidtheim was only the tip of the iceberg.[70] After all, witch-hunters like Möden searched for werewolves like hunting trophies. In 1636, he found another one in the devastating witch-hunts in Rheinbach.[71] Finally, more European peripheries have to be tested, like the Ardennes with the Terre of Saint-Hubert-en-Ardennes, where in the seventeenth century the abbots initiated a purification programme against adultery, demonic possession and witchcraft.

Notes

1. See: Jay M. Smith, *Monsters of the Gévaudan. The Making of a Beast* (Cambridge and London 2011).
2. Nicole Jacques-Chaquin, 'Nynauld, Bodin et les autres. Les enjeux d'une métamorphose textuelle', in: Jean De Nynauld, *De la lycanthropie, transformation et extase des sorciers, 1615*. Édition critique augmentée d'études sur les lyncanthropes et les loup-garous (Paris, 1990), 7–41.
3. *On the Inconstancy of Witches. Pierre de Lancre's Tableau de l'inconstance des mauvais anges et demons (1612)*, Gerhild Scholz-Williams (ed.) (Turnhout, 2006), 'Introduction', xxx.
4. Martin Rheinheimer, 'Wolf und Werwolfglaube. Die Ausrottung der Wölfe in Schleswig-Holstein', *Historische Anthropologie* 2 (1994), 399–422, esp. 416–417.
5. Stefan Donecker, 'Livland und seine Werwölfe. Ethnizität und Monstrosität an der europäischen Peripherie, 1550–1700', *Jahrbuch des baltischen Deutschtums* 56 (2009), 83–89; idem, 'The werewolves of Livonia: Lycanthropy and Shape-Changing in Scholarly Texts, 1550–1720', *Preternature: Critical and Historical Studies on the Preternatural* 1 (2012), 289–322; idem, 'The Lion, The Witch and the Walrus. Images of the Sorcerous North in the 16th and 17th centuries', *TRANS. Interent-Zeitschrift für Kulturwissenschaft* 17, 2010.
6. Willem de Blécourt, 'The Werewolf, the Witch, and the Warlock: Aspects of Gender in the Early Modern Period', in: Alison Rowlands (ed.), *Witchcraft and Masculinities in Early Modern Europe* (Basingstoke, 2009), 191–213, esp. 207–209.

7. Willem de Blécourt, 'A Journey to Hell: Reconsidering the Livonian "Werewolf"', *Magic, Ritual and Witchcraft* 2 (2007), 49–67; idem, 'The Werewolf, the Witch', 192–196; idem, 'Monstruos Theories: Werewolves and the Abuse of History', *Preternature: Critical and Historical Studies on the Preternatural*, 2/2 (2013), 188–212.
8. See: de Blécourt, 'A Journey to Hell'.
9. See: Stuart Clark, *Thinking with Demons: The Idea of Witchcraft in Early Modern Europe* (Oxford, 1997), 191–193.
10. See: Elmar Lorey, 'Wie der Werwolf unter die Hexen kam. Zur Genese der Werwolfprozesse' (http://www.elmar-lorey.de/werwolf/genese.htm), a first survey about demonological ideas about werewolvery and their relevance in the courtroom.
11. See the 'work-in-progress' sample of werewolf trials, which Elmar Lorey is presenting: http://www.elmar-lorey.de/Prozesse.htm.
12. De Blécourt, 'The Werewolf, the Witch', 207.
13. According to a Cologne chronicle, in 1445 a werewolf had bitten several people in Rheindorf near Bonn. Fourteen went to Saint Hubert in Cologne to get help, but two died; Guilelmus de Gouda, *Die Cronica van der hilliger Stat van Coellen* (Cologne, 1499), 309v.
14. Willem de Blécourt, 'Werwolfberichte. Der Fall Peter Stump und seine Folgen', in: Rita Voltmer (ed.), *Europäische Hexenforschung und Landesgeschichte – Methoden, Regionen, Vergleiche* (Trier, 2016, forthcoming).
15. For Institoris see the chapter by Johannes Dillinger in this book.
16. The first 'report' about witches, shifting into were-animals, comes from the chronicle of Hans Fründ, wherein information about one of the first witch-hunts in Switzerland (Sion) are given; Martine Ostorero, et al. (eds): *L'Imaginaire du sabbat. Edition critique des textes les plus anciens* (Lausanne, 1999), 36.
17. Rémy refers to the werewolf cases of Jean Maury and Pierron, le bouvier; see Nicolas Rémy, *La Démonolâtrie*. Texte établi et traduit à partir de l'édition de 1595 par Jean Boës (Nancy, 1997), 145, 151, 210, 218.
18. Robin Briggs, 'Dangerous Spirits. Shapeshifting, Apparitions, and Fantasy in Lorraine Witchcraft Trials', in: Kathryn A. Edwards (ed.), *Werewolves, Witches, and Wandering Spirits. Traditonal Belief and Folklore in Early Modern Europe* (Kirksville, 2002), 1–24.
19. Rémy, *Demonolatrie*, 209.
20. According to de Blécourt, werewolves were transgressing the boundary between human and animal, and thus, a third gender established itself; de Blécourt, 'The Werewolf, the Witch', 208.
21. Marianne Closson, *L'Imaginaire démoniaque en France (1550–1650). Genèse de la littérature fantastique*, (Geneva, 2000), 354, note 93.
22. Jonathan L. Pearl, *The crime of crimes. Demonology and Politics in France, 1560–1620* (Ontario, 1999); idem, 'Maldonado, Juan (1534–1583)', in: Richard M. Golden (ed.), *Encyclopedia of Witchcraft. The Western Tradition* (Santa Barbara, 2006), 710f.; Clark, *Thinking with Demons*, 14f., 535.
23. See *Traicté des Anges et Demons*, du R.P. Maldonat Jesuite. Mis en François, par maistre François de la Borie, grand Archidiacre et Chanoine à Perigueux (Rouen, 1616), 214r–216r. The first edition appeared in Paris in 1605. Maldonado preached the sermons in 1572.
24. See Rita Voltmer, 'Preaching Witchcraft? The Sermons of Johannes Geiler of Kaysersberg (1445–1510)', in: Louise Kallestrup and Raisa Maria Toivo (eds), *Heresy, Magic and Natural Philosophy* (London, 2015, forthcoming); idem.,

'Du discours à l'allegorie. Représentations de la superstition, de la magie et de la sorcellerie dans les sermons de Johannes Geiler de Kaysersberg', in: Antoine Follain and Maryse Simon (eds), *Sorcellerie savante et metalités populaires* (Strasbourg, 2013), 45–88.

25. Peter Maxwell-Stuart, *Martin del Rio. Investigations into Magic* (Manchester, 1998), 28f. On Delrio, see: Jan Machielsen, *Martin Delrio: Demonology and Scholarship in the Counter-Reformation* (Oxford, 2015).
26. Thomas Worcester, 'Jesuit dependence on the French monarchy', in: Thomas Worcester (ed.), *The Cambridge Companion to the Jesuits* (Cambridge, 2008), 104–119; Rainer Babel, 'Freigrafschaft Burgund / Franche-Comté – Freie Reichsstadt Besançon', in: Anton Schindling and Walter Ziegler (eds), *Die Territorien des Reichs im Zeitalter der Reformation und Konfessionalisierung. Land und Konfession 1500–1650*, VI (Münster, 1996), 198–223.
27. Rita Voltmer, '*...ce tant exécrable et détestable crime de sortilège*. Der "Bürgerkrieg" gegen Hexen und Hexenmeister im Herzogtum Luxemburg (16. und 17. Jahrhundert)', *Hémecht. Revue d'Histoire Luxembourgeoise. Zeitschrift für Luxemburger Geschichte* 56 (2004), 57–92, esp. 88
28. Henri Boguet, *An Examen of Witches*, Montague Summers (ed.) (London, 1929; reprint Mineola, 2009), vii.
29. Ostorero, *L'Imaginaire du sabbat*, 42.
30. Boguet, *Examen*, xiii–xv.
31. See the prefaces, adressed to Albert and Isabella, in the 1606 edition of Boguet's *Discours*; Voltmer, 'Bürgerkrieg', 89.
32. The witch-hunts in Saint Claude are part of the witchcraft prosecution in the Franche-Comté, where 795 witchcraft trials with 413 executions can be traced (mainly 1600–1660). All over Franche-Comté, about 17 people were executed as werewolves, one was lynched, another banned. The peak, with eight executions and one lynching, happened in Saint Claude. See: Brigitte Rochelandet, *Sorcières, Diables et Bûchers en Franche-Comté aux XVIe et XVII siècles* (Besançon, 1997); Caroline Oates, 'Metamorphosis and Lycanthropy in Franche-Comté, 1521–1643', in: Michel Feher (ed.), *Fragments for a History of the Human Body* (Cambridge, 1989), 305–363.
33. Rita Voltmer, 'Jesuiten und Kinderhexen. Thesen zur Entstehung, Rezeption und Verbreitung eines Verfolgungsmusters', in: Wolfgang Behringer and Claudia Opitz-Belakhal (eds), *Hexenkinder – Kinderbanden – Straßenkinder* (Bielefeld, forthcoming).
34. Rochelandet, *Sorcières*, 125–130.
35. Binsfeld, *Tractat*, 113–118.
36. Boguet, *Discours*, 132f.; Boguet, *Examen*, 153; Oates, 'Lycanthropy', 323f. (without mentioning the instruments).
37. Oates, 'Lycanthropy', 340. – In Luxembourg, likewise a Spanish-Habsbourgian province, the provincial gouvernment tried to control the extensive witch-hunts, which were orchestrated by witch-hunting committees in colaboration with minor lords. A comparision between the witchcraft policies in Luxembourg and Franche-Comté is forthcoming in one of my research studies.
38. De Lancre, *Inconstancy*, 270.
39. René Taveneaux (ed.), *L'université de Pont-à-Mousson et les problèmes de son temps* (Nancy, 1974); Joseph M. O'Keefe, 'The Pedagogy of Persuasion: The Culture of the University of Pont-à-Mousson', *Paedagogica Historica* 34 (1998), 421–442.

40. E. William Monter, *A Bewitched Duchy: Lorraine and its Dukes, 1477–1736* (Geneva, 2007).
41. See Rita Voltmer, 'Hexenpolitik im Saarraum? Zu Stand und Perspektiven landes- und kulturgeschichtlicher Hexenforschung in einer "passiven Geschichtslandschaft"', in: Brigitte Kasten (ed.), *Historische Blicke auf das Land an der Saar. 60 Jahre Kommission für Saarländische Landesgeschichte und Volksforschung* (Saarbrücken, 2012), 185–217.
42. William Monter, 'Witchcraft Trials in France', in: Brian P. Levack (ed.), *The Oxford Handbook of Witchcraft in Early Modern Europe and Colonial America* (Oxford, 2013), 218–231, esp. 230.
43. Clark, *Thinking with Demons*, 193.
44. Christa Tuczay, 'Animals', in: Golden (ed.), *Encyclopedia of Witchcraft*, 40–42.
45. Jonathan Schütz, *Johann Fischarts Dämonomanie: Übertragungs- und Argumentationsstrategien im dämonologischen Diskurs des späten 16. Jahrhunderts* (Berlin 2011), 246.
46. Michaela Valente, 'The Works of Bodin under the Lens of Roman Theologians and Inquisitors', in: Howell A. Lloyd (ed.), *The Reception of Bodin* (Leiden, 2013), 219–235, esp. 227.
47. Binsfeld, *Tractatus*, 113f.; Delrio, *Investigations*, 37.
48. In general see: Howell A. Lloyd (ed.), *The Reception of Bodin* (Leiden, 2013).
49. Rémy, *Demonolatrie*, 108f., 195, 209.
50. Lorey argues that trials against werewolves had already reached French and German courtrooms without any impact from Bodin's tract; de Blécourt contests it.
51. Cf. de Blécourt, 'Werwolfberichte'.
52. See: de Blécourt, 'Werwolfberichte'; De Blécourt, 'The werewolf, the witch', 196–199; Peter Kremer, *Der Werwolf von Bedburg. Versuch einer Rekonstruktion des Hexereiprozesses aus dem Jahr 1589* (Bonn 2006).
53. Jean Chapeauville, *Qui Gesta Pontificum Tungrensium, Traiectensium, et Leodiensium scripserunt, Auctores Praecipui : Ad seriem rerum & temporum collocati, ac in tomos distincti*, III (Lüttich, 1616), 557.
54. De Blécourt, 'The werewolf, the witch', 197f. – The note of Chapeauville, so far, was unknown in the research on the Stump case.
55. Also: Kremer, *Werwolf*, 77–81.
56. Jean Beauvois de Chauvincourt, *Discours de la Lycanthropie ou de la transformation des hommes en loups* (Paris, 1599), 2.
57. Claude Prieur, *Dialogue de la Lycanthropie ou transformation d'hommes en loups, vulguirement dits Loups-garous & si telle se peut faire* (Louvain, 1596), 38.
58. De Blécourt, 'The werewolf, the witch', 196, with note 27.
59. Idem, 200–202.
60. Idem, 205–207.
61. Francesco Maria Guazzo, *Compendium maleficarum*, (Milan 1626), 67 (first edition in 1608). In his English edition, Montague Summers misreads 'Bamberg' for 'Bedburg'.
62. The following considerations are based on my recent research, embedded in my large project on witch- hunts in the Rhine-Meuse-Area, notably in small ecclesiastical and secular princedoms and lordships as well as in cities. See a full publication list: www.uni-trier.de/index.php?id=45884.
63. Rita Voltmer, '"Hexendorf" Schmidtheim? Erste Überlegungen zu den Verfolgungen unter Reinhard d.J. und Bertram Beissel von Gymnich (1597–1635)', in: Arbeitskreis Kultur und Geschichte Gemeinde Dahlem (ed.), *Schmidtheim. 500 Jahre Beissel von Gymnich und Dorfgemeinde* (Dahlem, 2011), 387–403. The

trial records are preserved in the private noble archive of Frens (Dieter Kastner, *Die Urkunden des Archivs von Schloß Frens – Regesten*, II: 1566–1649 (Bonn, 2011).

64. Adolf Kettel, *Schmidtheim. Eine Herrschaft im Wandel der Jahrhunderte. Ein Beitrag zur Territorialgeschichte der Nordeifel mit genealogischen Stammtafeln und Quellen* (Dahlem, 2006), 65–69, 197–216.
65. Heribert Breiden, *Die Hexenprozesse der Grafschaft Blankenheim von 1589 bis 1643* (Bonn, 1954), 133–135.
66. Rita Voltmer, 'Zwischen Herrschaftskrise, Wirtschaftsdepression und Jesuitenpropaganda. Hexenverfolgungen in der Stadt Trier (15.–17. Jahrhundert)', *Jahrbuch für westdeutsche Landesgeschichte* 27 (2001), 37–107, esp. 97.
67. The lists are included in a sample of letters; Landeshauptarchiv Koblenz 29a, 490.
68. Walter Rummel, 'Möden', in: Golden (ed.), *Encyclopedia of Witchcraft*, 773f. Many thanks to Karin Trieschnigg, who is working on a basic biography of Möden and who has granted me insight into her manuscript.
69. Thomas P. Becker, '"Kein geringeres Verdienst vor Gott als in den weit entfernten Heidenländern": Die Geschichte der Jesuiten in Münstereifel', in: V. Johannes Mötsch / Martin Schoebel (eds), *Eiflia Sacra. Studien zu einer Klosterlandschaft*, II (Mainz, 1999), 391–410.
70. See note 62.
71. Hermann Löher, *Hochnötige unterthanige Wemütige Klage der frommen Unschültigen* (Amsterdam, 1676), 388–391.

8
The Werewolf in the Popular Culture of Early Modern Germany

Rolf Schulte

The werewolf of Trenthorst

In early March 1613, Joachim von Wettken, lord of the manor in Trenthorst, Holstein charges a woman with the serious crime of witchcraft and has her arrested. The accused is Gretge Hopperness, a woman who lives in a neighbouring village. Shortly beforehand, von Wettken had incarcerated a local healer, who he had accused of similar deeds and also charged with witchcraft; under torture the man had given the names of his alleged accomplices in his service of evil, among them Gretge Hopperness. Her special skills, he claimed, enabled her to transform into a wild, rapacious animal. The lord of the manor decides to confront the healer and Hopperness face to face with this accusation. The latter vehemently denies the charges, but von Wettken is still convinced that she is a playmate of the devil and decides to resort to the violent form of interrogation, to torture. Hopperness is unable to withstand the pain and confesses that she is able to perform magic and had caused the lord's cattle to perish as well as causing his wife to fall ill. She also maintains that she is able to become a wolf and that 'when it was so grim and cold she [had] bitten three wild deer to death'. The devil had enticed her to this act with a red apple, and to seal the pact with evil she had slept with him.[1] On 20 March 1613 von Wettken sentences Hopperness, the healer and two other women charged with the same (alleged) crimes to death by burning.

This is a classic example of a trial in the witch-hunt era. A woman is denounced as a witch, having performed maleficium, first as a human and then in animal guise, and hence as a werewolf; she is criminalised and executed. Yet it is uncertain whether the accused was even aware of the meaning of the term werewolf or of the specific crimes associated with it. In the confession extorted from her she admitted being a human-wolf and having breached the hunting privilege of the aristocracy by having killed and then eaten game in a time of need. However, there is no mention in her confession of the classic werewolf offences such as ferocious attacks on human beings; nor are any details of the transformation process given, such as the use of a special magic belt, an animal hide or an exceptional ointment.

Popular culture and the lack of sources

It was a lord of the manor who initiated the above case, a member of the upper stratum of northern German society who evidently believed in occult animal transformation and its necessarily aggressive ramifications. In large parts of northern and eastern Germany, ranging from east of the River Elbe up to the Baltic region, both the executive and the judiciary were the responsibility of the respective landlord of a large property, as was the case in Trenthorst. There were many ways in which the landlords could manipulate a situation to accuse people of, and subsequently punish them for, alleged infringements of norms; they also had means of pressurising their serfs into incriminating other people, so that the matter could then be pursued in the courts. In the trial of Gretge Hopperness the lord failed to heed the stipulations of the legislative but invoked in his verdict the locally valid 'Holstenrecht', a common law applicable to the territory of Holstein in the north of Germany[2] which had only been handed down verbally. He should, however, have adhered to the Emperor's Court Procedure Code applicable to large parts of the Holy Roman Empire; this made it mandatory to apply caution in dealing with such offences as well as in the use of torture.[3]

The records remaining from early modern trials such as that of Gretge Hopperness, often, although not always, represent the forms of thought and fantasies dominant in the minds of the persecutors and their learned culture, and not those of the persecuted. Statements by the accused regarding occult events were made in response to the question catalogues of the courts and were thus largely predictable. Court protocols, then, cannot be said to represent the opinions of popular culture since the interrogators' questions were designed, particularly when torture was applied, to extort the desired responses.

But what constituted these 'imagineries' in popular culture that are defined as 'a system of shared meanings, attitudes and values and the symbolic forms (performances, artifacts) in which they are expressed or embodied', as elements of an 'unofficial culture, the culture of the non-elite, the subordinate classes'?[4]

The early modern imagination and perception of the world knew a wide variety of workers of magic, one of which was the werewolf. We know this from extant church records, but also from secular texts such as legal statutes, poetry and literature as well as magic charms and spells: i.e., to say, from learned culture.

To what extent did these werewolf imaginaries of popular culture correspond to those of learned culture, of the culture which has been handed down to us today? Or in more concrete terms:

- To what extent was the werewolf part of 'white' or 'black' magic, a 'well-doer' or 'evil-doer'?
- To what extent was the figure different from that of other magic imaginaries?
- To what extent were there gender-specific categories of werewolf: he- or she-werewolves?

Reconstruction of the conceptions of popular culture poses methodological problems. The lack of other sources from the High and Late Middle Ages, as well as from parts of the Early Modern era, means that historians have to draw on learned chroniclers for a portrayal of popular culture. Such sources must necessarily be critically filtered; some researchers have thus expressed doubts regarding the validity of descriptions of popular belief based on learned sources, calling them speculative.[5]

Slander trials

The sources which provide us with the best insight into persons seen as werewolves in popular culture are those which have not been subjected to the influence of the authorities or legal structures, or to observations by learned culture. Of such sources, slander cases best meet these requirements, since the material gives a direct rendering of the conflicts as they were argued out among the ordinary people. Slander cases filed for prior accusations of werewolfism were the responsibility of the lower courts. The verdicts and the protocols of the lower courts or the so-called 'small magistrates courts' ('*kleine Magistratsgerichte*') can still be found in the penal records and the court accounts, reaching back as far as the mid-sixteenth century, a period in which there is otherwise little source material for Germany. Time and place are well documented so that the records can easily be assigned to the correct locality.

This contribution focuses on cases in north and north-west Germany, where records of many slander trials have survived and been conserved in the archives. The court authorities ensured that the penal records and court accounts were completed annually so that a classic serial source collection was established; however, although some of the records have survived in full, there are gaps in others. A record from 1622, in which the court scribe of the district of Trittau entered the following in the penal records, represents a typical example of a slander case record:

> Thies Biel, a farrier in Nahe, berated Hans Wrage, son of Marxen in the same village, as an old rogue and werewolf, but was unable to prove anything – 3 thalers (fine)[6]

The amount of detail given in the protocols varied depending on the scribe; very often the description was no longer than a few lines. The verdict, however, was generally given. Protocols of witnesses' statements are rarely to be found; so, the court's verdict is the only basis on which to evaluate the case. The sex of both plaintiff and accused is usually clear, but their social status is not normally mentioned, and information on the development of the conflict or the court proceedings is sparse:

> Johan Leistenschneider of Büderich struck Lueß of Waltringhausen in Johans Waßmans house. Lueß had earlier hit him, also scolded him, calling him a sorcerer, werewolf and witch's child. This prompted him to hit back. 8 Marks. Paid.[7]

In this case, the verdict tells us the actual names and villages of the two parties but gives no indication of their social standing. The writer notes the cause of the conflict but does not go into the details of the background to the case or mention whether the court even asked about this.

In the course of a trial the plaintiff presented his/her case, the accused defended him-/herself, and witnesses made their statements, sometimes under oath. The sources also report on so-called *'Beschicken'*, whereby prior to the case appearing in court the slander victim asked neighbours to seek out the slanderer and ask whether he or she upheld the accusation. If the latter then repeated the accusation in front of the emissaries, these neighbours would make statements to this effect in court.

The low courts consisted in the Districts of landed peasants led by a local magistrate, in the manorial districts of peasants and the lord of the manor, in towns of elected citizens, the reeve and one or more town councillors. After hearing the various parties the court declared its verdict. In some cases, the plaintiff was found guilty of slander and punished. In a very few cases when the court determined a charge to have been unfounded and the arguments of the opposing party to be convincing, the plaintiff was obliged to withdraw an accusation. For such unsubstantiated statements the low court sentenced the slanderer either to making an apology (*'Abbitte'*), a statement of honour or demanded the so-called *'Brüch(t)e'*, a fine.

The *'Abbitte'* consisted of a public apology by the slanderer to the unjustly insulted defendant. In a statement of honour he also withdrew the accusation, thus reinstating the honour of the person concerned. The statement of honour was of central significance for the plaintiff because the background to the case had threatened a loss of honour. Honour, however, played a central role as 'symbolic capital': in early modern times dishonourableness meant exclusion from the community, even to the point of a formal declaration of loss of social status and standing, whereby the stigma was transferred to family and even close friends. Thus early modern society exercised a certain pressure or obligation on every individual to demonstrate and preserve their honour or honourability towards their social environment. 'Symbolic capital' did not imply material possessions, economic capital or financial power; rather, the degree of honour was determined indirectly by means of self- and external assessment, and by the direct level of acceptance and respect shown by other individuals.[8] The main reason for the majority of slander trials, it seems from the protocols, was to defend the plaintiff's 'honourable' reputation and was thus not always voluntary: in some, though not all cases, not filing charges for slander could be considered tantamount to an admission of guilt.

Records of these slander cases can thus provide a solid and reliable source of information on the popular imagination of the werewolf. The slander trials are appropriate sources for the reconstruction of the general values and principles of the time and of the existence and prevalence of magic perceptions in popular culture so as to be able to assess the extent to which they were considered negative or even positive. On some levels, however, the validity of the slander verdicts is limited, particularly when no other sources, such as tax records, are available which would provide more detail with which to classify the opponents more accurately. The fact that the records have not all survived in full should also be taken into account when using these valuable sources. Since attempts are occasionally made, despite these limitations, to quantify important developments, any percentages given in the following should be considered simply as tendencies and as an attempt to provide approximate figures.

Werewolf imaginaries in Schleswig-Holstein

In the period 1580–1730, a total of 510 slander trials involving 615 people[9] were held as a consequence of prior occult activities in the two predominantly Protestant northern German duchies, Schleswig[10] and Holstein (Figure 8.1). This region knew a wide variety of words for allegedly magic activities: dragon, witch, witch-master, male witch, witch child, rogue, devil, she-devil, werewolf, cunning man/woman or soothsayer, weather cat, sorcerer or sorceress – all in the orthographical variants of the time.[11]

The largest number of cases (179) concerned people accused of being 'sorceresses/sorcerers'; this was followed by 'witch/male witch' (124 cases). These defamatory terms were not, however, used in parallel with one another, the term 'witchcraft' (*Hexerei*) abruptly superseded 'sorcery' (*Zauberei*) after 1640.[12] The number of werewolf accusations fell far behind these figures, occupying the courts in only 17 cases involving 20 people. In other words, only 3 percent of all magic offences were concerned with people who felt their honour had been insulted by another person using the word 'werewolf' in association with their behaviour. The population of Schleswig-Holstein in the seventeenth century[13] was about 400,000 to 500,000, and compared to this the small number of cases appears – despite

Figure 8.1 Schleswig and Holstein; fragment map from Abraham Ortelius, *Daniae Regni Typus, Cornelius Antoniades Descripsit Cum Privilegio* (Antwerp, 1584). In the early modern era the duchy of Schleswig belonged to the kingdom of Denmark, the duchy of Holstein to the Holy Roman Empire of the German Nations. Both territories were politically bound together by treaties. Germany was only united in 1871

the incompleteness of the material – to be very low, if not insignificant. Yet the existence of these cases demonstrates definitively that the werewolf figure was existent in popular culture.

The records of some of the early slander trials have survived since the sixteenth century; almost a third of these early cases document insults related to magic as an attack requiring a court action but do not yet list the werewolf as a derogatory term.[14] The first werewolf accusation occurred more or less parallel to the outbreak of the Thirty Years War (1618–1648), which devastated large parts of Central Europe, including Schleswig and Holstein. The majority of slander cases are found in court records immediately after the end of the war in the middle of the seventeenth century. After this, werewolf slander accusations occur only sporadically and irregularly. As late as 1748, however, a court scribe noted that a man was quoted in court as having described another as 'a wolf who wanted to bite my wife';[15] he was subsequently fined for this insult.

The idea of metamorphosis of humans into animals, then, played a subordinate role in the minds of the people in this land between the Baltic and the North Seas. This notion first appeared with the outbreak of the Thirty Years War and was most widespread in the years immediately following the war; it does not, however, appear to have been as widespread as other magic imaginaries. In the north, as elsewhere, this aggressive human-animal figure only appears to have spread, or perhaps become active, in popular culture after the signal trial of the peasant Peter Stump, which took place near Cologne in 1589; with time, tales of the alleged deeds of such human wolves spread to other regions.[16]

Nevertheless, the 17 slander cases demonstrate clearly that the notion of a human wolf was not a positive one and that the werewolf was by no means considered a figure with humane or even supportive characteristics. In half of the cases the accusers combined the werewolf accusation with other terminology which implicates breaches of norms and insults to people's honour, such as thief, '*Schelm*' (rogue), traitor, and, most frequently, magic. Whereas most of these terms do not require any further explanation, the meaning behind the word '*Schelm*' only becomes clear from its etymology. Admittedly, a '*Schelm*' was considered a general villain or evil-doer[17] in early modern times but the semantic background of the word points directly to death, epidemics and carcasses, showing that there was an association between the werewolf and death.[18] Although half of the werewolf insults were combined with other insults from the realm of magic, the other half remained alone, without any further additions. Sexual connotations – as known from other European regions[19] – are non-existent in Schleswig and Holstein.

Nevertheless, the werewolf in popular culture may have been both a positive and a negative figure, although the former would not, of course, have appeared in the slander cases. It is, however, possible that peasants and craftsmen made this division into 'good' and 'bad' metamorphoses. The fact that the majority of false werewolf accusations are listed in the sources without any further qualifying or descriptive terms, is an argument against such a division into the salutary and ominous forms of the man-wolf being. No derogatory prefixes such as '*Aas-*' (literally: carrion), '*Schand-*'

(literally: shame, disgrace, ignominy) or '*Galgen-*' (literally gallows) or attributive adjectives (bad, black, evil, dishonourable, wicked) such as those found in other terms of abuse are recorded here. Only twice did an accuser use an additional term, the word 'old', in order to abuse and dishonour another person.[20] Anyone who used the expression 'you werewolf' as a consciously controlled term of abuse in public or as an insult in an outbreak of anger was obviously clearly understood. This implies that it is hardly likely that the werewolf was thought of as an ambivalent figure from the magic realm.

Quantity, that is the number of instances in which the term 'werewolf' occurs, can be used as an indicator of how widespread this occult figure was. However, these numbers tell us little about the characteristics or the quality and degree of menace which popular culture associated with the werewolf. The historical sources provide little information on this point as the verdicts often consist only of one phrase and an abbreviated form of the sentence, e.g.:

> Heinrich Eggersen insulted Johan Barenstecher, calling him a werewolf – [fined] 1 Reichsthaler[21]
>
> Asmus Kacken widow of Heilshoop for insulting Marx Schölen, who lives in the same village, calling him a traitor, werewolf and suchlike – [fined] 3 Reichsthaler[22]

In only two cases is it still possible to reconstruct in any detail the conflict which preceded the court case. In the first case, in 1651, a woman, who was already suspected of being a witch and had been seen very early one morning in a woodland glade, was accused by her neighbours of being a werewolf. They accused her of maleficium which she was also able to perform in the guise of a wolf.[23] This background to the case shows that animal metamorphosis had already penetrated the witchcraft image and that seemingly conspicuous behaviour was integrated into this image as a confirmation of destructive magic activity. In another case, which came before the court in 1696, the geese of a peasant couple ate up all the cabbage in a neighbour's garden; the neighbour then attempted in vain to chase the geese away. The fact that he was unsuccessful was interpreted by the owner of the geese as an attempt to actually take possession of them, and she berated him, calling him a werewolf. The latter considered this such an affront to his honour that he took her to court and won the case; his neighbour had to pay a fine of one Reichsthaler.[24] Here, evidently, the werewolf accusation was expressed in connection with what was taken to be illegal appropriation of the geese. These two cases do not, however, provide a sufficiently solid basis on which to draw a concrete profile of the werewolf in the popular culture of Schleswig-Holstein.

The fines laid down by the peasants' courts could give some indication of how the werewolf was assessed in comparison with other magic figures.

However, with a few exceptions, the fines imposed were between one and four Reichsthaler or Mark Lübsch (depending on the usual currency of the respective region); on average, then, the same as those imposed for the term 'witch' or 'sorcerer/sorceress'.[25] This shows that there was no difference between the witch and the werewolf in terms of their estimated threat potential. Nor did the alleged ability to become a wolf make a witch potentially more harmful.

Certain terms used in the slander cases imply particular kinds of magical activities. The term 'dragon', for example, was used in the context of unlawful gain;[26] the sources remain vague and indistinct regarding the characteristics of people vilified as werewolves.

The small statistical sample size – there are only 17 specifically named magic offences – means that it is not possible to make any conclusive statements. The contours of the werewolf figure are nonetheless unmistakable. The werewolf was primarily considered to be a male being: 69 percent of the werewolf slander cases concerned men. Despite reservations regarding the representativeness of these numbers, they do show that such magic offences were associated largely, but not exclusively with the male sex, with women representing almost a third of the werewolf cases. The primarily male contours of the werewolf in popular culture in Schleswig and Holstein become clearer when the various ways of performing harmful magic are regarded from a gender-specific angle.

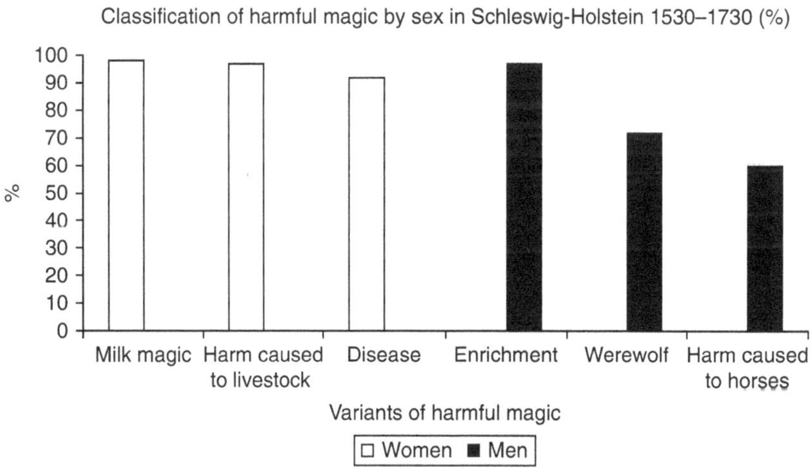

Of the cases relating to the use of magical means to sour milk and butter, 98 percent were ascribed to women; 97 percent of cases where harm was caused to livestock, and 92 percent of cases in which magic was worked to cause sickness were ascribed to women. Men, and very few women, were charged with unlawful gain; 72 percent of cases involving magical violent aggression in the form of a werewolf, and 60 percent of cases involving harm caused to horses were ascribed to men.

This evaluation clarifies the associations in popular culture of specific offences with either men or women. This must, however, be seen in the context of the gender roles in early modern times. In this traditionally agricultural society certain areas of work were the stronghold of either men or women. Livestock and dairy work, caring for the sick, and the preparation of meals were the realms of women. The division of labour even went as far as placing a strict taboo on men doing the milking. The male work areas included caring for the draught animals, such as the horses.[27] The gender-specific association of criminal occult offences correlates with the traditional gender-specific domains: the position of men and women, then, played an important role in the emergence of accusations of alleged magical aggression.[28]

Northern German popular culture incorporated the werewolf into the realm of occult and harm-bringing figures although this semi-animal, semi-human creature was not yet coherently integrated into comprehensive witchcraft paradigms. Insofar as the small number of sources allows any general conclusions at all, the werewolf figure appears to have been considered, at least partially, as not belonging to this paradigm. What is more, none of the werewolf slander cases led to either of the opponents being charged with witchcraft, a charge which could have ended in a death sentence.

Werewolf cases only occurred in certain more densely forested regions of Schleswig and Holstein: in the district of Eiderstedt, which is surrounded by the North Sea and for which the records are still complete for the years 1591–1696, not a single werewolf was sighted or indicted. No werewolves appear in the marshy, almost treeless areas along the coasts. In the densely forested district of Trittau, on the other hand, the records for 1608–1705 (which are also still available in full) show occurrences of the werewolf: 6 percent of all lawsuits are slander cases resulting from honour-threatening werewolf insults. An extensive forest covered, and still covers today, large areas of this district,[29] and it is possible to draw generalisations from this: the majority of cases occurred in administrative districts in which the forest, the traditional habitat and retreat of wolves, covered a large area. Werewolf accusations were largely made, for example, in the districts of Trittau, Ratzeburg or Eutin in the east and in the county of Rantzau in the south of the territory, all of which have large, continuously wooded areas. As research has determined,[30] imagined harm is a function of natural space and an appreciation of its resources: the mere fact that Schleswig and Holstein, as sparsely wooded regions surrounded by the sea, did not present ideal ecological prerequisites for large populations of wolves, meant that the werewolf imaginary was unable to gain a real foothold. The fact that the cases which did occur were concentrated in and shortly after the Thirty Years War could be related to wolves' tendency to feed on human and animal corpses; the military events of 1618–1648 may increasingly have led the animals to leave their habitual habitats and move into areas with human settlements. This temporal concentration of werewolf accusations may, however, also have to do with the increased concentration of source materials found for the period after the Thirty Years War. Nevertheless, accusations of werewolfism, humans transforming into wild, devouring animals, remained the exception in both the slander and the witch trials in the region between the North and the Baltic seas.[31]

The fact that the werewolf notion was not widespread in Schleswig and Holstein has to do, not only with objective but also subjective factors. The witch paradigm, with which the werewolf was doubtless associated in early modern times, was concentrated among large percentages of the rural population in Schleswig and Holstein (unlike other German-speaking regions) on the maleficium element. Learned and cumulative demonology, with its comprehensive ideas of the witches' Sabbat and their consequences, was unable to take a firm hold.[32] In the north, the werewolf was considered to be just *one* variant of harmful magic, a figure which does not entirely fit into the cumulative magical paradigm. The werewolf had not yet shed its singularity, and, in contrast to other regions of the Holy Roman Empire, popular culture in Schleswig-Holstein generally considered werewolfism to be a clearly delineated offence; the werewolf image remained distinct from that of the witch and the all-embracing, destructive and organised crime which popular culture associated with witchcraft.

Werewolf imaginaries in Westphalia

The district and town of Werl lie in north-western Germany and in the early modern era belonged to the Duchy of Westphalia (Figure 8.2). The town was one of the four capitals of Westphalia and thus had a role model function. In the course of counter-Reformation campaigns at the beginning of the seventeenth century, for example, all citizens of the town were obliged to become Catholics.[33]

Large numbers of records were lost in central Europe as a result of the Thirty Years War (1618–1648), which laid waste to many regions and cost one third of the population their lives. In Werl, however, many files containing slander trials have survived; of these the records for the period 1597–1682 are complete. This was a time when the citizens of the district and the town frequently took one another to court: 1238 slander trials are recorded for this period and area the population of the town was around 3000–3500 in the second half of the seventeenth century). Approximately 140 of these involved magic offences.[34] These charges and trials are indicative of severe tensions and conflicts in the face-to-face society of the town in the early modern period.

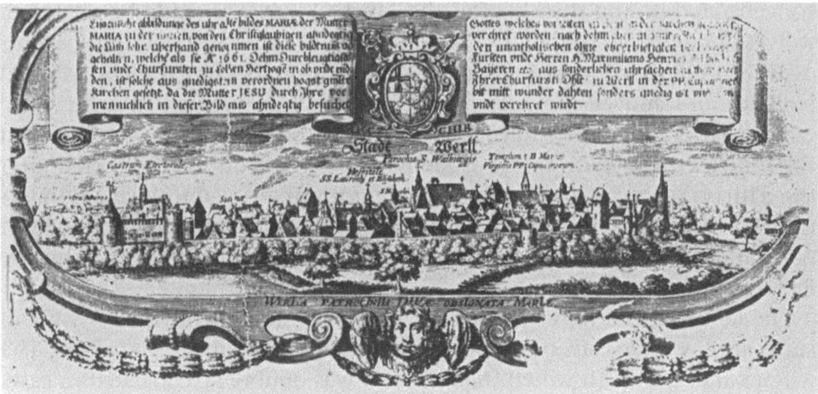

Figure 8.2 View of the town of Werl, 1661 by an unknown artist; Wendelin Leidinger: *Documenta Werlensia* (Werl, 1972). The town of Werl was situated on a central trade route and in the early modern period evolved into an important town in Westphalia. Its main trade resulted from salt mining but also from many craft shops and farms, which were protected by the town walls. wikimedia commons)

Although the records for Werl show a wide variety of accusations and insults related to alleged magical activities, unlike Schleswig and Holstein most of these were to do with witchcraft or sorcery. In every other case, the plaintiffs were objecting to having been called a witch, sorceress, sorcerer or male witch and demanding that the matter be rectified in court.

Werewolf followed as the next most frequent accusation, comprising 17 percent of all slander cases. In 24 court trials involving 28 people the plaintiffs attempted to have their honour restored. The notion that werewolves existed was obviously alive in public life and first arose earlier than in Schleswig or Holstein. In 1609 two town councillors insulted a baker 'as a cheese thief and werewolf'; between 1620 and 1640 the number of accusations increased significantly; and the last case was registered by the magistrates court in 1680.[35]

The frequency of these defamations in comparison with Schleswig and Holstein is all the more surprising when one considers that the Werl area did not provide a good ecological basis for real wolf populations either. This relatively densely populated lowland area lies in a sparsely forested region of Westphalia at some distance from the large forests of Sauerland and the German uplands. Nevertheless, the records report that during and after the Thirty Years War wolves did occasionally penetrate into residential areas.[36]

Unlike Schleswig and Holstein, maleficium barely played a role in the defamations: in Werl, the accusations quickly became generalised, using the terms 'witch' or 'sorcerer' and implying that the accused were well acquainted with magic in general.[37] Such generalisations also applied to the werewolf cases. Only in one case, notably the only one involving a woman, did the accusation take the form of a classic werewolf offence – 'tore apart and devoured a donkey'.[38] In these sources, the werewolf appears as an abstract figure in a figurative sense, a long way from its classic features as a harmful, wild, aggressive and devouring wolf-metamorphosing human prone to attack people as well as animals.

The Westphalian cases are different from those in Schleswig and Holstein in that only in a minority of cases is the werewolf mentioned alone. The accusation, as an attempt to disparage or even demean an opponent, was mostly combined with other expressions aimed at offending their honour. The following entries from 1630 and 1621, respectively, are representative for the town and district of Werl in which the term werewolf was often used in connection with that of sorcerer:

> Wilhelm Rive called Claß Konertz a sorcerer and werewolf and used otherwise grossly offensive language towards him, whereupon the latter abused him as a rogue and thief and otherwise severely insulted him ...[39]
>
> Kordt Hinrich admitted insulting the Drost [highest ranked administrative official] of Lodtringhausen as a sorcerer and werewolf – 6 Reichsmark (fine). Paid[40]

The combinations used in Werl were limited. The terms 'rogue' or 'thief', and the original 'Calvinist rogue' or 'Galgenschwingel'[41] (hanged man, criminal hanging from the gallows, especially as a term of abuse 'bad fellow, ripe for the gallows'). The majority, however, combined the animal metamorphosis insult with the assertion that the person mastered the general art of magic. This combination could potentially end in a witchcraft trial, since incriminating a fellow citizen as a dangerous criminal in pact with the devil could be fatal for the accused. Insults in Werl were often cumulative and aimed to defame a person as generally dishonest and evil. In the presence of uninvolved witnesses, complicated often verifiably long-standing disputes were reduced, and thus simplified, to a general suspicion. At the same time, the accusers were able to use the opportunity to present themselves, in contrast to their opponents, as citizens committed to the norms and values of society. In this part of Westphalia, then, the werewolf in popular culture was a dark, occult and inauspicious figure, and was very much as negatively charged as its counterpart in Schleswig and Holstein.

Even more than in the far north of Germany, the werewolf in Westphalia was male; of the 28 people accused of being werewolves, only one was a woman. Whereas it was widespread for women to use 'witch' as a term of abuse amongst each other,[42] the term 'werewolf' was used almost exclusively in male disputes;[43] with regard to both plaintiff and defendant, werewolf accusations persisted in an almost exclusively male domain. This exclusiveness can be interpreted in various ways: it may be connected with so-called honour conflicts and represent a way of dealing with social conflicts in which men from different strata of society sought to demonstrate their own social prestige and stabilise their own social status by attempting to deny others their honour.[44]

Indeed, some of the cases did involve politically and socially privileged merchants interacting with less privileged artisans from the town,[45] but these numbers are too low to verify any attempts at a theory. Werl was a town permeated by social conflict,[46] so that indirect conflicts of rank seem a likely explanation; however, the absence in the werewolf slander trials of any members of the clearly defined upper social stratum of the town, the salt entrepreneurs, shows, even without looking into other possible reasons, that this can at best partially explain the phenomenon. On the contrary, the werewolf insult occurred in cases in which male opponents turned to physical violence in their conflicts with one another. Those trial records which actually report the background to the slander case, tell of violence – even though there were differences in the versions rendered in court by the two opponents. The courts capitulated in the face of widely differing accounts, abandoning attempts to establish the truth and occasionally sentencing both parties.

In 1623 a citizen of Werl accused a Johan Wulner of being a werewolf; Wulner had previously intimidated people with a halberd (a combined spear and battle-axe), threatening to make cripples out of them.[47] In the same year, two citizens became involved in a fight with one another, in the course of which one of them accused the other and the other's father of being a 'sorcerer and werewolf'.[48] Seven years later Wilhelm Rive accused Claß Konertz of being an animal-metamorphosing sorcerer, with the result that the latter pursued him with a loaded gun wanting to kill him. Even in court both were still so enraged that they threatened to kill each other in public. Konertz was already known to the court for his violence: among other things, he had broken a woman's arm in the course of a dispute.[49] In 1637 Christian Armedes insulted Hermann Hurrelmann: he called him a werewolf, pulled his beard, tried to throttle him and then pulled out a knife in order to kill him. Passers-by prevented this from happening. Armedes defended himself in court saying that Hurrelmann had murdered his (Armedes') sister.[50] Three years later a borrowed tub was the cause of a dispute which arose between two neighbours, who fought out the conflict with heavy wooden clubs, causing each other severe injuries. Here, too, the term werewolf was used as an insult.[51] In 1640, Blasius Menne, intending to have his honour restored, took a citizen of Werl to court who had called him a werewolf. In the course of the trial it became clear that Menne had first knocked down the citizen, kicked him and hit him on the head with a piece of wood, inflicting a serious wound.[52] A few years later Goddert Kampmann accused another citizen of Werl of calling him a werewolf. In court it emerged that Kampmann had attacked the defendant with a fork and tried to injure his genitals. The werewolf insults were made, it was claimed, before and during the physical fight. The court sentenced both opponents: one for slander, the other for bodily harm.[53] In a verbal dispute which occurred in 1680, Aleff (=Adolf) Billecke took hold of a shovel with which he broke his neighbour's arm; she subsequently swore at him, calling him a werewolf, whereupon he took her to court claiming his reputation had been damaged.[54]

The werewolf underwent a particular development in Werl, becoming a kind of secondary variant of the sorcerer figure, when the conflicts were not carried out indirectly via an alleged anonymous and invisible magic act but directly in the form of visible physical aggression. The frequency of werewolf accusations in Werl, as compared with Schleswig and Holstein, has to do with the difference in intensity of the people's belief in witches and gradually varying magic imageries. Numerous sources testify to a virulent belief in witches in this part of Westphalia and demonstrate that the majority of the population shared the notions of the works of the devil and his agents on earth, the witches and male witches. In the minds of the populace, these people truly had made pacts with the devil and represented a very real danger, with the result that a mentality developed in the seventeenth century which was permeated by a particular fear of occult forces.[55] Westphalia was one of the hotspots of the central European witch-hunts, and Werl was a faithful reflection of these developments: both the Council and the people themselves were extremely hard on anyone who was alleged to have performed witchcraft. Even though the loss of records has meant that it is not possible to obtain a precise number for the witch trials, this town is clearly one of the areas in which the persecution rate was high.[56] Well-known witch commissioners, such as Christian Kleinsorge and Caspar Reinhartz, were recruited from Werl; they had the Prince's mandate to hunt out witches, and left behind them a bloody trail throughout the whole of the Westphalia-Rhineland area.[57] In 1607 Kleinsorge was appointed to the post of General Witch Commissioner for the Prince-Elector, a post he held together with the Mayor of Werl, Jobst von Hoxar. After his renowned and consequential activities as a witch-hunter Reinhartz was elected several times as Mayor of Werl; the notorious witch-hunter Heinrich von Schultheiß also worked in the town.[58] Not by chance: mentally and socially rooted in the town, these witch-hunters advocated a witch theory, which no longer worked on the premise that witches caused harm to individuals, but that witchcraft represented organised crime aimed at harming the whole of society. It was no longer maleficium itself which was the focus, but the disengagement of individual Christians from their faith and their engagement in a subversive community of devil's advocates in which all moral standards were turned upside down. The danger which the alleged witches represented no longer lay in their ability to work harm on people or livestock, but in their participation in a global conspiracy which was imagined to be the goal of a secret society and its clandestine operations. What is more, men – as sorcerers – played an important role in this demonology and, like the female witches, had to be removed from the face of the earth.[59] Those people who held such convictions spread their ideas of witchcraft throughout the town and area and obviously found considerable support among the population. Around the year 1590 pamphlets describing the alleged deeds of the werewolf Peter Stump were produced and disseminated in the vicinity of Cologne and these may well have reached Werl and triggered people's sensitisation to the supposed activities of these half-man, half-wolf creatures.[60]

The concept of men as werewolves – whether or not it existed prior to this point must be left open here – took a good hold on popular culture in Werl. Only thus is it possible to explain the fact that, unlike in Schleswig and Holstein, accusations of werewolfism which became known to the courts did not remain without consequences. Admittedly, no werewolf accusation actually ended directly in a witch trial; however, such insults clearly had a major negative impact on the reputation of some of the accused, starting up or intensifying rumours about them, and five of the people originally insulted as werewolves were later indicted for witchcraft – and executed.[61]

Not an adversary of witches but not a congenial character either...

The slander trials verify that the notion of a person able to metamorphose into a wolf existed in north and north-west German popular culture in the early modern era. The werewolf was not the invention of literary writers or learned demonologists. However, the actual contours and characteristics varied considerably from one region to another. The werewolf in the north of Germany was not a congenial character; he was, without doubt, an evildoer, contaminated by the diabolical. He was not like the Italian *benandanti,* the Hungarian *taltós* or the Croatian *Kersniki* who, though magical and sometimes metamorphosing figures, were nonetheless benevolent beings, who fought the threatening witches to secure the villagers their harvests.[62] But, as long as the dark side of werewolfism remained outside the witchcraft discourse and the judicial system of elite culture, popular culture let the werewolf be and saw no reason for persecution. As soon as he was identified by elite culture as a wolf-metamorphosing witch, however, he could be legally prosecuted and persecuted – often to end up at the stake in one of the frequently burning fires of the time. But the werewolf lived on in the tales of the rural population.[63] Not until the films and trivial literature of the twentieth century did the werewolf acquire new, occasionally even amicable, character traits.

(Translated by Linda Froome-Döring)

Notes

1. Statement made by Gretge Hopperness recorded in the court proceedings 1613, Archives of the City of Lübeck, Altes Senatsarchiv, Asa Interna Landgüter XI, Trenthorst, Fasc. 5.
2. Germany as such has only been in existence since 1871; in the early modern era many of the states belonged to the Holy Roman Empire of German nations, a

pre-modern, supranational state, which included large parts of Central Europe and in which numerous languages were spoken. For the sake of simplicity, the geographical term 'Germany' is used here.

3. F.C. Schroeder (ed.), *Die peinliche Gerichtsordnung Kaiser Karls V. und des Heiligen Römischen Reichs von 1532* (Stuttgart, 2000) § 44, 45–47, 52, 58, 69, 109, 219. The imperial laws applied to large areas of the Holy Roman Empire of German Nations, and thus also for the Duchy of Holstein.
4. Peter Burke, *Popular Culture in Early Modern Europe*, 3d ed. (Farnham, 2009) Prologue, xiii.
5. See Burke, *Popular Culture*, 7–19, 50–51 for a discussion of the concept popular culture, where he also deals with the criticisms made of his arguments since the first edition of the book, 103–118.
6. Verdict passed by the lower court (*Niedergericht*) Trittau 1622, Landesarchiv Schleswig-Holstein AR 111 (1622/23).
7. Verdict passed by the magistrates court (*Magistratsgericht*) in Werl 1655, printed in: Werner Kohn and Heinrich Josef Deisting (eds), 'Brüchtenregister der Stadt Werl [Fines Register for the town of Werl] 1597–1671', *Beiträge zur westfälischen Familienforschung* 50, Nr. 952 (1992), 246.
8. On the concept of honour in early modern times see: Gudrun Gersmann, '"Gehe hin und verthetige dich' – Injurienklagen als Mittel der Abwehr von Hexereiverdächtigungen – ein Fallbeispiel aus dem Fürstbistum Münster', in: Sibylle Backmann e.a. (eds), *Ehrkonzepte in der Frühen Neuzeit. Identitäten und Abgrenzungen* (Berlin, 1998), 237–270; Ralf-Peter Fuchs, *Um die Ehre. Westfälische Beleidigungsprozesse vor dem Reichskammergericht 1525–1805* (Paderborn: Schöningh 1999), 20–30, Paul Münch, *Lebensformen in der Frühen Neuzeit* (Berlin, 1998), 66–68; Rainer Walz, *Hexenglaube und magische Kommunikation im Dorf der Frühen Neuzeit* (Paderborn, 1993), 422–425.
9. Landesarchiv Schleswig-Holstein (LAS): Court protocols Eutin, Abt. 275, Nr. 456–60 (1624–1712); (1723–1754); Reinbek, Abt. 111, Nr. 253 (1615/16); Court accounts (AR) Ahrensbök, Abt. 108 (1638–1753); Bordesholm, AR 106 (1616–1663); Cismar, AR 107 (1646–1682); Eiderstedt, Abt. 163 (1591–1696); Eutin, Abt. 275, Nr. 584–627 (1663–1711); Fehmarn, Abt. 180, AR 87 (1581); Flensburg, Abt. 167 (1692–1696); Gottorf, Abt. 168 (1597–1768); Hütten, Abt. 168 (1665–1758); Husum, Abt. 163 (1633–1637); Kiel, AR 106 (1613–1655); Neumünster, AR 105 (1712, 1721); Norderdrittenteil Dithmarschen, AR 101 (1560); Oldenburg i.H., Abt. 107, Nr. 1001–1007 (1640–1681); Pinneberg, AR 112 (1641–1701); Grafschaft Rantzau, AR 113 (1651–1656); Reinbek, AR 111 (1593–1703), AR Reinfeld AR 109 (1640–1728); Rendsburg, AR 104 (1585–1671); Schwabstedt, Abt. 162 (1580–1681); Segeberg, AR 110 (1560–1663); Steinburg, AR 103 (1606–1648); Süderdithmarschen, AR 102 (1585–1616); Tondern, Abt. 161 (1615–1692) (1643/44); Tremsbüttel, AR 111 (1674–1690); Trittau, AR 111 (1608–1705); Apenrade, State Archives Copenhagen, Film–No. 43156, 43161 (1616–1650); Brüchteregister Wilster, Town Archives Wilster, Man. 11 b, Nr. 273 (1597, 1612); Court book Kiel, City Archives Kiel, XXII/l, Court book 1611, Court protocols Oldenburg i.H., Town Archives Oldenburg i.H., A 1 12/4 (1624–1683), Court protocol Plön, Town Archives Plön, Altarchiv, Nr. 189 (1674), Court protocol Plön, Abt. 142, Nr. 1–12 (1729–1767), Nr. 30–32 (1650–1692), Court protocol Rixdorf, Abt. 125.20, Nr. 8 (1644); Estate protocol Niendorf, Rundhof, Estate archive Rundhof, B VI 1 (1686, 1692/93); Court fines protocol, no call no. District Archives Ratzeburg (1723–1752); Case register Wilster, Town Archives Wilster, Man. III b, Nr. 396–397 (1594–1620). The material was transcribed and filed as part of a research project in the Institute of Ethnology/European Ethnology at the University of Kiel in Germany. Many thanks to Nils Hansen, who again enabled me to access this very valuable material.

10. The Duchy of Schleswig did not belong to the Holy Roman Empire of German Nations in Early Modern times but to the Kingdom of Denmark despite strong legal ties to the Duchy of Holstein. However, slander cases from Schleswig have been used here in order to enlarge the statistical sample and its statistical significance; for the sake of simplicity data from small territories such as the County of Pinneberg are treated as part of the Duchy of Holstein.
11. See also the lists in: Karl S. Kramer, 'Hohnsprake, Wrakworte, Nachschnack und Ungebühr. Ehrenhändel in holsteinischen Quellen', *Kieler Blätter zur Volkskunde*, 16 (1984) pp. 63–85.
12. Kirsten Sander, *Aberglauben im Spiegel schleswig-holsteinischer Quellen des 16. bis 18.Jahrhunderts* (Neumünster, 1991), 32; Karl S. Kramer, *Volksleben in Holstein (1550–1800), Eine Volkskunde aufgrund archivalischer Quellen* (Kiel, 1987), 283.
13. Karl Wegemann, 'Die Volkszahl Schleswig-Holsteins seit dem in Mittelalter', *Zeitschrift der Gesellschaft für Schleswig-Holsteinische Geschichte* 47 (1917), 66.
14. First mention of the werewolf: LAS AR Trittau 111, 1622/23. For a quantification of slander cases in Schleswig and Holstein see: Rolf Schulte, *Man as Witch, Male Witches in Central Europe* (Basingstoke, 2009), 168–171.
15. Notification 1748, LAS AR Ratzeburg, no call number.
16. Willem de Blécourt, 'Werwolfberichte. Der Fall Peter Stumpf und seine Folge', in: Rita Voltmer (ed.), *Europäische Hexenforschung und Landesgeschichte – Methoden, Regionen, Vergleiche* (Trier, 2016, forthcoming).
17. Fuchs, *Um die Ehre*, 106.
18. Friedrich Kluge, *Etymologisches Wörterbuch der deutschen Sprache* (Berlin, 1999), 716; Lutz Röhrich, *Das große Lexikon der sprichwörtlichen Redensarten*, III (Freiburg, 1992), 1316; Kramer, 'Hohnsprake', 60. Martin Rheinheimer, 'Die Angst vor dem Wolf. Werwolfglaube, Wolfsglaube und. Ausrottung der Wölfe in Schleswig-Holstein', *Fabula. Zeitschrift für Erzählforschung* 36 (1995), 36.
19. The following works discuss the connections between werewolf and sexuality: Willem de Blécourt, 'The Werewolf, the Witch, and the Warlock: Aspects of Gender in the Early Modern Period', in: Alison Rowlands (ed.), *Witchcraft and Masculinities in Early Modern Europe* (Basingstoke, 2009), 204–206; Elmar M. Lorey, 'Das Werwolfstereotyp als instabile Variante im Hexenprozeß', chapt. 3. 6; s. http://www.elmar-lorey.de/Stereotyp.htm (accessed on 31 July 2011), first print publication: *Nassauische Annalen* 112 (2001), 135–176.
20. Verdicts LAS AR 112 Pinneberg 1663/64, LAS AR 111 Segeberg, 1656/57.
21. Verdict LAS AR 112 Pinneberg, 1660/61.
22. Verdict LAS AR. 109 Reinfeld, 1715.
23. LAS Abt. 275 Eutin, Nr. 457.
24. LAS AR 112 Pinneberg, 1696/97.
25. The fines varied within central Europe depending on the territory.
26. Sander, *Aberglauben*, 67.
27. Michael Mitterauer, 'Geschlechtsspezifische Arbeitsteilung und Geschlechterrollen in ländlichen Gegenden Mitteleuropas', in: Jochen Martin and Renate Zoepffel (eds), *Historische Anthropologie. Aufgaben, Rollen und Räume von Frau und Mann*, II (Freiburg/München, 1989), 819–909.
28. Heide Wunder, 'Hexenprozesse im Herzogtum Preußen während des 16. Jahrhunderts', in: Christian Degn, Hartmut Lehmann and Dagmar Unverhau (eds), *Hexenprozesse. Deutsche und skandinavische Beiträge* (Neumünster, 1983), 189.
29. On forest cover in the early modern era see: Walter Hase, 'Abriß der Wald- und Forstgeschichte Schleswig-Holsteins im letzten Jahrtausend', *Schriften des Naturwissenschaftlichen Vereins Schleswig-Holsteins*, 53 (1983), 90–8.

30. All witchcraft research profits from looking beyond the narrow European context and incorporating other ethnological insights. For an initial and fundamental discussion of the connections between magic offences, social structure and society's resources see Monica Hunter Wilson, 'Witch-Beliefs and Social Structure', in: Max Marwick (ed.), *Witchcraft and Sorcery: Selected Readings* (Harmondsworth, 1970/82), 252–263.
31. Werewolf accusations were only made in five out of a total of 892 witch trials in the sixteenth to eighteenth centuries, see Rolf Schulte, *Hexenverfolgung in Schleswig-Holstein* (Heide, 2001), listing of all victims of the witchcraft trials in Schleswig and Holstein from 1530–1735, years: 1613, 1647, 1668, 1686, 1687.
32. Schulte, *Hexenverfolgung*, 84–93; Sander, *Aberglauben*, 69–73.
33. Maria Elisabeth Grüter, '"Unruhiger Geist" – Politik und Religion im 16.Jh', in: Amalie Roher and Hans Jürgen Zacher (eds), *Werl. Geschichte einer westfälischen Stadt*, I (Paderborn, 1994), 363, 370, 378. Hans Josef Deisting and Annegret Karsten, 'Aspekte der Sozial- und Wirtschaftsgeschichte des 17. und 18.Jhs.', in: Roher and Zacher, *Werl*, 505.
34. For the period 1597–1671: Werner Kohn and Heinrich Josef Deisting (eds), 'Brüchtenregister der Stadt Werl 1597–1671', *Beiträge zur westfälischen Familienforschung* 50 (1992); for 1671–1680: Hans Stodt, 'Hexenwahn und Zaubereiprozesse in Stadt und Amt Werl im 17. Jahrhundert (1628–1630, 1642/43, 1660). Ein Beitrag zur Sozialgeschichte Werls des 17. Jahrhunderts', *Mitteilungen der Werler Arbeitsgemeinschaft für Familienforschung* 24 (2003), 402–422. The records are only missing for a marginal part of the district where the Count of Plettenberg presided over the courts.
35. Stodt, 'Hexenwahn', 349, 380.
36. Rainer Decker, *Die Hexen und ihre* Henker (Freiburg, 1994), 137.
37. The court cases mention maleficium affecting, butter, milk, the weather, livestock and potency; however, these only played a minor role. See the lists in Kohn and Deisting, 'Brüchtenregister', 228, no. 852; 255, no. 1004; Stodt, 'Hexenwahn', 362, 364, 381, 388, 412.
38. Verdict 1658, see: Kohn and Deisting, 'Brüchtenregister', 380, no. 995; Ratsprotokoll no. 141 nach Stodt, 'Hexenwahn', 381.
39. Verdict 1630, see: Kohn and Deisting, 'Brüchtenregister', 199, no. 701.
40. Verdict 1621, see: Kohn and Deisting, 'Brüchtenregister', 170, no. 505.
41. Verdict 1650, see: Kohn and Deisting, 'Brüchtenregister', 231, no. 866.
42. For Great Britain: Peter Rushton, 'Women, Witchcraft and Slander in Early Modern England. Cases from the Church Courts of Durham 1560–1675', *Northern History* 18 (1982), 116–132.
43. 83 percent of the insults were made by men and addressed men.
44. Fuchs, *Um die Ehre*, 190–194.
45. 1637: 'Brüchtenregister', 202, no. 723; 1640: Stodt, 'Hexenwahn', 361; 1640: Brüchtenregister, 209, no. 759; 1658: Stodt, 'Hexenwahn', 367: 1658: 'Brüchtenregister', 246, no. 952; 1658: Stodt, 'Hexenwahn', 381.
46. Michael Hecht, *Patriziatsbildung als kommunikativer Prozess: die Salzstädte Lüneburg, Halle und Werl im Spätmittelalter und Früher Neuzeit* (Köln: Böhlau 2010), 203–211.
47. Verdict 1623, see: Kohn and Deisting 'Brüchtenregister', 159, no. 428; Stodt, 'Hexenwahn', 353, 356.
48. Verdict 1623, see: Kohn and Deisting, 'Brüchtenregister', 174, no. 523.

49. Verdict ca. 1630, see: Kohn and Deisting, 'Brüchtenregister', 199, no. 70; for the background on the basis of other sources, see: Stodt, 'Hexenwahn', 361.
50. Verdict 1637, see: Kohn and Deisting, 'Brüchtenregister', 202, no. 723; Stodt, 'Hexenwahn', 362.
51. Hans Stodt, '"Alß die Zauberschen gerichtet" – Hexenwahn und Zaubereiprozesse in Stadt und Amt Werl', *Mitteilungen der Werler Arbeitsgemeinschaft für Familienforschung* 20 (1999), 64–5.
52. Verdict ca. 1640, see: Kohn and Deisting, 'Brüchtenregister', 209, no. 759.
53. Verdict 1646, see: Kohn and Deisting, 'Brüchtengister', 222–223, no. 830.
54. Stodt, 'Hexenwahn', 412.
55. Rudolf Fidler, 'Quellen zur Hexenverfolgung in Werl/Westfalen', in: idem, *Rosenkranzaltar und Scheiterhaufen, Das Rosenkranzretabel zu Werl/Westfalen (1631) im Wirkfeld von Konfessionspolitik, Marienfrömmigkeit und Hexenglaube* (Köln 2002), 129–136.
56. Stodt, '"Alß die Zauberschen gerichtet"', 45, 59, 65, 74; Gerhard Schormann, *Hexenprozesse in Nordwestdeutschland* (Hildesheim, 1977), 106.
57. Tanja Gawlich, 'Der Hexenkommissar Heinrich von Schultheiß und die Hexenverfolgungen im Herzogtum Westfalen', in: Harm Klueting (ed.), *Das kurkölnische Herzogtum Westfalen von den Anfängen der kölnischen Herrschaft im südlichen Westfalen bis zur Säkularisation 1803* (Münster, 2009), 304–305.
58. Peter A. Heuser, 'Die kurkölnischen Hexenprozesse des 16.und 17.Jahrhunderts in geschlechtergeschichtlicher Perspektive', in: Ingrid Ahrendt-Schulte et al. (eds), *Geschlecht, Magie und Hexenverfolgung* (Bielefeld, 2002), 141, 174; Stodt, 'Hexenwahn', 351; Hans Stodt, 'Hexenwahn und Zaubereiprozesse in Stadt und Amt Werl im 17. Jahrhundert (1628–1630, 1642–1643, 1660). Ein Beitrag zur Sozialgeschichte Werls des 17. Jahrhunderts', *Mitteilungen der Werler Arbeitsgemeinschaft für Familienforschung* 21 (2000), 199.
59. For information on the demonology of a witch commissioner in Westphalia: Schulte, *Man as Witch*, 136–138.
60. Willem de Blécourt, 'Werwolfberichte. Der Fall Peter Stumpf und seine Folge' (2016, forthcoming).
61. Those executed were: Eboltz, Stodt, 'Hexenwahn', 349; the 'Drost' of Lüttringen (as well as his wife) Brüchtenregister, 246, no. 952 and a man known as 'kurzer [short] Heinrich', 'Brüchtenregister', 170, no. 505, also Stodt, 'Hexenwahn', 339; the villager Westhoff, 'Brüchtenregister', 191, no. 647, Stodt, 'Hexenwahn', 339; and the merchant Menne, 'Brüchtenregister', 209, no. 759, Stodt, 'Hexenwahn', 367.
62. Carlo Ginzburg, *The Night Battles. Witchcraft and Agrarian Cults in the Sixteenth and Seventeenth Centuries* (Baltimore, 1983); Gábor Klaniszay, *The Uses of Supernatural Power: The Transformation of Popular Religion in Medieval and Early Modern Europe* (Princeton, 1990).
63. Willem de Blécourt, '"I Would Have Eaten You Too". Werewolf Legends in the Flemish, Dutch and German Area', *Folklore* 118 (2007), pp. 23–43; Elmar M. Lorey, 'Besichtigung der Restbestände. Was nach Ende der Hexenprozesse im 20. Jahrhundert schließlich vom Werwolf übrig blieb in Volkserzählung, Literatur und Film', http://www.elmar-lorey.de/werwolf/Literatur.htm (accessed 3 August 2011).

Interlude: The Shepherd of Wolves

Like the wolf-riders, leaders, masters or shepherds of wolves cannot precisely be seen as werewolves either. They aided the understanding of human, or even supernatural, authority behind the actions of wolves. In Eastern European tradition especially, saints, spirits, God or even wolves fulfilled this role.[1] The *Wolfsbanner* (banner of wolves), who is the human version of the wolf herdsman, has been considered a 'weakened form' of the werewolf (Lorey, following Austrian folklorist Fritz Byloff).[2] This supposes the priority of the werewolf, whereas the relation may very well have been the other way around: in some (rare) cases wolf banners morphed into werewolves.[3]

The image above is by Maurice Sand and published in his mother's, *Légendes rustiques*.[4] The question remains, though, whether this is a genuine piece of French folklore, or an appropriation from Eastern Europe. WdB

Notes

1. Mirjam Mencej, 'The Role of Legend in Constructing Annual Cycle', *Folklore: Electronic Journal of Folklore* 32 (2006), 99–128.
2. Elmar Lorey presents a list of trials of these figures at the end of his list of werewolf trials, see: www.elmar-lorey.de/prozesse.htm.
3. Lorey's trial examples are mainly based on: Martin Scheutz, 'Bettler-Werwolf-Galeerensträfling. Die Lungauer "Werwölfe" des Jahres 1717–18 und ihr Prozeß', *Salzburg Archiv* 27 (2001), 221–68, esp. 246–7.
4. George Sand, *Légendes rustiques* (Paris, 1858), between pp. 28 and 29.

9
Estonian Werewolf History

Merili Metsvahi

Werewolves in the works of sixteenth- and seventeenth-century writers

In a 1550 reprint of his *Cosmographia* Sebastian Münster wrote on the belief in werewolves shared by the inhabitants of Livonia[1]: 'In this land there are many sorcerers and witch-women, who adhere to the erroneous belief – which they have often confessed before court – that they become wolves, roam about, and cause harm to all they encounter. Afterwards they transform back into human shape. Such people are called werewolves' (*warwölff*).[2] This was the first time learned westerners were informed that Livonia was a home for werewolves. The first edition of *Cosmographia*, which appeared in print in 1544, did not contain a section on werewolves, as Münster had no information on the topic then. The information was forwarded to him by Johannes Hasentötder, a Hessian scholar who had visited the Eastern coast of the Baltic region in 1547–1548.[3]

In 1555 *Historia de gentibus septentrionalibus*, by the Swedish priest Olaus Magnus, was published in Rome; it describes the activities of werewolves in Livonia at greater length. The inhabitants of Livonia were said to suffer more due to werewolves than real wolves. Around Christmas time the werewolves gather in Prussia, Livonia and Lithuania and harm animals as well as people. They break down doors, make their way into cellars and drink up all the beer they can find there. Magnus also writes on how the shape-shifting abilities can be acquired. To become a werewolf, the werewolf-to-be has to share a mug of beer with an experienced werewolf who has to recite a certain spell while doing this. Magnus also writes about a big annual gathering of werewolves at which several thousands of werewolves come together in some ruins on the border between Lithuania and Curonia. There they compete at jumping over the wall. Those who cannot cope with this task – for instance, because they have grown too fat – are beaten with a whip by their leader. It is not only local peasants but also members of the nobility who gather there. Magnus also retells three tales of werewolves. Firstly, there is a story about a nobleman and his peasant companions staying out in the woods at night; a peasant offers to bring the company some meat, turns into a wolf, goes off and fetches a sheep. The second story is of a Livonian noblewoman who is arguing with a serf on whether a human being can turn into a wolf. The serf wants to prove this: he goes into a cellar and soon a wolf appears in the same place. Dogs attack the wolf and rip out one of its eyes. When the peasant returns to the lady on the following day, one of his eyes is missing. The third tale speaks of a Prussian duke who did not believe in witchcraft of this kind and made a man accused of lycanthropy demonstrate this, later having him burned at the stake for having so gravely trespassed the laws of God and of people.[4]

As there were very few descriptions of the Northern regions of Europe published at the time, these two works turned out to be extraordinarily influential in Western Europe.[5] According to the Austrian historian Stefan Donecker, during the following 150 years no writing on Livonia by learned Europeans would fail to mention the local inhabitants' ability of animal transformation.[6] By the middle of the seventeenth century the knowledge of Baltic werewolves was so common that they were mentioned side by side with the lynx and hares in characterisations of Livonian fauna in the travel writing of early modernity.[7]

It is worth mentioning that it was foreign writers rather than local Baltic Germans who were particularly prone to convey sensational stories of wild werewolves and their horrible deeds.[8] An overwhelming majority of the authors associated the ability to turn into a wolf with the native peoples in Livonia. It was with the purpose of characterising them as unknown strangers that comparisons were needed that would prove evocative for the reader. Educated Western Europeans knew about the image of the werewolf, first and foremost due to the authors of classical antiquity.[9] It was already back then that the wolf was associated with both conceptual and bodily wildness.[10] Christianity made the image of the wolf even more negative.[11] Thus, the (were)wolf turned out to be a productive symbol that helped the intellectuals of early modernity visualise the representatives of the eastern Baltic region in all their barbarity and paganism.[12] On the basis of numerous sources it is known that Estonians were rather inert when it came to converting to Christianity and they did not abandon their pre-Christian customs and habits easily.[13] One of the reasons for the religious inertia was the great distance between the upper classes, whose language belonged to another family of languages, and the peasants, who were serfs: it was difficult for the peasants to understand the Christian doctrine, which was conveyed in a foreign language.[14] Seen from the perspective of the upper classes the distance contributed to the exoticisation of the peasants.

Donecker portrays the connection between werewolves and war as one of the aspects explaining the reasons for Livonia becoming an El Dorado of werewolves in early modernity. Today's historians have established that werewolf beliefs are more widespread during wars and in generally restless times.[15] As the boundary lines between the supernatural and natural worlds did not run in the same places in the minds of the people of early modernity and today, and as wolves who behaved unusually were connected with werewolves anyway at the time,[16] it is not surprising that a major role in the image of the werewolf becoming established and associated with Livonia was played by the Livonian war of 1558–1583, which had the most devastating consequences. Even historians of the time recorded the dire behaviour of wolves during the war and made note of the extraordinary numbers of wolves in the sixteenth and seventeenth centuries.[17] A military motif appearing as the leader of werewolves equipped with a whip could be found already in Olaus Magnus' description of werewolves.

That, in the following centuries, there were great numbers of wolves (as well as bears) in Estonia and Livonia, is also confirmed by travel accounts of the time: the topographic-ethnographic writings of local pastors and calendar articles.[18] It was only in the second half of the seventeenth century that landowners were made to hunt wolves and bears and their numbers began to be reduced systematically.[19] It is known that, in 1827 for instance, 935 wolves were killed in the Riga Guberniya, while wolves killed about 138000 domestic animals there within a year.[20]

Thus, the recurrence of the werewolf motif in connection with Livonia can be considered a sociopolitical means that helped to fashion the image of the inhabitants of the periphery of Europe. Firstly, it suggested that these strangers tending towards paganism had to be educated, brought to the Christian community (from among the wolves). Secondly, the werewolf image spoke of danger: If peasants are attributed magical abilities they are presumably capable of damaging the social order, despite their status as serfs who have been excluded from political power.[21] Consequently, preventive disciplinary means should be used.[22]

It is difficult to determine to what degree the motifs used by the authors of early modernity, in their werewolf descriptions, originate from the oral tradition circulating in Livonia or from the authors of antiquity and the Middle Ages. What is certain is that the readers of the time highly valued references to reputable old writings. The fact that Herodotus,[23] the 'Father of History', had written about the Neuroi of Scythia was more important than a thorough explanation of how the Neurians' ability to transform into wolves occasionally could be related to the corresponding ability of the Livonian peasants.

However, several passages from the texts, describing the Livonian werewolves of the sixteenth and seventeenth centuries, refer to the use of motifs from popular discourse concerning werewolves – rather than to compilations of earlier written sources.

The werewolf in the protocols of witch trials

In addition to the descriptions found in chronicles, travel writings, sermon transcripts and other writings, which describe the local conditions already mentioned above, data about the belief in werewolves in the seventeenth century can also be found in the protocols of witch trials. According to the preserved data a little more than 30 people were accused of practicing lycanthropy at witch trials that took place on the territory of contemporary Estonia. Yet, when we consider the scarce scope of the preserved material and its laconic nature, this is still a relatively large amount.[24]

The first witch trial for which data has been preserved[25] and which involves an accusation of being a werewolf, took place in Tallinn in 1615: two women (one of them a widow) were accused of practicing transformation into a wolf and killing cattle in this shape.[26] A somewhat more thorough protocol comes from the witch trial that took place in Kiviloo, Harju region, in 1617, in which the accused, Alit (Alheit), confesses to having been a werewolf ten years earlier and doing great harm. She also claims that, together with two more witches, she killed two oxen, two horses and other cattle and ate them. To the question of if she knew whose animals they were, the woman answered, laughing: 'Does a wolf ever ask who the animals belong to before it goes and kills them?' When questions were asked about eating the animals, Alit answered that they laid the wolfskins aside before boiling the meat, and that they'd boil the meat in their own manner.[27] The same protocol says that the testimony of another accused person (the wife of Arends) coincides with that of Alit in all aspects, the only difference being that the second accused claimed that she had also killed and eaten her own animals. When the third accused, Anna Wosell, was questioned after torture about the time and place of transforming into a werewolf and killing the animals, Anna answered that this had happened a week before Whitsun in the village of Jakob of Lunden that was called Pergel. They also killed two pigs in their own village. All the accused – there were three more in addition to those mentioned above – were sentenced to be burned at the stake.[28]

The main pattern of the trials involving an accusation of lycanthropy on the Estonian territory in the first half of the seventeenth century is the following: at first the accused deny their guilt, but after being tested by water and/or torture admit, among other sins, to having turned into werewolves and killing cattle, after which they are burned at the stake. From mid-century on the trials did not end with the death of the accused so often.[29] For instance, in 1650 a trial was held in Pühalepa on the island of Hiiumaa. Among other accusations, the plaintiff claimed that his flock of sheep had been attacked by wolves acting like werewolves. The accused denied his connection with the whole incident and attempted to explain the accidents that had befallen the plaintiff with natural reasons and coincidences. Due to the lack of evidence, the accused was exempted from torture and further interrogation, but as this was insisted upon by the accusers he was exiled from the neighbourhood and it was promised that should there arrive any more accusations concerning the same person, he would be called back and torture would be applied.[30]

Several protocols of the trials show considerable differences in the statements made by the accused in front of the judges before and after the torture. Thus, in 1640 Henn of Nossi claims at a session of the provincial court of Pärnu that he has only been healing people and not harmed them. The healing power was acquired from an old man. He also reveals the words of the spells he uses for healing in front of the court. After torture, however, he admits to knowing how to cause harm and knows what spells should be used in this case, and among other things also admits that he has put on a wolf skin and caused great harm as a wolf. Together with his accuser, Henn of Kiewase, he had gone to a hilltop, to which they had been invited by the Devil; they had taken a calf as a gift for the Devil so as not to provoke his wrath. The Devil had ordered him to harm the lord of the manor Wolmer von Schlippenbach, but he had not done so.[31]

In becoming a werewolf the same method is often used that occurs in becoming a healer or a witch: an older practitioner transfers his or her skills to a younger. Thus, a woman called Maret claims at a trial held in 1636 that the art of turning into a wolf was taught to her by a woman from Kurtna called Magdaleena, who had invited her into the forest and given her a sweet root to eat, after which both of them were turned into wolves.[32] On other occasions, becoming a wolf occurs in drinking and toasting together. The motif of becoming a wolf with the help of putting on a wolf skin is widespread. On several occasions the skin is said to have been hidden in a field under a stone; on one occasion the skin was obtained from an old man dressed in black who hid it in a mouse-hole by the stream.[33] The court materials also show that turning into a werewolf could be put upon one as a curse. Thus, Mikk of Karja, after ordeal by water, confessed in a 1633 Pärnu trial that three years earlier he had hit a Lapp drinking companion who had not paid for his beer. The other man had told him: 'This revenge I will have on you. During three years you will have neither peace nor rest at home. Whenever the Devil calls you, you will have to go', upon which he had started going out as a werewolf.[34]

The court protocols show that very often several people went out as werewolves together. According to the testimony of Greeta, the daughter of Thomas of Titza, in 1696, the largest 'guild of werewolves' mentioned in the trial protocols was active, in Vastsemõisa, Suure-Jaani parish, where 11 people from the villages of the neighbourhood practiced lycanthropy.[35]

I have mentioned here only those protocols of witch trials that took place in the territory of contemporary Estonia. As I do not have complete data on the trials that occurred in present-day Latvian territory, I have left these out of the discussion. All in all it can be claimed that the belief in werewolves was in full bloom in the present-day Estonian territory up to the end of the seventeenth century. This can be concluded, not only on the basis of witness' testimonies, but also from other court materials. Thus, in 1695 a case of a woman from Rapla, who had died as a result of a gunshot, was studied at an inquest in Nabala. A suspicion was spreading that the woman had been a werewolf and the question arose if she could be buried in the churchyard. After the corpse had been examined by the royal surgeon, the court decided that the woman was to be buried in a separate place, not in the churchyard, and the pastor had to make the incident known from the pulpit to warn people against such godless persons.[36]

Folk tales from the nineteenth and twentieth centuries

The richest source for studying the belief in werewolves in Estonia is the texts on the topic stored at the Estonian Folklore Archives. Most have never been published and are stored in manuscript, a small number in audio and manuscript, and a tiny number only as audio files. As regards genre, the largest part of the texts could be classified as legends; in addition, there are fairy tales belonging to the Aarne Thompson type 409,[37] reports of belief and some occasional memorates. The material on werewolves was recorded from the beginning of the nineteenth century to the early twenty-first.

In addition, there are legends about werewolves in printed sources, both in Estonian as well as in German translation. The first printed folk tales on the topic of werewolves can by found in a travelogue by Christian Hieronymos Justus von Schlegel (1755–1842) *Reisen im mehrere russische Gouvernements in den Jahren 1801, 1807 und 1815. Fünftes Bändchen. Ausflug nach Ehstland im Junius 1807.*[38] Von Schlegel was a man of wide knowledge, who had studied at the *Gymnasium* in Weimar, where Johann Karl August Musäus,[39] one of the first authors of literary German 'folk' tales, was among his teachers. In the years 1780–1782 von Schlegel worked in Estonia as a private tutor, and married a local Baltic German girl in 1783. As Estonian folk culture was one of von Schlegel's favourite topics, and his writings on the topic are reliable,[40] it may be believed that the Estonian tales of werewolves he translated into German do derive from the mouths of the peasants.

It is remarkable that as many as three of the four legends provided by von Schlegel (all in all the travelogue contains six folk tales, two of which are fairy-tales) deal with the topic of animal transformation.[41] The first is a story of how in one winter there were more wolves around in the parishes of Torma and Laiuse than usual, but they killed only people. It was said that the wolves had come from Russia, across the ice of Lake Peipsi; they were more terrifying than ordinary wolves as they were not afraid of people. It was difficult to catch the wolves. Finally, a German hunter decided to shoot them. He loaded two guns, one with an ordinary bullet and the other with a silver one, claiming that the latter could be used to shoot the Devil himself. He went out and saw a pack of wolves, but it was not sufficiently close to him. In the end a wolf with two black spots on its breast came near him to attack him. At first he took a shot using one gun, but to no avail. He reloaded the gun and took another shot, but still to no avail. The wolf was coming closer and closer. Then the hunter took the other gun, shot it and hit the wolf. The wolf still attempted to jump at him, but fell down dead. The dead wolf was so heavy that it was impossible to move it. Thus, the hunter invited people there so that they could skin the wolf on spot. It was a great surprise when the wolf skin revealed a dead woman who had transformed herself into a wolf with the help of witchcraft (*durch Teufelkünste*).

The other tale was about a wild bear in Simuna parish, which would steal cattle from the cowsheds every night. In the end, a cowherd set up an iron trap to catch it, but it was useless. Subsequently, he shot a burst of shots into the bear's body, upon which the bear released a heifer it had just stolen and disappeared, roaring. The brave cowherd followed it at a distance, unnoticed. The man saw the bear enter a cabin and listened at the door as the bear warned its children against choosing its path: now his last hour had come and he was in the hands of the Devil. The cowherd entered the house, saw a bearskin on the floor and a man upon the bed. He made the man confess that it was he who transformed into the bear with the help of the Devil. After the man's death his name was made known to everybody and his family was put to shame. His cabin was burned and his children imprisoned. The site of the cabin is still known today.

The third tale is of a peasant who was driving through the forest when he heard voices calling up to him: 'Take us along!' He looked around but saw no one. In the evening he heard the voices again. Now he crossed himself, muttered a quick prayer, and went towards the sound. He reached a wooden house in which a woman was sitting with her child. 'Why did you cross yourself?' she asked, 'otherwise you would have been ours.' Her eyes were gleaming red like the eyes of a wolf and, at the same moment, the woman and the child did turn into wolves and ran off into the forest.[42]

Von Schlegel's folklore collection contained more folk tales than published in print. This is shown by the manuscript texts found in the collection of the Literary Society of Estonia in the Estonian Folklore Archives (EFA).[43] I have already written on von Schlegel's fairy-tale that contains the motif of the werewolf and was probably also recorded in 1807.[44] It is interesting that another – but Estonian language – version of the same fairy-tale also dates back to the first quarter of the nineteenth century, although the exact date is not known.[45] As the recording of Estonian fairy-tales on a considerable scale started as late as the 1870s, while wide-scale collection began only in the final decade of the nineteenth century, these are very early texts in the history of collecting folklore in Estonia.

Most of the werewolf legends were written down in the first half of the twentieth century, particularly from the time directly after the foundation of the EFA in 1927. The great number of texts reflects the intensity of folklore collection during this period in comparison with other periods. In addition, it is important to note that the EFA contains more materials related to folk belief than other world archives of a similar type.[46] Firstly, this is due to Jakob Hurt, the great collector of folklore who, at the end of the nineteenth century encouraged his correspondents[47] also to collect 'old faith' among other materials;[48] and secondly, this is due to Oskar Loorits, the founder of the EFA and its first director, who considered folk belief a topic with plenty of 'gaps' in the archives that should be filled during field expeditions when questioning village people.[49]

The fact the the EFA employees specifically attempted to collect material on werewolves in the years immediately following the founding of the EFA can be seen from brief (so-called belief) reports recorded in 1930–1931, of which some examples can be given here:

> *Werewolf – unknown (Who knows what kind of an animal it might have been?)*[50]
>
> *Werewolf – this is a wolf as well.*[51]
>
> *People made of wolf.*[52]
>
> *Werewolf, real wolf – there is no difference anyway.*[53]

The texts rather demonstrate the fieldwork methods adopted in the early times of the EFA, and say less about folk belief. However, the numerousness of short reports of this kind does not reduce the significance and reliability of the werewolf legends or even memorates recorded from other people during the same period.[54]

It is clear that, over the ages, communities and societies have consisted of people with different mental worlds. People with similar thinking (in other words, personality structures) form channels through which stories of belief move from the consciousness of one person to that of another. If the tale meets a person who is not capable of receiving it, the channel ends in an impasse, or joins another channel in case modifications arise. If, however, a particularly good narrator and a number of persons with similar mental worlds belong to the links of the channel, it can start producing numerous new branches with the help of new retellers.[55] If a storyteller is masterful with words and evokes mental pictures in the mind's eye of the listener, a sympathetic listener will relate the story very closely to the experiences of his or her own life. The resources needed to make sense of the world, that arise in a listener's consciousness, are very similar to the resources we use to understand the world through personal experience.[56]

This explains the why the teller of a legend suddenly switches to using the first person in the singular or plural.[57] We can analyse the werewolf legend of Ann Pilberg (recorded in 1933) from this theoretical point of view (Figure 9.1). The elderly female informant of the folklorist Richard Viidalepp connected a werewolf story belonging to fairy tale type AT 409 with events from her own life: as a child she and her mother had seen a wolf come out of the woods, take off its wolf skin, put it onto a stone and, in the shape of a local woman, suckle her human baby upon another stone. She willingly showed the same stones to Viidalepp, who took pictures of these and added the photos to the werewolf tale told by Pilberg in the EFA volume.[58]

The *conduit* theory can explain the existence of very different world views side by side – for instance, the scepticism of Herodotus regarding the existence of werewolves, and the conviction of Jean Bodin, nearly 2000 years later, and of Montague Summers as much as 2400 years later, that humans indeed can turn themselves into wolves.[59] In order to understand the case of the werewolf Thies applying *conduit*-theory and the ideas of experiential

Figure 9.1 The story teller Ann Pilberg from Karilatsi (South-Estonia) in 1933, ERA, photo 192. (Reproduced with permission of the Estonian Folklore Archives, Estonian Literary Museum (*Eesti Kirjandusmuuseum*), Tartu)

resources will not suffice[60]; rather, in addition to that, the ingenious old man should be approached as a bricoleur,[61] who could make inventive use of experiential resources gained when living through life events and listening to tales, as well as employ his storytelling skills according to the requirements set by the occasion. In the same way Chonrad Stoeckhlin described by Wolfgang Behringer used motifs of folk belief in order to justify his role as a magical specialist, Thies joined his narrating skills, familiar motifs of folk belief, and his reputation with the villagers as a healer in a way that allows him to be truly called 'a virtuoso of folk culture', to use Behringer's terms.[62]

I myself have recorded the tales of a final link in a channel of werewolf legends and delivered the recordings to the EFA. Namely, in addition to several other folk tales, Ksenia Müürsepp,[63] a 90-year-old woman of Seto origin, told the folklorists who visited her in 2001 the legends of 'Merchant into a Werewolf'[64] and 'Bridal party into Werewolves', and claimed that werewolves could be recognised by the fact that the hair below the neck was white.

A glance into the development history of Estonian werewolf legends

If Estonians adopted the werewolf concept from Germans,[65] then one of the factors that made such a transfer, and the ensuing rapid spreading of the idea, possible is the fact that the image of animal transformation had been familiar in local folk belief for a long time. For instance, this is proved by an old folk song called The Song of the Fish Maiden in Estonia. The song is about a brother who goes out to fish, brings a fish home and, when starting to cut it up, discovers beads, a chaplet, a brooch, an apron or other attributes of a maiden.[66] Antti Aarne, who has written a comparative study of the Finnish-Karelian and Estonian versions of the song, claims that the Estonian and Karelian versions have a common origin. The song has been included in the Finnish national epic, the Kalevala. In the Kalevala, Väinamöinen goes down to the sea and catches a fish. At the moment when he intends to cut the fish up, it jumps back into the sea and announces it is Joukahainen's sister whom Joukahainen had promised to Väinamöinen in marriage, but who would not marry him and went to live in the sea as a fish instead.[67]

The motif of finding objects belonging to people under the fish's skin is similar to the motif of finding human beings or objects belonging to humans under a wolf's hide when skinning the wolf. The motif, that can be found in Estonian werewolf legends with different plots, can also be found in a folk tale published by von Schlegel. It only rarely appears in werewolf legends from the rest of Europe. In the few cases when a human being is found in a wolf skin in the traditions of other peoples, it is just the human body and not jewellery or accessories.[68] In Estonian folklore, however, it is very common that, either pieces of jewellery (most often a brooch or beads) or an element of clothing (a silk scarf, a ribbon around the neck or the like), is found under the skin of the wolf. The plot is not always fully developed; occasionally just short belief reports can be found, such as the following:

> Once a wolf who had come to the village was killed and beads were found around her neck.[69]

Finding the objects is very often accompanied by abandoning the plan of skinning and burying the werewolf. Thus, for instance, in the 1890s Jakob Hurt recorded a tale in the Seto region of how a village near the edge of a forest was pestered by a werewolf. The werewolf would chase children and snatch at them with its teeth. It killed nobody but kept the whole village in fear. In the end the wolf was shot with a gun. When the skinning started, a large bridal brooch and beads were found under the skin. Since then no one in the village dared to kill any more wolves.[70]

It seems that, although nineteenth-century calendars encouraged peasants to kill wolves and monetary rewards were offered for this,[71] the peasants were not particularly eager to do this, apparently partly from fear that they could unwittingly kill a human who had transformed himself or herself into a wolf,[72] partly because wolves were respected and it was believed that they were necessary for something. The wolf's right to its prey was recognised – it was believed that wolves have been allotted their share from each farm that they have to get. When a sheep was killed, the first vertebra from the neck was thrown into the forest to the wolf. Wolves stealing dogs from farmyards in the winter was not disapproved of, as it was believed that wolves needed to eat dog's meat before the start of the rutting season or else they would not be able to conceive.[73] It was believed that, when wolves killed some of the cattle in the herd it was good for the herd, for it contributed to the cattle's fertility.[74] At the same time, writers who aimed at enlightening people admonished the peasants for having too reverent attitudes towards wolves.[75] The wolf's significance, and its special status in comparison with other animals, is testified by the Estonian language having more euphemisms for 'wolf' than for any other animal, and by the folklore connected with the wolf, which is richer than that of any other wild animal.[76] The wolf would chase and kill evil spirits, thus rescuing humans many a time.[77] It is possible that the positive characteristics of the wolf and the special attitude towards it are a continuation of one-time totemic ideas.

A particularly interesting detail about werewolf beliefs in Estonia is the fact that both the texts from the EFA, as well as the materials of witch trials, show that women would transform themselves into wolves somewhat more often than men. I believe that an important reason for the popularity of female werewolves was the fact that, before the Christianisation of Estonia in the thirteenth century the position of women was stronger here than in any other part of Europe.[78] The transition into a patrilineal and patrilocal society brought along such great changes in the status of women that following the rules of the new social order would have been impossible without compensating for the status of the subjugated in genres of folklore. Stories of and beliefs about werewolves were – at least from the sixteenth century – one of the so-called safety valves that offered women a possibility to experience their selves in an alternative manner.

An example of how a tale of werewolves reached Estonia, most likely mediated by Germans, and was modified to fit the world picture of the local peasants, and meet their expectations according to the specifics of their environment, is the tale of two girls on their way to a manor.[79] There are 35 different versions of the tale, collected in various places in Estonia stored and at the EFA. The first published version comes from J.B. Holzmayer's work *Osiliana*, dating from 1873.[80] The following example comes from a manuscript that was recorded on Saaremaa, Estonia's largest island, as was Holzmayer's version:

> Once two girls from Räägi village were going to the manor of Reo to do their statute labour. Their road went through the manor's paddock where the foals of the manor were pasturing. One girl said to the other: 'Look at how nice and fat the foals are!' The other took a look but did not answer anything right then; afterwards she said: 'Why are we in such a hurry; we'll come to the manor too early, let us rather lay down for a while, I so want to sleep!' The first girl agreed and they laid down to sleep. The first acted as if she had fallen asleep and was snoring rather loudly, the other stayed awake, listened for a while and, thinking that the other was asleep, stood up and went behind a bush. There she turned three somersaults and became a wolf, ran to the foals, killed the nicest and fattest of them, drank its blood and ate some of the raw meat, also storing some of it in her box. The first girl did not dare to watch any more, came away from behind the bush and laid down to sleep again. Soon the other girl came back to the first one in the shape of a wolf and listened to whether she was still asleep. The first girl acted as if she was sleeping heavily and guessed that if the other understood she was awake, she would surely kill her. The other, having understood that there was nothing to be worried about, again turned three somersaults and was a human being again. She went to the first one, woke her and they started to walk towards the manor together.
>
> At the manor, the second girl offered the first one some meat from her box. The first one had gained courage and told her: 'I don't want your raw foal's meat, you can eat it yourself if you like it.' The other girl became angry, took a piece of the meat and threw it into the eyes of the first one, saying: 'See for yourself if it is raw!' The piece of meat, however, was so hot that the face of the first girl was burnt.[81]

The book *Deutsche Sagen* (1816), by the Brothers Grimm, includes a legend with a similar plot that is still relatively different from the Estonian version,[82] the most significant differences being that the German legend has male characters and its ending lacks the motif of the meat becoming baked on its own.

The self-baking meat motif becomes less puzzling when the werewolf is treated as someone who transgresses boundaries and is capable of moving between the supernatural and natural worlds. The supernatural world (sphere) is where various resources come from and exists in a place that is difficult to access.[83] This imperceptible sphere, which remained beyond the activities of the ordinary peasant, became more perceptible through Thies, and other talented virtuosi of folk culture, who evoked it in front of their listeners with the tales they told and performed, claiming that they were the mediators and transgressors of the boundaries on which the welfare of the ordinary world depended. The accused of several seventeenth-century Latvian witch trials claimed that their souls had visited Hell, taking the souls of dead cattle with them to be revived there.[84] The Latvian peasant Thies also claimed, before a seventeenth century court, that the abundance of the harvest of the following year depended on his and other werewolves' visits to Hell on St. Lucy's Day.[85]

It seems that the crossing or meeting places of the two spheres displayed a specific potential. Thus, the most often occurring way for a wolf to change back into a human in the legend type 'Farmhand does as his master'[86] is to rub one's body against the corner of the pigsty.[87] As a rule, the pigsty used to be the building on the farmyard that was furthest away from the house, thus a place suitable for returning from the sphere of the wild. Whenever an element of the quotidian world of humans entered the wild sphere, a radical change took place there: if a wolf was given some bread in the depth of the forest, it changed back into a human being;[88] if a rebuking human voice interrupted the main wolfish activities of a werewolf,[89] the meat of an animal just killed suddenly became baked on its own accord. In addition to the legend of the two girls on their way to the manor, the motif also often occurs in the EFA as a separate belief record, or part of a lengthier description of a werewolf:

> When a werewolf kills cattle without anyone seeing this, it must eat the meat raw; if it is seen and shooed away, the meat will bake on its own and therefore the saying goes: when a wolf is shooed away, meat is being baked for the werewolf.[90]
>
> As soon as the home wolf's teeth touch an animal whom it wants to kill it said to become as if cooked and not at all raw any more.[91]

In the last example the mere fact of the wolf killing the animal not being an ordinary one, but a human wolf connected with the quotidian sphere is sufficient for the meat to become baked. It is possible that the phrase 'the one upon the stove'[92] – an old name for werewolf – is connected with the werewolf's supernatural ability to heat meat. Alit's testimony, given before a court in 1617, and Thies' testimony from 1692,[93] according to which werewolves do not eat raw meat but bake or cook it before eating, seem to refer to the same belief complex. Both of the accused take up the topic without being asked directly about it.

Concluding remarks

Individuals considered capable of moving between two spheres certainly existed, even before the witch trials.[94] A change that occurred in the sixteenth century at the latest[95] was the idea of a werewolf,[96] imported from abroad, that facilitated connection with the people moving between spheres with wolves. It is not known if the connection was first created by external observers, or by the local inhabitants. Either way, the concept proved to be most productive, and mutually profitable: writers would win more readers with their exotic descriptions, that may also have had socio-political aims; eloquent locals boasted of their supernatural talents, both in front of their own kind and local Germans, as well as those who had arrived from abroad.[97]

At the same time it is interesting that Germans, on the one hand, and native local inhabitants, on the other, had widely differing conceptions of the werewolf.[98] One of the important reasons for this is the difference in associations connected with the wolf. It is possible that the idea of the werewolf, introduced by demonologists, was easily adopted by the peoples of the eastern Baltic in the sixteenth century, specifically because it offered something positive in comparison with the entirely negative image of the witch spread by demonologists.[99] The clever and eloquent witches who were brought to court – particularly those who were active as healers and mediators between the above-mentioned two spheres – did not have to deny all of their useful activities, but could in a way introduce the topic in their testimonies.

In the testimonies of the witch trials, and in the legends stored at the EFA, the motif of wolf transformers killing only smaller animals occurs quite often. If killing an animal would increase the fertility of the cattle it apparently made sense to kill smaller animals. No reports of people being attacked can be found in any materials of witch trials containing accusations of lycanthropy. The Estonian materials at the EFA only extremely rarely contain the motif of a werewolf killing a human being. The text introduced above about a werewolf chasing children in a Seto village (recorded by Jakob Hurt) stands out as an exception among the EFA materials. It is much more common for the werewolf to not harm people, but only domestic animals. The mild behaviour of Estonian werewolves towards people is all the more remarkable if we take into account that there are relatively many reports of either rabid or man-eating wolves with disturbed behaviour from the seventeenth to the nineteenth centuries.[100]

The above suggests that one and the same term can refer to rather different phenomena in different cultures. It can also be concluded, on the basis of the above, that folk tales and concepts adopted from other peoples can develop into fascinating modifications when travelling into new environments and finding fertile ground for growth there, becoming so rooted in tradition that they are soon accepted as something inherent to the people.

Notes

This research was supported by the European Union through the European Regional Development Fund (Center of Excellence, CECT) and by the institutional research funding IUT (2-43) of the Estonian Ministry of Education and Research.

1. The name of Livonia came into use in the beginning of the 13th century to denote 'the new Christian colony which covered broadly the territories of modern Estonia and Latvia and was inhabited by diverse ethnic groups' (Marek Tamm, 'Inventing Livonia: The Name and Fame of a New Christian Colony on the Medieval Baltic Frontier', *Zeitschrift für Ostmitteleuropa-Forschung* 60 (2), 2011, 186–209, 196). The meaning of 'Livonia' has changed over time. In the 18th and 19th century, Livonia was the region that was located in contemporary South-Estonia and the Northern part of Latvia.
2. Stefan Donecker. 'The Werewolves of Livonia: Lycanthropy and Shape-Changing in Scholarly Texts, 1550–1720', *Preternature, Critical and Historical Studies on the Preternatural* 1 (2012), 289–322, 295.
3. Donecker, 'The werewolves…', 295.
4. 'Des Olaus Magnus Bericht über die livländischen Werwölfe. Uebersetzt von Eduard Pabst', *Archiv für die Geschichte Liv-, Esth- und Curlands* 4 (1845). The most complete description of Livonian werewolves by Olaus Magnus occurs in the first edition of his work that appeared in 1555; it has been considerably shortened in the later editions. Walter Anderson, 'Uus töö balti libahundiprotsesside kohta', *Ajalooline Ajakir*i nr 4 (1927), 151–4.
5. For instance, 44 editions of Münster's *Cosmographia* appeared in all European languages during the next 100 years. Eduard Laugaste, *Eesti rahvaluuleteaduse ajalugu. Valitud tekste ja pilte* (Tallinn, 1963), 290; Donecker, 'The werewolves…', 301.
6. Stefan Donecker, 'The Lion, the Witch and the Walrus. Images of the Sorcerous North in the 16th and 17th centuries', *Internet-Zeitschrift für Kulturwissenschaften*. 17 (February 2010). http://www.inst.at/trans/17Nr/4-5/5-5_donecker.htm (accessed on 18 January 2013).
7. Donecker, 'The werewolves…', 302–303.
8. Donecker, 'The werewolves…', 299. Balthasar Russow, who is of local origin, does write about the atypical behaviour of real wolves during the Livonian war, but is silent upon the topic of werewolves, Donecker, 'The werewolves…', 301.
9. See, e.g., Wilhelm Hertz, *Der Werwolf: Beiträge zur Sagengeschichte* (Stuttgart, 1862), 32–36. Leander Petzoldt, *Kleines Lexicon der Dämonen und Elementargeister* (München, 1990), 181.
10. Aleksander Pluskowski, *Wolves and the Wilderness in the Middle Ages* (Woodbridge, 2006), 12, 71.

11. Pluskowski, *Wolves and the Wilderness in the Middle Ages*, 84; A. Erler, 'Friedlosigkeit und Werwolfglaube', *Paideuma*, 1. 7 (1940), 304–317, 307. On the opposition of the wolf and the lamb as the opposition of the Christian and the un-Christian in Martin Luther see, e.g., Utz Anhalt, *Der Werwolf.Ausgewählte Aspekte einer Figur der europäischen Mythengeschichte unter besonderer Berücksichtigung der Tollwut*, Magisterarbeit (Hannover, 1999), http://www.historicum.net/themen/hexenforschung/thementexte/magisterarbeiten/art/Der_Werwolf_Au/ (accessed on 22 January 2013).
12. That the use of the wolf in the sense of the un-Christian was widespread is proved by an extract from August Lercheimer's (= Hermann Witekind's) *Christlich Bedenken vnd Erinnering von Zauberey* (Heidelberg, 1585) in which he writes of his own contact with a Livonian 'werewolf', calling a local peasant who considers himself to be a werewolf, a madman who knows as much of God as a wolf does, 26–28.
13. A source that proves this is provided by visitiation protocols, see, e.g., P. Pedakmäe, Nõiausust Taaniaegsel Saaremaal, *Saaremaa muuseum. Kaheaastaraamat 1997–1998* (Kuressaare, 1999), 18–25. Christian Kelch writes in his chronicle that appeared in 1695, and is significant in Estonian and Livonian historiography, that the following proverb is known among Livonian congregations: 'I must needs pay the clergyman, but know little of God and his word'. H. Kruus. *Eesti ajaloo lugemik II: valitud lugemispalad Eesti ajaloo alalt 1561–1721. a.* (Tartu, 1926), 50; see also J. Kahk, 'Rristiusk, teadus ja nõiaprotsessid XVII sajandil', in: J. Kivimäe (ed.), *Religiooni ja ateismi ajaloost eestis. Artiklite kogumik*, III (Tallinn, 1987), 146–171, 158–162.
14. The first translation of the entire Bible into Estonian appeared in 1739, the first translation of the New Testament into South Estonian appeared in 1686 and into Latvian in 1685. In 1582–1625 most of Latvia and South Estonia belonged under Polish and Lithuanian Roman Catholic power. Jesuit preachers would use even more Estonian and Latvian than Lutheran pastors. It is possible that it was due to the activities of the Jesuits that Lutherans, who wanted to win the peasants' souls back to their church, that the local languages started to be used more. Rein Taagepera, 'Albert, Martin and Peter Too: Their Roles in Creating the Estonian and Latvian Nations', *Journal of Baltic Studies* 42 (2011), 125–141, 132.
15. Donecker, 'The werewolves...', 311. See also W.M.S. Russell and Claire Russell, 'The Social Biology of Werewolves', in: J.R. Porter, W.M.S. Russell (eds), *Animals in Folklore* (Ipswich, 1978), 158–159. Occasionally it is impossible to distinguish between wolves who behaved atypically and werewolves in Early Modern sources, Donecker, 'The werewolves...', 319.
16. Hertz, *Der Werwolf: Beitrag zur Sagengeschichte*, 18; Donecker, 'The werewolves...', 319.
17. See, e.g., the description of Adam Olearius of how cattle, if not shut up close to the farms, would be presently taken away by the wolves, and that one had to carry a heavy stick to ward off wolves when walking on village roads; see *Archiv für Geschichte*, 330. See also Tiina Vähi, 'Kirjalikud teated eesti libahundipärimuse kohta kuni 20. sajandi alguseni', *Õpetatud Eesti Seltsi aastaraamat* (2010), 242–267, 253–254.
18. Ilmar Rootsi, *Tuli susi soovikusta* (Tartu, 2005), 52–54.
19. Rootsi, *Tuli susi soovikusta*, 228.
20. 'Huntidest', in: *Eesti-ma-rahwa Kalender ehk Täht Ramat 1829 aasta peäle* (Tallinn, 1829).

21. During Early Modern times Estonia and Livonia used to belong to the regions in Europe in which serfdom was the most advanced, Katrin Kello, *Isikliku sõltuvuse piirid ja tunnused 18. sajandi Liivi- ja Eestimaal päruskuuluvuse teket või muutumist käsitlevate kohtuotsuste põhjal.* Magistritöö (Tartu, 2003), 4.
22. Donecker, 'The werewolves…', 313.
23. Donecker, 'The werewolves…', 299; Jean Bodin, *Von ausgelasnen wütigen Teuffelsheer* (Strassburg, 1591), 123, on the importance of authoritative texts for the writers of later times see also M. Tamm, Liivimaa leiutamine…, 212–213.
24. See Maia Madar, 'Nõiaprotsessid Eestis XVI sajandist XIX sajandini', in: Jüri Kivimäe (ed.), *Religiooni ja ateismi ajaloost Eestis. Artiklite kogumik III* (Tallinn, 1987), 124–145.
25. Only after completing this article I learned about a trial that took place in Reval/Tallinn in 1596 that contains accusations towards different persons being wolves and also information about a witch trial two years earlier against two peasant women who were accused of changing themselves into wolves. The trial record from 1596 is obscure and misses the end. It has been published in: Paul Johansen and Heinz von Zur Mühlen, *Deutsch und Undeutsch im mittelalterlichen und frühneuzeitlichen Reval* (Köln, 1973), 465–469.
26. Otto von Riesemann, 'Hexen und Zauberer in Reval 1615–1618', *Beiträge zur Kunde Ehst-, Liv- und Kurlands*, II. 2 (1877), 325–343.
27. A similar motif can be found in the materials of the court case of Thies. See also H. von Bruiningk, 'Der Werwolf in Livland und das letzte im Wendenschen Landgericht und Dörptschen Hofgericht i. J. 1692 deshalb stattgehabte Strafverfahren', *Mitteilungen aus der livländischen Geschichte.* Band 22, 3. Heft (1925), 163–220, 205.
28. 'Ein Hexenprocess', *Das Inland: Eine Wochenschrift für Liv-, Esth- und Kurlands Geschichte, Geographie, Statistik und Litteratur* 5 (1840), 341–344.
29. Actually, there is a great regional variation in the statistics of death sentences. For instance, the provincial court in Tartu sentenced quite a few witches to be executed as late as the 1680s, and the last execution of a witch known to history and based on the decision of the Tartu provincial court occurred in 1699. Madar, 'Nõiaprotsessid Eestis…', 129–130.
30. Rudolf Winkler, 'Über Hexenwahn und Hexenprozesse in Estland während der Schwedenherrschaft', *Baltische Monatsschrift*, 67 (1909), 321–355.
31. Villem Uuspuu, 'Nõiaprotsesse Pärnu Maakohtu Arhiivist kuni 1642', *Usuteaduseline Ajakiri* Nr. ¾ (1937), 119–21, also published in the web-based journal http://www.folklore.ee/tagused/nr22/uuspuu.pdf (accessed on 16 January 2013).
32. Winkler, 'Über Hexenwahn…', 337.
33. M.F. von Toll. 'Zur geschichte der Hexenprocesse. Auszug aus dem Protocoll des Wier- und Jerwschen Manngerichts', *Das Inland. Eine Wochenschrift für Liv-, Esth- und Kurland's Geschichte, Geographie, Statistik und Literatur* 17 (1839), 257–263.
34. Kruus, *Eesti ajaloo lugemik*, 78; Winkler, 'Über Hexenwahn…', 335–336.
35. Madar, 'Nõiaprotsessid Eestis…', 138–139.
36. Winkler. 'Über Hexenwahn…', 348.
37. On why I prefer referring to the Aarne-Thompson Catalogue (Antti Aarne and Stith Thompson, *The Types of the Folktales. A Classification and Bibliography.* Helsinki 1961) rather than Uther's Catalogue (Hans-Jörg Uther, *The Types of International Folktales. A Classification and Bibliography.* Helsinki, 2004); See also Merili Metsvahi, 'Die Frau als Werwölfin (AT 409) in der estnischen Volkstradition', in:

Willem de Blécourt and Christa A. Tuczay (eds), *Tierverwandlungen: Codierungen und Diskurse* (Tübingen, 2011), 193–219, esp. 193–194 and idem, 'The Woman as Wolf (AT 409). Some Interpretations of a Very Estonian Folk Tale', *Journal of Ethnology and Folkloristics* 7.2 (2013), 65–92.

38. Folk tales can be defined in different ways and also the stories retold by Olaus Magnus could be considered as folk tales in case of a different approach. I am proceeding from the aims and definitions provided by the author. Von Schlegel has six folk tales printed in his travelogue in order to show Estonian folk tales to his audiences and point at the difference between the tales of different peoples. Speaking of tales of animal transformation, he himself uses the expression 'fairy-tale-like stories' (*märchenhafte Erzählungen*), *Reisen im mehrere russische Gouvernements in den Jahren 1801, 1807 and 1815*. Fünftes Bändchen. Ausflug nach Ehstland im Junius 1807 (Meiningen, 1830), 179–181.
39. J. G. v. Bunge (ed.), *Archiv für die Geschichte Liv-, Esth- und Curlands*, 96, 106. Von Schlegel himself wrote the following about his teacher Musäus*: selten ist ein Lehrer, der in Seine Fache sich seiner Pflicht so brav entledigte, als Musäus.* v. Bunge (ed.), *Archiv für die Geschichte Liv-, Esth- und Curlands*, 96. Musäus is said to have been a person of very nice character. Noticeable facts about him worth mentioning in this context include his hospitality as he often received houseguests from Livonia in his Weimar house and also invited common people (children in the street, an old soldier, old women) to come and visit him, asking them to tell him folk tales (for a small fee). Johann Karl August Musäus, August von Kotzebue (ed.), *Nachgelassene Schriften* (Mannheim, 1803), 14.
40. v. Bunge (ed.), *Archiv für die Geschichte Liv-, Esth- und Curlands*, 105–106. The fact that von Schlegel published many Estonian *regi* songs in his travelogues also shows that he was interested in folklore and considered it important to publish authentic popular texts.
41. The genre called 'legend' (*Sage*) developed into an independent category only after the collection *Deutsche Sagen* (1816–1818) by the Brothers Grimm.
42. v. Schlegel. *Reisen in mehrere russische Gouvernements...*, 181–183.
43. See also Kärri Toomeos-Orglaan, 'Vanimate imemuinasjuttude kirjapanekust, nende koopiatest ja trükiversioonidest', in: M. Hiiemäe, K. Labi, J. Oras (eds), *Aega otsimas*. Pro Folkloristica XII (Tartu, 2005), 151–164.
44. See Metsvahi, 'Die Frau als Werwölfin (AT 409) in der estnischen Volkstradition...', 202–203.
45. The version has also appeared in print: R. Järv, M. Kaasik, K. Toomeos-Orglaan (eds.), *Eesti muinasjutud I: 1. Imemuinasjutud* (Tartu, 2009), 348–351; see also the comment on the fairy-tale on p. 559.
46. According to Mall Hiiemäe 'the fields that Hurt had called *customs* and *habits* and *beliefs*,/.../ gained even more attention than the main kinds of folklore?? at the end of the 1920s and in the 1930s. M. Hiiemäe, 'Eesti Rahvaluule Arhiivi rajaleidmised', in: Tiiu Jaago (ed.) *Pärimus ja tõlgendus. Artikleid folkloristika ja etnoloogia teooriate, meetodite ning uurimispraktika alalt* (Tartu, 2003), 50–60, 51. See also p. 52 and p. 54 where it is pointed out that the Finnish, Latvian and Lithuanian folklore archives contain less material on beliefs.
47. On Hurt having had a very wide network of correspondents all over Estonia, see e.g., Tiiu Jaago, 'Jakob Hurt: The Birth of Estonian-language Folklore Research', in: Kristin Kuutma and Tiiu Jaago (eds), *Studies in Estonian Folkloristics and Ethnology. A Reader and Reflexive History* (Tartu, 2005), 45–64.
48. See Laugaste, *Eesti rahvaluuleteaduse ajalugu*, 213.

49. On Oskar Loorits, see Ergo-Hart Västrik, 'Oskar Loorits: Byzantine Cultural Relations and Practical Application of Folklore Archives', in: Kuutma and Jaago (eds), *Studies in Estonian Folkloristics and Ethnology. A Reader and Reflexive History,* 203–215. Among other things, Loorits is the author of *Grundzüge des estnischen Volksglaubens*. I–V (Lund, 1949–1960), the most extensive general survey of Estonian folk belief.
50. The Estonian Folklore Archives (EFA) ERA II 25, 73 (30) < Järva-Madise khk., Albu v., Aravete k. – Richard Viidebaum < Liisa Liiv, 71 a. (1930).
51. EFA, ERA II 28, 485 (34) < Viru-Nigula khk., Kalvi v., Männiku k., Süvadi t. < Viru-Nigula khk., Kalvi v., Lahe k. – Rudolf Põldmäe < Maarja Süvadi, 86 a. (1930).
52. EFA, ERA II 28, 612 (44) < Viru-Nigula khk., Kunda-Malla v., Kunda k., Mäepealt s. – Rudolf Põldmäe < Leena Mikker, 62 a. (1930).
53. EFA, ERA II 34, 53 (1) < Paistu khk., Heimtali v., Rõõsa t. – Oskar Loorits < the Grünbergs, farm owners (1931).
54. At the same time it should be added that there are more legends meant for entertainment than legends related to beliefs in the narrower sense among the werewolf legends to be found at the EFA. By the early twentieth century the werewolf was not so much an object of living folk belief any more, but rather a character in fictional tales that were being retold first and foremost due to their narrative value. On the topic of it being easier for folklore collectors from towns whose social background is different to access fictional tales than rumours and other real-life stories see, e.g., Merili Metsvahi, 'Contemporary Fieldwork and Archival Recordings', in Ulrika Wolf-Knuts and Anniki Kaivola-Bregenhoj (eds), *Pathways. Approaches to the Study and Teaching of Folklore* (Turku, 2001), 99–108.
55. Linda Dégh, 'Conduit-Theorie', *Enzyklopädie des Märchens*, III (1981), 124–126.
56. Donald Braid, 'Personal Narrative and Experiental Meaning', *Journal of American Folklore* 109 (1996), 5–30.
57. Will-Erich Peuckert, *Sagen: Geburt und Antwort der mythischen Welt. Europäische Sagen* (Berlin, 1965), 15.
58. Merili Metsvahi, 'Muistend kui tegelikkus', in: M. Hiiemäe, J. Oras (eds), *Maa-alused*. Pro Folkloristica VII (Tartu, 2000), 44–59, 54–57.
59. Bodin, *Von ausgelasnen wütigen Teuffelsheer,* 118–29; W.M.S. Russell and Claire Russell, 'The Social Biology…', 155–156, the Russells also describe a case in which belief in werewolves manifested itself in places even as late as in 1925 – it was then that a French police officer shot a boy whom he considered to be a werewolf, 55.
60. See Merili Metsvahi, 'Werwolfprozesse in Estland und Livland in 17. Jahrhundert. Zusammenstöße zwischen der Realität von Richtern und von Bauern', in: Jürgen Beyer and Reet Hiiemäe (eds), *Folklore als Tatsachenbericht* (Tartu, 2001), 175–184; Willem de Blécourt, 'A Journey to Hell: Reconsidering the Livonian "Werewolf", *Magic, Ritual, and Witchcraft,* 2 (2007), 49–67.
61. Wolfgang Behringer, *Chonrad Stoeckhlin und die Nachtschar. Eine Geschichte aus frühen Neuzeit* (München, 1994), 160.
62. Behringer, *Chonrad Stoeckhlin…*, 168.
63. On this excellent informant see in more detail Merili Metsvahi, *Indiviid, mälu ja loovus: Ksenia Müürsepa mõttemaailm folkloristi pilgu läbi*. Dissertationes folkloristicae Universitatis Tartuensis 10 (Tartu, 2007).
64. Antti Aarne, *Estnische Märchen- und Sagenvarianten*. FFC 25 (Hamina, 1919), 131, no. 74. It is a tale found in several places in Europe about a werewolf who has been transformed into a wolf against his will and becomes human again when it receives

some bread on a knife handed to it by a man sitting by a campfire. The man who gave the wolf bread goes into town and sees his knife in a shop window, enters the shop and learns that it was the merchant who was the wolf. The merchant thanks the man and gives him a reward (e.g., allows him to have free goods from his store throughout his life). On this tale, see also Felix Oinas, 'Kitzbergi draama 'Libahunt' ja selle rahvaluuleline tagapõhi, Kalevipoeg kütkeis ja muid esseid rahvaluulest, mütoloogiast ja kirjandusest' (Toronto, 1979), 77–89, 85.

65. The idea of the image of the werewolf being borrowed from Germans was expressed already as early as in the first quarter of the nineteenth century – see Vähi, 'Kirjalikud teated…', 258. Loorits claimed that the earliest layer in the Estonian belief in werewolves could date back to the Viking era, but he was certain that werewolf belief flourished in Estonia in the sixteenth and seventeenth centuries. Loorits has also suggested that the idea of transformation generally appears to be of secondary importance in Estonian folk belief and the only mythological being representing it in a somewhat more advanced form is the werewolf: Loorits, *Grundzüge des estnischen Volksglaubens* I, 306, 311.
66. See Merili Metsvahi, 'Muundumised libahundimuistendeis', in: Mall Hiiemäe and others (eds), *Lohetapja*. Pro Folkloristica VI (Tartu, 1999), 111–122.
67. *Kalevala. Soome rahva eepos* (Tartu, 1939), 36–38. See also Antti Aarne, *Das Lied vom Angeln der Jungfrau Vellamos: eine vergleichende Untersuchung*. FFC 48 (Helsinki, 1923), 3.
68. Rudolph Leubuscher, *Ueber die Wehrwölfe und Thierverwandlungen im Mittelalter. Ein Beitrag zur Geschichte der Psychologie* (Halle, 1850), 35; Norbertas Velius, *Lithuanian Mythological Tales* (Vilnius, 1998), 234 (The Killed Wolf-Man).
69. EFA, E, StK 22, 107 (6) < Setu – H. Jänes 1924.
70. EFA, H II 60, 721 (1) < Vastseliina < Setu – J. Sandra 1897.
71. *Maarahwa Kalender ehk Täht-ramat 1826 aasta peäle, pärrast Jesuse Kristuse sündimist* (Tartu, 1826), 23.
72. In 1822 O.W. Masing retells a tale of people in Tartumaa not daring to shoot two wolves who had come to the village, taking them for a couple of werewolves; Tiina Vähi, 'Kirjalikud teated…', 257; at the EFA a story can be found about there having been a law in Saaremaa according to which anyone who could shoot 100 wolves would have been given a pension to the end of his life; a man had killed 99 wolves but as the last one he shot was a human wolf, he did not gain the award, EFA, EKmS 4°5, 388/9 < Karja khk – W. Mägi, 1889.
73. The generalisations are based on nineteenth-twentieth-century materials from the EFA, Rootsi, *Tuli susi soovikustu*, 432–434.
74. Rootsi, *Tuli susi soovikusta*, 435.
75. C. Körber, *Lomisse ramat* (Tartu, 1850), 39–40.
76. Rootsi, *Tuli susi soovikusta* (Tartu, 2005), 431.
77. Examples of this, as well as other folk tales in which the wolf is a positive character can be found in, e.g., Oskar Loorits, *Endis-Eesti elu-olu II. Lugemispalu metsaelust ja jahindusest*. ERA Toimetused 21 (15) (Tartu, 2004 (1st ed. 1941)), 305–313 (the electronic version can be found at http://www.folklore.ee/rl/pubte/ee/eluolu/elu2/198.html).
78. This is certainly connected with the matriliear social system that was prevailing in Estonia at the end of the first millennium and probably also at the beginning of the second millennium: M. Mägi, 'Abielu, kristianiseerimine ja akulturatsioon. Perekondliku korralduse varasemast ajaloost Eestis', *Ariadne Lõng: nais- ja meesuuringute ajakiri*. 1–2 (2009), 76–101. See also Merili Metsvahi, 'Estonian Werewolf

Legends Collected from the Island of Saaremaa' in: Hannah Priest (ed.), *She-Wolf: A Cultural History of Female Werewolves* (Manchester, 2015), 24–40.

79. As regards catalogues of folk tales a description of this type can be found only in the catalogue of Livonian folk tales, Oskar Loorits, *Livische Märchen- und Sagenvarianten*. FFC 66 (Helsinki, 1926), no. 166. On this legend type more can be read in Merili Metsvahi, 'Neiu hundiks. Libahundimuistendi „Teine puhkajatest murrab looma' analüüs', *Ariadne Lõng. Nais- ja meesuuringute ajakiri* 1/2 (2001), 39–51.
80. Jean Baptiste Holzmayer, *Osiliana. Erinnerungen aus dem heidnischen Göttercultus und die Gebräuche verschiedener Art, gesammelt unter den Insel-Esten* (Tartu, 1873); see also Metsvahi, 'Estonian Werewolf Legends Collected from the Island of Saaremaa'.
81. EFA, H III 25, 261/3 (9) < Valjala khk., Lõune v. – J. Evert < K. Nuut 1895.
82. Brüder Grimm, *Deutsche Sagen*, Bd. 1 (München, 1993), No 213. In other German versions of the same tale the characters are also male, mostly farmhands, see, e.g., P. Zaunert (ed.), *Westfälische Sagen* (Jena, 1927), 259–260.
83. According to Thies and the other accused in Latvian witch trials, Hell was located somewhere far away, in a swampy region or a swamp, across the sea or underground, von Bruiningk, 'Der Werwolf in Livland…', 204.
84. The data from witch trials that took place on the territory of today's Latvia in 1608, 1637 and 1647, K. Straubergs, 'Zur Jenseitstopographie', *ARV. Journal of Scandinavian Folklore* 13 (1957), 56–110, 87–88. Estonian werewolf legends do not include the motifs of the souls of dead animals being revived in Hell or there being an enmity between werewolves and witches.
85. von Bruiningk, 'Der Werwolf in Livland…', 205–207.
86. Antti Aarne, *Estnische Märchen- und Sagenvarianten* (Hamina, 1918), 131–132, no. 75.
87. See Metsvahi, 'Muundumised libahundimuistendeis', 120. The most widespread way of transforming into a wolf involved magic movements (first and foremost rolling about on the ground, somersaults, cartwheels, walking around a stone, a tree or a bush – 43 of the 63 versions of the given tale include different magic movements).
88. This is the most widespread motif in Estonian werewolf legends and belief reports concerning werewolves. In his testimony, Thies claimed that when werewolves had eaten meat and were full they transformed into people again without having to use *bread* (my emphasis, M.M.), von Bruiningk, 'Der Werwolf in Livland…', 205.
89. Killing animals is the most typical activity of werewolves on the basis of the witch trials as well as the EFA materials.
90. EFA, H, R 2, 10 (52) < Põlva – J. Hurt, 1866.
91. EFA, H, Jõgever 1, 31/2 (105) < Kaarma – J. Jõgever, 1888.
92. See, e.g., F.J. Wiedemann, *Estnisch-Deutsches Wörterbuch* (St. Peterburg, 1893), 11.
93. Von Bruiningk, 'Der Werwolf in Livland…', 205.
94. Few written sources have been preserved from earlier times. One of the few reports that has been associated with shamanism derives from the Chronicle of Novgorod (1071) and speaks of a man of Novgorod who came to the land of the Chuds (= Estonia) to receive divination from a wise man. The body of the wise man was lying rigidly, while devils were at work at it. Vt. E. Laugaste, *Eesti rahvaluuleteaduse ajalugu…*, 11–14.

95. The dating is based on the quotation from S. Münster introduced at the beginning of the chapter according to which the accused admitted being guilty of lycanthropy also in sixteenth-century trials.
96. The werewolf as a character does not occur in older genres, such as the *regi* song.
97. Donecker describes an interesting occurrence in an inn in Curonia in 1637 described by Christian Kortholt as an experience of an acquaintance of his. A German travelling in Curonia went into an inn in Dublen together with local Germans. There were Latvian peasants sitting nearby and one of them came up to their table with a friendly face, raised its stein and said in the way of a toast, 'To you, sir, as it is to me'. The travelling German did not know any Latvian, yet was about to say something similar in return. The local Germans jumped up and told him to be silent. They beat up the Latvian peasant and threw him out. When the surprised traveller asked why they had treated the friendly man like this, they answered that had he blessed the peasant's drink, he would have been forced to run around as a werewolf at night, Donecker, 'The werewolves…', 305; on the basis of the experience of an anonymous traveller visiting Curonia in 1719 Johann Kanold writes: '(…) many peasants liked to boast that they were werewolves, knowing that such tall tales would strike other villagers with terror and wonder', Donecker, 'The werewolves…', 311.
98. It is worth mentioning at this point that the Estonian words for the werewolf *libahunt* and *soend* were not used as yet during the witch trials; also, a great number of different words denoting werewolves can be found in the EFA texts, which is partly explained by different Estonian dialects having different words for the concept. Very often simply *hun*t ('wolf') is used where there would be 'wer(e)wolf' in German or English. Von Schlegel, writing in German, only uses the word *Wolf.* Also Knüpffer who recorded the earliest werewolf fairy-tale in Estonia in the first quarter of the nineteenth century only uses *Wolf/hunt.* Loorits has noted that in the tale type AT 409 the concept of a werewolf does not appear, but the woman is simply turned into a wolf. Loorits, *Grundzüge des estnischen Volksglaubens* I, 312. Mere *hunt* appearing instead of the concept of *libahunt* often occurs also in case of other plots.
99. Even today, *noid*, the Estonian word for 'witch', has more positive connotations than the English 'witch' or the German 'Hexe', bearing also the meaning of a healer.
100. Rootsi, *Tuli susi soovikusta*, 102–227.

10
The Werewolf in Nineteenth-Century Denmark

Michèle Simonsen

In 1798 the Danish priest Joachim Junge, in an ethnographic description of his parishioners' ways of life, gave a rational explanation to the extraordinary accounts of 'goat-feeted' and 'one eyed' people who were supposed to live near the polar circle. The goat-feeted people may have been so named because they climbed up mountains with a goat's speed, probably with the help of skis of a sort; whereas the 'one-eyed' probably wore a kind of travelling cloak with only a narrow slit for the eyes; and the 'wolfsmen' could be just Nordic people entirely dressed in animal skins as protection against the polar winter. In the same way, he explains the fact that pregnant women, afraid of werewolves, always take a young boy with them if they have to go out after dusk, by the superstition's positive function: it prevented them from falling down, hampered as they were in their walk when heavily pregnant.[1] To discover a more satisfying reason, and one that related more to the people who said those kinds of things, as to why pregnant women had to be careful about werewolves, we will have to look at the legends of about a century later.[2]

Werewolves occur mostly in two more or less overlapping kinds of narrative: legends, which focus on the story, and memorates which deal with pieces of reminiscence. In the Northern parts of Sweden, Norway and Finland, werebears often replace werewolves, since bears were a frequent plague there (both wolves and bears disappeared from Sweden around 1860). In *Sundfjord* it was said that when there was an abundance of wolves, it was an omen that there would soon be a war; while in *Hardaland*, that werewolves were especially numerous in time of war. Swedes believed that during the war against Russia in 1808–1809, Russians had deliberately turned Swedish prisoners into werewolves and turned them loose to plunder the Swedish countryside.[3]

Shape-shifting can happen either by an act of will or by compulsion, but there are clear regional variations. In the northern part of the Nordic countries, as well as in the Baltics, humans can turn into werewolves by putting on a magic belt, sometimes made of wolfskin. Some narratives from

Figure 10.1 Evald Tang Kristensen, the shoulder bag contained his notation utensils. Photo: Mart. Christensen, with permission of the Danish Folklore Archives (*Dansk Folkemindesamling*), Copenhagen.

Norway claim that one could leave such a belt on the entrance of a house, and whoever stepped over it was in turn transformed into a werewolf. It is worth noting that, in accounts from the north of Scandinavia werewolves could decimate domestic animals but very rarely attacked human beings.

Werewolves in legend

During most of the nineteenth century hardly any werewolf legends were recorded in Denmark. This changed with the activities of the schoolteacher and folklorist Evald Tang Kristensen (Figure 10.1).[4]

Some of the Danish legends and memorates about werewolves have a different content. There humans also turn into werewolves deliberately, with the help of a magic belt, usually in order to indulge in eating raw meat, or for revenge. The following narrative emphasises the eating:[5]

> There was a man and his servant, who went on the heath and collected turf…The man says to the servant: 'let us have a good sleep today, turfdiggers deserve it'. This man had a quarrel with his neighbour. This neighbour had a foal, which stood next to its foal nearby. The man got up and went away, but the servant could nor see where and thought he had resumed his work. Then he sees a huge wolf come and spring on the foal…The servant grabs it and gets the wolf off the foal, but the foal was dead.…then he sees the man come from the left side of the hill. 'Where have you been?' he asks. 'I have laid down for myself, on the other side of the hill'. 'This is a lie,' says the servant. 'You are a werewolf and a devil, and you have torn the neighbour's foal apart'. According to the legend, as soon as it is made public, they can no longer shift shape.[6]

In most of Denmark, however, as well as in the south of Sweden (Scania, Blekinge and Halland, which used to be Danish provinces), we meet quite a different tradition altogether, concerning both the ways a human being turns into a werewolf and the ways he can become human again.[7] The Danish tradition is especially well documented, thanks to Tang Kristensen, certainly one of the best and most prolific folklorists of Western Europe at the end of the nineteenth and beginning of the twentieth centuries. He has published a seven-volume collection of Danish legends, partly collected by himself (about half of them), and partly by others.[8] The legends about werewolves in *Danske Sagn* cover the whole of Denmark, from Jutland to Zealand. Most narratives share a common semantic core. Here shape-shifting is an involuntary phenomenon, a curse. The werewolf is the victim of a magical ritual practiced by his mother. Girls and women who wanted to escape labour pains during childbirth could take the afterbirth from a mare – i.e., the placenta in which the foal had been wrapped – and stretch it between four poles. They would then take all their clothes off and crawl through the stretched placenta, completely naked. This secret ritual would ensure that they would give birth without pain. But the child thus born would become a 'mare' if it was a girl (a female supernatural being that rides on men's chests during their sleep and oppresses them, and which can also be found in the English word 'nightmare'), or a werewolf if it was a boy. This is the only way a man becomes a werewolf in Danish tradition. All narratives but one agree on that. The one exception states that it is a man who, for unknown reasons, wants to become a werewolf and practices the same secret ritual. There is no example of a spell being due to being the seventh child of a seventh child and no magic belt is used, as in the North of Sweden or in Germany.[9]

When it comes to the means by which a werewolf can be saved from the curse, the Danish narratives fall into two groups. In the first, the werewolf tries to attack a pregnant woman. If he succeeds in eating the unborn child or drinking his blood, he ceases to be a werewolf and becomes a normal human being.

> If a werewoolf can get hold of an uborn child's heart and eat it, he can be saved from the curse.[10]
>
> Once a farmer whose wife was pregnant was coming home after his day's work. The door was locked and he could not easily come in, but he saw that the earth was dug out under the door. When he came into the room, he saw his wife lying dead on her bed, her tummy opened. So the poor man understood what had happened and he went around the house and saw the werewolf on the manure heap eating the baby.[11]

In the second group of stories, the werewolf is saved from the curse if someone says aloud to him 'You are a werewolf!' while he is in his human shape. In most cases, this gives rise to quite an elaborate narrative. For example, a man feels that he is going to turn momentarily into a werewolf while he is travelling alone with a woman, usually his girlfriend or his wife, through a wood or a heath. He tells her that he has to leave her for a while on the call of nature. She must not be afraid if, in the meantime, a beast attacks her: all she has to do is defend herself with her apron. While the man is away, a wolf comes and attacks her – she fends it off with her apron, and the beast only runs away when the garment is completely in tatters. Shortly after her boyfriend comes back. The girl notices that he has blue and white threads between his teeth, and she understands that it was him who attacked her. When she exclaims, 'But you are a werewolf!', he thanks her. For, just by making this statement aloud, she has saved him from the curse.[12] However, according to some narratives she has to say it three times, otherwise she will become a werewolf instead of him.

These two groups of stories are quite different. In both cases the curse is broken, but in the first group salvation takes place at the cost of two murders: that of a pregnant woman and that of an unborn child. Whereas, in the second group the werewolf is saved by a word at the modest cost of a spoilt apron. And yet, these two different outcomes, the tragic one and the happy ending, are very closely related at the semantic level. Both start with a woman who wants to give birth without pain, as animals supposedly do. So she crawls naked through a mare's placenta, thus imitating the birth of an animal. This is considered by the community as an unacceptable social disorder, as can be seen from the following narrative:

> This happened in the days when there still were werewolves, and my father did not see it himself, but he has heard it from his father, who had heard it from an old man from Magleby, who had seen it with his own eyes. So this old man from Magleby he was driving towards the wood one evening, and he saw some ten young girls who had stretched out the placenta from a mare, and three of them had already crawled through it. So he came near them with his whip and he chased them away in front of him to the village and through the main street, and they were all stark naked. [13]

We might think that the forbidden ritual shows the clash between folk magic and the official Christian authorities, trying to enforce the biblical punishment: 'In sorrow thou shalt bring forth children' (*Genesis* 3:16). But I think it more likely that it is to do with the more widely spread prohibition against giving birth secretly, away from the community. Women who need most imperatively to give birth without pain are unmarried girls who hide their pregnancy, try to give birth secretly and then get rid of the unwanted child. Already in 1798 Joachim Junge claimed that many a girl who had given birth out of wedlock, and killed her new-born child, used the common superstition to blame the murder on some werewolf.[14]

As a consequence of this imbalance, she gives birth to a hybrid child, half-human and half-animal. A werewolf looks and behaves like a normal human being most of the time, but when reaching puberty, he turns now and again into a wolf and savagely attacks his fellow human beings. In order to restore this imbalance and become a full-blooded human being, the werewolf has then two possibilities: either to attack a woman's womb literally or symbolically.

In the first group of stories, the werewolf must eat the heart of an unborn child: to incorporate the child, so to speak, in order to become the human foetus which, as a result of his mother's mistake, he had never had the chance to be. To achieve this, he must destroy a mother's womb, and inflict upon it a much greater violation than the mild violence of labour pains from which his own mother escaped.

In the second group of stories, the woman's apron plays a key role. Most narratives insist on that. One narrative says: 'Do not use a fork, just use your apron'. Another states that the girl first tries to defend herself with her scarf, but this does not help. Only when she uses her apron does the wolf run away. The apron here is clearly a symbol of the womb; both a metaphoric symbol (the apron looks like a womb, can be used to wrap and carry things), and a metonymic symbol (the apron is in contact with the womb, covers the womb). In this second group of narratives, then, the werewolf does not have to destroy a woman's womb literally. Attacking its symbolic substitute, the woman's apron, is just as efficient. Here the werewolf is able to perform a symbolic action: i.e., a human action. At least, this symbolic action is efficient enough to neutralise this one sudden incident of 'werewolfishness' and save the girl. But it is not enough to deliver the werewolf permanently from the curse.

That is where the second part of the narrative comes in: the girl's cry, 'But you are a werewolf!' This suggests that the curse that dwells upon the werewolf is not just due to his dual nature, at the same time human and animal, but to the fact that his double nature is not acknowledged by his community. Admittedly, some men are suspected of being werewolves, because of their looks or their asocial behaviour: the animal is suspected in the human. But the threads of apron between the man's teeth allow the girl conversely to detect the human in the animal. 'You are a werewolf!' As we know from *Genesis*, the ability to name things and animals is a human prerogative which gives man power over them, and by acknowledging openly the werewolf's double nature, the woman neutralises his animalism.

Thus, considered as a whole, Danish narratives about werewolves make up an elaborate, close-knit and very coherent semantic system. But what do they tell us about folk beliefs?

Danish werewolves between narratives and beliefs

Scholars in the past have been quick to infer belief in werewolves from narratives about werewolves. But 'saying isn't believing', to quote the title of a book about legends from Newfoundland.[15] Ballads, like folktales, are works of fiction and are recognised as such by the narrator and his/her community. Although they often claim the ballad to be 'true', more thorough investigation has shown that what they usually mean is that the story gives a true picture of the human predicament, not that the episodes are literally true. Therefore, we cannot infer anything about folk beliefs about werewolves from either the ballads *Jomfru I Ulveham* or *Varulven*.[16]

Legends, on the other hand, have long been said by folklorists to express folk beliefs, in opposition to tales, which were told as fictitious stories. But, as most modern folklorists agree, this is a truth with many modifications. It is more accurate to say that legends are told in such a way as to be credible, because the question of their veracity is relevant; unlike a tale, which is fictive and known to be fictive. But legends use the rhetoric of truth, not to state an article of faith, but to test it. In their natural context, legends acquire meaning through the discussion from which they arise. But we have almost no contextual information and do not know the exact circumstances in which these narratives were produced. Even those collected by ETK himself have very likely arisen as a response to his questions during his many house visits, and not as part of a conversation within the community. We do not know what triggered these narratives off, nor how they were commented on.[17]

One could claim that the traditional belief in werewolves is proved by the ritual. Women and girls have been caught trying to crawl through a mare's afterbirth. Understandably enough, childbirth in the nineteenth century was extremely painful and dangerous. But strictly speaking, this only proves the belief in the efficiency of the ritual for the mother, not necessarily the belief in its consequences for the child. And anyway, I suspect that, even then the ritual's efficiency was understood as symbolic rather than pragmatic. Besides, not everyone fully trusted it, as can be seen from the following narrative:

> Three girls from Forlev had heard that if they crawled naked through a foal's skin three times, they would never get (proper) children, for they would be werewolves, and they felt like trying. They went on the heath with a mare's afterbirth. But a lad who had overheard them told about it to a couple of men, who at once chased the girls home with their whips, so the girls never found out whether this was old wives' gossip or not.[18]

Legends have an ambiguous status on the sliding scale from assertive speech act (daily conversation) to expressive speech act (artistic performance). Tales, marvellous tales in particular, are staged and told through self-conscious performances by renowned storytellers, and marked by notable features: opening and closing formulas, special mimics, use of voice, etc. The performance of tales leads to aesthetic value judgments and commentary. Legends, on the contrary, are more often than not phatic speech acts: acts of communication which have less meaning by themselves than that they are triggered off by the preceding speech act and produce a subsequent speech act (an approval, a question, a protest, etc.). In a way, the more elaborate a legend is as a narrative, the less we can infer about the 'belief' it is supposed to express, and vice versa.

Those legends reproduced in *Danske Sagn*, which were collected by Ewald Tang Kristensen himself, are written in a manner much closer to the informants' words than was customary at the time. They display much stylistic variation, which enables us to examine for most informants the interplay between collective mental representations and individual aesthetic preoccupations. The legends about werewolves range from a few phrases to one-and-a-half pages. They include 'dites', 'fabulates', and all degrees of 'memorates', to use Carl-Wilhelm von Sydow's terminology.[19]

Let us have a look at the following text:

> Werewolves used to attack pregnant women to open their womb and free themselves from the curse. For that they had to eat a male foetus. There was a farm hand, not far from here, Weaver George, he was suspected to be a werewolf.
>
> Girls can avoid child-labour pains by crawling naked through a foal's placenta.[20]

It is not very elaborate and shows no sign of aesthetic intention. It is in fact hardly coherent, and would not make sense at all if read in isolation. The informant apparently just states at random all that comes to his mind when asked about werewolves. There is no apparent connection between the first and last sentences, between the existence and actions of werewolves and the ritual practised by pregnant girls. We only understand the logical connection because we know all the other werewolf stories in ETK's collection; and I assume the informant understands it because it is common knowledge in his community. Hence, the logical connection between both statements need not be formulated. In some way, this elliptic text may be closer to the beliefs of the informant than more elaborate and polished texts. Maybe most legends have never been told from beginning to end as a full-blooded narrative. Maybe they live mostly an underground life, as a pool of motifs and collective mental representations and a simple allusion in the conversation suffices to make them efficient.

At the other end of the scale we have the following narrative: told by Jens Kristensen, a clog-maker from Himmerland whom Tang Kristensen regarded as one of his best informants.[21]

> My father told me that once there were two women who had taken their clothes off and who had crawled through the placenta of a foal which had been stretched between four pales. They were pregnant, and they were naked, so that their children became werewolves. But most of the time they looked like human beings. When they had their wolf shape, they would attack the first being they met on their way, whoever that was, and tear him to pieces. A farmhand from hereabouts had gone on the heath to collect some heather together with a maid, and she was his sweetheart. Once they came on the heath, the lad got off the cart and said to his girlfriend: 'If a wolf comes here, I beg you, don't hit it with your fork!' – 'What shall I defend myself with, then?' – Well, she could untie her apron and hit it with it. He had hardly gone away before a wolf turned up and tried to attack her. But she took hold of her apron and hit it with it. The wolf set its teeth in the apron and tore it into a thousand pieces, then went away. Shortly after the lad came back. 'Why have you

been away so long? A wolf came and tried to tear me to pieces!' The young man started laughing. The girl's apron was blue with white stripes. As he was laughing, the girl saw that he had white and blue threads between his teeth. 'So it was you who attacked me! You still have threads between your teeth!' And since that day, he stopped being a werewolf, because his secret had been revealed.[22]

This is an elegant narrative, which alternates skilfully between direct dialogue and third person narration, between description and commentary. Except for the initial 'My father told me that…', it is told as a fabulate, with none of the rhetorical tools used by memorates to enhance credibility. It does not seem much concerned with convincing the audience of its 'truth': 'My father told me' may well be a playful manipulation with the tools of memorates. This narrative bears the mark of a well-polished performance, and I assume that, in a social context, it would evoke appreciation and pleasure rather than discussions about its content; it might also lead to some other legend well told by another gifted storyteller.

Rather than a cognitive statement, belief is an emotional experience. Belief is not assertive, but performative. It does not state a general law, like a scientific law or a religious act of faith, but it influences our actions and our discourse. And it does so in a concrete, particular situation. With all these Nordic legends about werewolves, what matters is not so much whether informants and their community 'believed' in werewolves – in their existence, in their origin, or in their peculiar road to salvation. What matters is that they felt the need to talk about them. They were concerned about the necessity of painful childbirth, about the animalism of mankind, about the power of human speech and of symbolic actions, and they needed to weave around these concerns all sorts of meaningful narratives.

Notes

1. Joachim Junge, *Den Nordsiellandske landalmues Charakter, Skikke, Meeninger og Sprog* [The Character, Customs, Opinions and Language of the Peasants of Northern Zeland] (Copenhagen, 1798) (ed.) Ellekilde (1915), 234. [233–236 varulve].
2. Earlier on this topic: Michèle Simonsen, 'La variabilité dans les légendes: les récits danois sur les loups-garous', in: Veronika Görög-Karady (ed.), *D'un conte à l'autre – la variabilité dans la littérature orale* (Paris, 1984), 181–186. The present chapter is a reworked version of a paper given at the ISFNR conference in Athens in 2010, see also: Michèle Simonsen, 'Danish Werewolves Between Beliefs and Narratives', *Fabula* 51 (2010), 225–34.
3. Cf. Ella Odstedt, *Varulven i svensk folktradition* (Uppsala, 1943).
4. About him: Bengt Holbek, *Interpretation of Fairy Tales. Danish Folklore in a European Perspective* (Helsinki 1987), 69–87; Timothy Tangherlini, *Danish Folktales, Legends, and Other Stories* (Seattle and London 2013), 36–39 and more extensively in the Danish Folklore Nexus.

5. This concerns one of the most popular stories in Germany, The Hungry Farmhand, see: Willem de Blécourt, '"I Would Have Eaten You Too": Werewolf Legends in the Flemish, Dutch and German Area', *Folklore* 118 (2007), 23–43.
6. ETK, *Danske Sagn* (Arhus 1893), no. 32, told by Niels Kristian (Pilmanden) Jensen in Fredbjærg. Cf. no. 15, from Hårup. See also: Bengt Holbek and Iørn Piø, *Fabeldyr og sagnfolk* (Copenhagen 1967), 165–176.
7. An overview in: Bengt af Klintberg, *The Types of the Swedish Folk Legend* (Helsinki, 2010), 306–310.
8. *Danske Sagn som har lydt i folkemund. Udelukkende efter utrykte kilder* (*Danish Legends as told by the folk, exclusively from unpublished sources*) (Arhus, 1892–1901). The 1980 edition (Copenhagen) has an additional index volume.
9. Cf. the references under 'Gallus' and 'Geburt' in the *Handwörterbuch des deutschen Aberglaubens* III (Berlin, 1931), cols 281, 409, 411.
10. Sent by E.F. Madsen, Magleby, Zealand, *Danske Sagn* II (1893), no. 33, 34.
11. Sent by teacher Anton Nielsen, *Danske Sagn* II (1893), no. 35.
12. An example of this legend: *Danske Sagn* no. 18 from Testrup, see also the story by 'Bitte Jens' Kristensen, below. The type is particularly popular in Norway, see: Reidar Thoralf Christiansen, *The Migratory Legends: A Proposed List of Types with a Systematic Catalgue of the Norwegian Variants* (Helsinki, 1958), no. 4005.
13. E. Trier, Magleby, Lendemark. Tang Kristensen, *Danske Sagn som har lydt i Folkemunde*, II (Copenhagen, 1928), 152.
14. *Den Nordsiellandske landalmues*, 236.
15. Gary Reginald Butler, *Saying isn't Believing: Conversational Narrative and the Discourse of Tradition in a French-Newfoundland Community* (St. John's, 1990).
16. Svend Grundtvig,, Axel Olrik, H. Grüner-Nielsen, Erik Dal et al. *Danmarks gamle Folkeviser*, 12 vol. (Copenhagen, 1853–1976), types DGF 54 and 55; *DgF* IV, 890–1, version E, sung by Jens Isakson, Viborg, collected by Evald Tang Kristensen.
17. Tang Kristensen left fieldwork notes, including information about most of his informants in his four-volume autobiographical *Minder og Oplevelser* (Viborg, 1923–1928). See also: Timothy Tangherlini, *Interpreting Legend: Danish Storytellers and their Repertoires* (New York, 1994).
18. Christian Rasmussen, teacher; Evald Tang Kristensen, *Danske sagn* II, no. 13.
19. Carl-Wilhelm von Sydow, *Selected Papers on Folklore* (Copenhagen, 1948), 73–76, 106–126.
20. Told by Jørgen Jørgensen, coachman, Flintinge, *Danske sagn* II, 1928, 148.
21. About the narrator: Tangherlini, *Danish Folktales* (as in note 4), 64–91.
22. Told by Jens Kristensen, from Ersted, Tang Kristensen, *Danske sagn* II, no. 14; cf. the translation in Tangherlini, *Danish Folktales*, 78–79.

11

Dead Bodies and Transformations: Werewolves in Some South Slavic Folk Traditions

Maja Pasarić

South Slavic folklore and ethnographic sources from the end of the nineteenth and the beginning of the twentieth centuries abound with folk narratives about various supernatural beings, thus providing a fruitful ground for a range of mythological, folklorist and ethnological approaches to the subject. Certain aspects of the above-mentioned cultural forms alongside folk customs can be linked with traditional culture's conceptions about death and afterlife, the transformational potential of the human body and animal symbolism. In such terms, accounts about werewolves are no exception – even more so, they seem to provide to the mentioned inquiries quite well.

Narratives about werewolves are well known and widely spread in South Slavic folk cultures. However, they differ from the same cultural practices from other European folk traditions homogenised around the lycanthropic character of a living man, as they mostly stand for notions about dead people coming back from their graves while taking on various physical features, very often those of an animal. I will begin by briefly sketching the phenomena of werewolves and their focal traits followed by glances into the bodily characteristics of werewolves, the transformational potential of the human body and the practices directed towards the dead human body.

The desired task will be carried out with the help of some South Slavic ethnographic and folklore material, such as oral narratives and customs. The focus will be on Croatian, Serbian, Bosnian and Montenegrin sources, the majority of which date from the end of the nineteenth and the beginning of the twentieth centuries, when such folklore and ethnographic data were extensively gathered.

The mentioned data about werewolves from some South Slavic folk traditions will prove to be valuable grounds for looking into different notions about the boundaries of human bodies and identities regarding bodies' ability to change after death, the otherness of the transformed body and the practices that are thought to prevent such changes. These questions also touch on concepts of cultural and social norms, the liminal potential of death and traditional cultures' notions about the borders of the human world and the ways they can be transgressed.

Werewolves and vampires – a syncretism of terms

Records about werewolves in South Slavic folklore and ethnographic sources are characterised by a slight amount of dispersion. The confusion often arises together with a wide variety of names used to describe the identical being, whether in one and the same or in different South Slavic areas, as well as with the interlacement of werewolves with other supernatural beings. Foremost, looking at the ethnographic and folklore sources at hand, it is interesting to notice how the word werewolf, *vukodlak* and its diverse South and other Slavic versions (for example, *vukozlak, volkolak, vovkulak, vlkodlak,* etc.) is considered to derive from the Slavic *vьlk, 'wolf' and the South Slavic *dlaka*, 'hair, fleece', and have first and foremost marked a living person that can occasionally turn into a wolf, in whose shape he or she attacks people and animals,[1] assimilated a different meaning which can be recognised in the nineteenth and mid-twentieth century in South Slavic folk traditions where it essentially refers to the so-called walking dead. Scarce relics of the beliefs about a living person that can turn into a wolf and back, whether marked with the term werewolf or not, are, however, still observable in some South Slavic ethnographic and folklore data.[2]

While narratives about the lycanthropic character of a human being can be traced, in their written form, back to antiquity,[3] and were further elaborated within the rich but complex Middle Age mythology fused with different cultural influences, it has been suggested that, somewhere in the late Middle Ages, in certain parts of Europe, and especially amongst the Slavic people, the werewolf lost his wolf-like characteristics. Bloodthirstiness and the desire for evil endeavour remained, though, and blended with another supernatural being, the vampire.[4]

Various data testify to the complex and not always thoroughly clear relations between werewolves and vampires. As brought forward by Irena Benyovsky records of South Slavic beliefs from the sixteenth and the first half of the seventeenth centuries speak about the *souls* who died under the curse of a werewolf and were therefore destined to return to the world of the living as vampires.[5] Records of Serbian beliefs dating from the first half of the twentieth century from Southeast Montenegro and Central-West Serbia note that every vampire had to take up the shape of a wolf for a certain amount of time.[6] Finally, the syncretism of terms used to define vampires and werewolves can be found in reference works such as the 1970s *Yugoslav Academy Dictionary*, in which the word 'vampire' is linked with 'werewolf', the latter being defined as a person that returns after death.[7]

For several centuries already, the term werewolf has been mostly used to describe the *demonic dead,* dead people that return from their graves. Not only did the werewolf and vampire merge together in meaning across the South Slavic territory, but ethnological records from the first half of the twentieth century suggest that these words can be treated as synonyms.[8] This notion has, thus far, been generally accepted and is referred to in recent folklorist, ethnological and cultural anthropological scholarly works. Even most recent fieldwork data from the Dalmatian hinterland in Croatia, collected by Luka Šešo, show that, among the different variants of the term werewolf, the term vampire turns up in one village although it is noted by the informant that the expression was used only by older people who are already dead and that the term vampire is no longer in use.[9] According to available sources, it seems that the word werewolf more often or even primarily appears in Croatian ethnographical and folklore data, while the word vampire is more widely used in the areas east of Croatia, especially among the Serbs.[10] For the purpose of this contribution these two words will be treated as synonyms.

Those who have previously studied the reason why these two expressions amalgamated in meaning did not reach a final conclusion. According to some, the belief in vampires developed from narratives about werewolves,

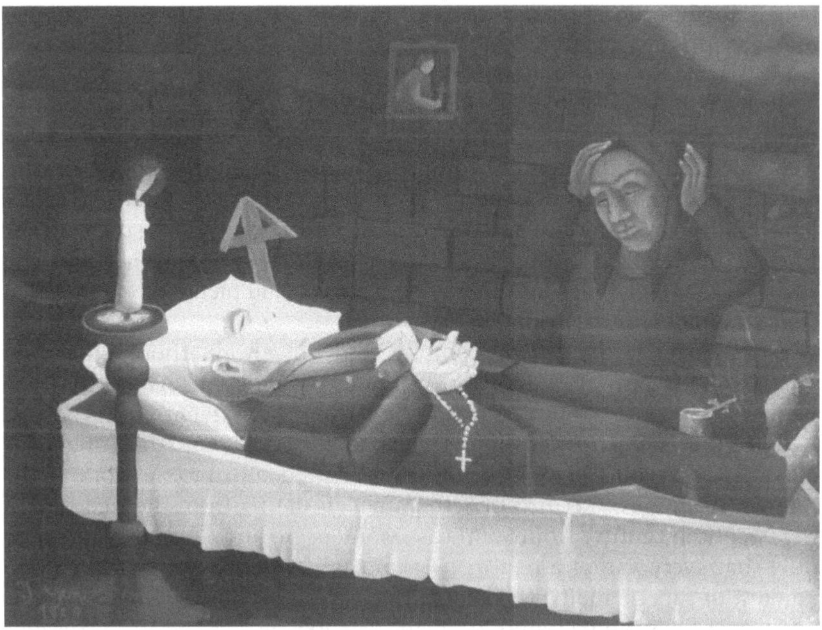

Figure 11.1 *Mourning* (1958), oil on glass, by Ivan Generalić (1914–1992), reproduced with permission of his son Goran Generalić; www.generalic.com

while others doubted that the characteristics of a vampire could arise from those of a primordial werewolf.[11] The first seems to draw near to Gábor Klaniczay's opinion, who considered that the vampire incorporates different elements of various beings diverse in origin: such as, the spirits of the dead, nightly mountain spirits of the Germanic world, Ancient *strix* and Slavic *moras* and werewolves.[12] Some authors even supposed that the term vampire replaced the term werewolf as a taboo pronoun in an attempt to bewilder evil forces,[13] which would imply that these two words had the exact same meaning. Although it is thought that, together with the newly acquired meaning, the werewolf lost some of its previous animal features it had not lost the possibility to transform its bodily appearance. One of the werewolf's main features as a supernatural being is the ability to shape-shift. This led some to search for an explanation of the werewolf–vampire syncretism in their simultaneous links with conceptions about witches who also display shape-shifting characteristics. The close link between werewolves and witches has further been attested to in some South Slavic notions about male witches becoming werewolves after death. In areas such as Istria in Croatia, the relationship between these two beings becomes even more complex since notions about werewolves and male witches even amalgamated together under the same name, *kudlak*, used for defining them both. More details about the mentioned issues were investigated in the works of Maja Bošković-Stulli and Luka Šešo.[14]

To a lesser extent, South Slavic werewolves found themselves as the themes of other inquiries as well. Devoting his interest to beliefs about supernatural beings in the Dalmatian hinterland in Croatia, Šešo included beliefs about werewolves in his research, centring his work on new fieldwork data and the contemporary population of the area.[15] Historical approaches to vampires or werewolves in Dalmatia based on eighteenth-century documents have been proposed by Benyovsky and Lovorka Čoralić et al.[16] Relations between the wolf, werewolf and vampire have been discussed by Pieter Plas as part of a wider attempt to re-examine the thanatological meaning of the wolf in Western South Slavic traditional culture.[17] Although dating back to the mid-twentieth century, broader insight into South Slavic notions about vampires offered by Tihomir Đorđević is still taken as a relevant source,[18] while more recent investigations devoted to vampires in South Slavic beliefs, particularly in Serbian myth and literature, have been completed by Dagmar Burkhart and Ana Radin.[19]

If one was briefly to sketch the most common traits of werewolves in South Slavic folk traditions, noting that a person primarily becomes a werewolf after death, although in some cases a werewolf's identity is acknowledged through birth (by being born in a caul, for example) might serve as a start. Werewolves' time of action is linked with the nocturnal chronotope while some accounts relate their extensive appearance to wintertime as well. The malevolent actions of werewolves in the neighbourhoods they return to are comprised of making noises, frightening, scaring and attacking people or animals, sucking their blood on some occasions, and even killing them or snitching them to graves. Sometimes werewolves simply return to their families and continue the relationships with their wives. The reasons for their reappearance are commonly unsettled accounts with the members of their family or the wider community. They ask for redemption and the community is usually requested to provide for their deliverance; after which they stop appearing. Nevertheless, different magical and/or religious practices have been extensively used for protection against werewolves and as prevention from becoming one; the final means of fighting a werewolf was by stabbing his heart or piercing his body, most often with a hawthorn stick. Narratives and cultural practices that evolve around werewolves in the South Slavic folk tradition foster investigations of a plenitude of questions. However, this contribution will focus on notions about the transformational potential of the human body and changes an individual can undergo after death in the process of becoming a werewolf, as well as the specificities of physical appearance.

Dead bodies living on

Since beliefs about werewolves are also beliefs about dead people getting out of their graves, dead bodies living on, they prompt different questions about corporeality and the physical form. Some of them are related to notions about the body image of werewolves, while others refer to dead human bodies which manifest the possibility of the specific transformation. Werewolves are bodily beings and are usually imagined in different physical forms.[20] It is believed that the body of the deceased who is prone to turning into a werewolf does not decompose after death but only transforms in a specific way. The physical existence of werewolves in South Slavic folk traditions is also related to notions about the soul which, for specific reasons, does not leave the body after death but stays with it for a certain amount of time and revives it into different bodily shapes.[21] According to Dušan Bandić, beliefs about werewolves (or vampires) in traditional Serbian culture are closely linked with beliefs about the bodies of dead people that continue to live on, although in a slightly different form, and actually stand for relics of an animatistic worldview.[22]

The human body itself is known to be imbued with different cultural concepts and ideas. As Åsa Larsson points out, 'social sciences and humanities have long realised that the living body is used and manipulated, moulded and embellished in ways that both communicate and create social identities, and make them appear biologically given'.[23] Since the way bodies are perceived and imagined differs across cultures and their inner socialising processes, it has been suggested that, even though we acknowledge the materiality of the body, we have no other way of reading it but according to different interpretative schemes. For example, Nicole Sault notes that, despite its physical existence, our perception of the body is a culturally constructed process: the only way we can perceive it is in terms of a body image.[24] Laura Stark outlines the term body image as a theoretical conceptualisation of the way in which persons conceive their corporeal presence particularly in regard to the presence of other persons or personalised beings.[25]

However, bodies also change through time, together with the identities inscribed in them, and new social practices are brought forward in the face of death whose liminal characteristics have been widely acknowledged. Some refer to the liminal character of the dead body as a forceful materialisation of the crisis of death; others highlight the fact that the liminality of death experience brings us the knowledge that the human world and the meanings attributed to it, as well as our bodies, are not as solid as they appear.[26] Having death as the final *rite de passage*, a transgression that ultimately affects the human body as its recognisable trait, the etiology of werewolves in the South Slavic folk tradition points to questions about change and transformations of body matters. Interesting notions about changes that the body undergoes over time and the role of the corporeality of the body in affecting identity change or mutation which are closely related to the liminal potential of death have been brought forward by Dušan Borić.

Borić highlighted the significance of focusing on the Otherness of the body matter itself that regularly destabilises the form and of acknowledging and exploring the intercession between the material resistance of the body and the process of subjectification – ways of sustaining and/or inventing stable images of the self that are always already socially mediated.[27] The author stresses the fact that in such a way one deals not only with the questions of social construction and individual subjectification in regard to the body, previously much discussed in the domain of body theory, but also with the corporeal Otherness of the body that triggers the change of form.[28]

To explore the concept of body change Borić addresses the issue of body metamorphosis. According to him, by focusing on the processes of body metamorphosis one achieves a necessary balance in the treatment of exteriority, surface, body aesthetic and culturally-imposed norms shared by a collective agency on the one hand, and individual embodiments and the biographical significance of the individual lives on the other.[29] The materiality with which this author is concerned, is 'the body's radical alterity which requires constant attempts at social taming, control and mastery in order to overcome the threatening movement of matter'.[30]

Aside from the notion that the human body can potentially change its form through death and continue to exist afterwards, narratives and other practices evolving around werewolves in South Slavic folk cultures point to different ways of dealing with the dead human body in order to master or prevent the possible change of the deceased. Such practices are particularly interesting if we acknowledge that the way communities deal with the physical remains of people who were until recently part of their daily life, and the practices they implement upon them, highlight deep convictions and ideological norms of what it is to be human.[31]

Werewolves and their physical characteristics

As already noted, a werewolf is usually imagined in different physical forms. When not appearing in the shape of an animal, werewolves commonly demonstrate themselves as human beings, or in a form that resembles the human body but is clearly marked as different. This is not surprising if we perceive the body as the most common denominator of human experience – our understandings of the world, ourselves and others are framed through the concepts of our body.[32] Therefore, it is interesting to see how the body image of werewolves as threatening and malevolent beings is constructed.

Although more often than not people become werewolves through death as an act of ultimate transgression, some accounts, however, speak of werewolves actually being born and recognised by the society through their specific bodily characteristics. Several narratives state that children born from a relationship with a werewolf, being werewolves themselves, have big, red, pointed teeth, or refer to their bodies being 'wet and slimy as an excretion'.[33] The same children cause their mothers breasts bleeding in attempts of breastfeeding.[34] Leaking or bleeding bodies could, by following Julia Kristeva's notions, be seen as abject or intolerable, opposed to desirable and wholy integrated bodies. These bleeding bodies are situated outside the boundaries of social propriety, suggesting their ambiguous limits and ultimately their death.[35] Werewolves are also mentioned as being born in a membrane or caul.[36] The notion of being born in a caul, which itself is considered to hold supernatural powers, blurs werewolves with other supernatural beings, this time with witches and moras.

The appearance of werewolves that get out of their graves and return to their communities in the shape of human beings, is marked by certain recognisable features. They are described as walking around dressed all in white, barefoot, being wrapped in blankets or carrying a shroud over the shoulder.[37] However, more often their physical shape differs from the usual human form in some particular traits. Their bodies are blown up bodies, made from skin filled only by blood with no flesh and bones. A story that illustrates this more vividly, while also linking animals to the genealogy of werewolves, has been written down in the village of Poljice in Croatia, describing a white ox and a black pig that visit the graves at midnight. The ox uses its horns to move the tombstones and the pig sniffs the earth in order to find out whether the deceased is a potential werewolf. When they come across such a deceased, they pick him up from the grave, shake away his bones, blow up his body and turn him into a werewolf.[38]

Such narratives offer an image of a body that looks like a bellows and does not contain other bodily structures except blood. According to some accounts, if joints and blood vessels would possibly start to grow inside the bellows they would make the werewolf look like a living human being and at the same time grant him eternal life and resistance to death.[39] As has been mentioned, bones are usually thought to be absent from werewolves' bodies, but in some narratives they are the only things their bodies are made of, while meat and even the skin are absent.[40] Compared to a human's, the blood of a werewolf is considered to be thick, dark or even black.[41] The eyes can be large and bloody and so can its teeth, which may also be black and sooty, or shiny and made from iron.[42] In search of other distinctive werewolves' bodily characteristics one finds that their bodies can be extremely tall or very cold, their faces red and they might even have a rotten head.[43]

Blood, bones and flesh, as integral parts of the physical structure of a human body, are some of the features through which the body image of werewolves is expressed. The bodily characteristics of werewolves also highlight the difference between the corporeal reality of these supernatural beings and living humans. With the exception of being born in a caul, it is usually the lack of one or more bodily substances that marks the dissimilarity. As stated in one of the accounts, growth of certain vital body parts would almost make them human. Other cases testify to different structures of the bodily constituents, not to mention the fact that a werewolf's body is sometimes comprised of non-bodily substances, such as iron. Another insight into the werewolves' body image is offered through the open boundaries of their bodies as the excretion appears on the surface of a child werewolf's body. Also, because of his pointed teeth, the child's interaction with his mother during breastfeeding causes the opening of her body boundaries to a greater extent than usual, as her breast starts to bleed. Bloody eyes and teeth might as well point to the internal substances which attempt to transgress the body's boundaries. Together with the images of rotten flesh, or a body comprised only of bones, the above-mentioned traits speak about the abject characteristics, *sensu* Kristeva, of the werewolves' bodies as well. One interesting account that links werewolves with the transgression of body boundaries is noted in Crnorečki region in Serbia, where it was believed that a sick person who developed wounds in the lower back area, or around the hips due to lying down for a long time, will become a werewolf after death.[44] All discussed features make werewolves distinguishable from living humans through their constructed body image. This way their otherness is also constructed. However, the otherness of a werewolf's body is the feared and undesired otherness that almost any human being can potentially experience after death.

Theriomorphic traits of werewolves

Another interesting aspect of werewolves' bodily characteristics is their ability to appear in the form of an animal. Accounts that describe such manifestations speak of a wide range of animals whose shape werewolves can take, such as a hen, a goose, a pig, a goat, a sheep, a horse, a donkey, an ox, a bull, a cat, a dog, a frog and others.[45] It is interesting that the vast majority of animals whose shape a werewolf appropriates, are domestic animals that live in everyday close contact with humans. Lycanthropic traits of the above-mentioned transformation are scarce, although they can be found in rare beliefs, which describe a werewolf as a creature i.e., hairy like a dog with bloody eyes and teeth and fire coming out of its mouth or, on the other hand, as a human being covered with wolf hair.[46]

The image of the transformation of werewolves that first manifest themselves in human form and then into an animal is nicely illustrated in a narrative noted in the village of Otok, in Croatia, where such an occasion was witnessed:

> He went to the graveyard. When it was time for the dead to come out he set down so that he could see from which grave someone would come out and he noticed a man standing on one grave and taking off his suit. When he took it off, the man hangs the suit on a cross, turns into a sow and leaves for the village...When it was one o'clock after midnight the sow came back from the village and turns back into a man...[47]

This account might also indicate that the act of undressing and leaving the clothes behind symbolically stands for leaving behind a humanlike shape that now easily, although not permanently, changes into a teriomorphic one.[48]

In animal form werewolves' physical and acoustic capacities are usually the same as the animal's whose form they take. However, sometimes their appearance has a specific trait, such as being very large as in the case of *'dog large as a donkey'*[49] or being of black colour with a shroud placed over the body.[50] In other cases, special characteristics are linked with their previous lives as human beings. One example from Otok speaks of a werewolf appearing as a horse, neighing and running through the village although his legs were chained, indicating that the man had previously stolen a chained horse from a villager.[51] Sometimes animals whose shape werewolves take display traits of anthropomorphism. Conceptions about the anthropomorphism of animals are not unknown in South Slavic folk culture. Usually it is thought that domestic animals sometimes take on certain human characteristics, such as the ability to talk and communicate among each other, as it was also believed that animals have feelings like human beings.[52] In the case of werewolves, their anthropomorphism is usually expressed in the animals' ability to speak and communicate with living humans but also through certain forms of physical activity that are otherwise not a distinction of animal behaviour, such as hitting objects with a stick.[53]

Werewolves easily go through different bodily shapes, from a human form or just a bellows to animal and vice versa. Such changes might indicate that their bodily matter cannot be considered stable, but constantly fluid. Sometimes changes from one physical form to another are triggered through contact with a living human being or they happen in the moment of a werewolf's death. The former has been testified by a few accounts that describe a werewolf in a human shape turning into a dead black dog after being killed, or a werewolf in the form of a horse that, once shot, changes back into the form of a human being.[54]

As we have seen, the image of werewolves' bodies is constructed through traits of animality as well. Aside from the discussed teriomorphic characteristics that mark the otherness of werewolves' corporeality as supernatural beings, the same is also noted through the ability of these supernatural beings to easily move from one physical shape to another. It is in the realm of the supernatural that such changes occur easily and *naturally*. The animal form does not destabilise the threatening identity of werewolves but rather becomes an important part of it.

Crossing the boundaries

Finally, who are these individuals that possibly undergo such changes of their bodily matters through death, whose bodies continue to exist in one or the other previously discussed form? It seems that notions about people who can potentially turn into werewolves imply a strong link with the moral and social code of the community. Beliefs that sinful and bad people usually undergo such transformations are widely present in South Slavic folk traditions.[55] The unmoral deeds of a person during his lifetime and proneness to aggressive acts or arguments appear to be among the prime reasons that keep the dead unsettled and urge them to return from their graves. Thieves and cones are especially often pertinent to such kinds of goings-on after death. Even in places where it is believed that people of certain religious confessions – for example, the Orthodox in the Bukovica region in Croatia – do not become werewolves, a person who has stolen a ploughshare from a field is, however, going to become one and will return from his grave with a ploughshare attached to his leg.[56] In addition, one's theft can even influence the form in which the werewolf will reappear.[57] Interesting accounts of unmoral and socially unappreciated performances are noted in some parts of Montenegro where parsimony, mercenary and imposture are also mentioned as characteristics of sinful people that lead to the same, already mentioned consequences.[58]

Some accounts about werewolves seem to involve notions of fragmentation of one's lifecycle. This seems to be suggested in a narrative from Croatia about a man who lived with a woman for whom he cared and had children with but still remained unmarried to, even though he was encouraged to do so once he fell ill and others recognised his imminent death, which was possibly the reason why he returned afterwards as a werewolf.[59] Other similar records usually speak of young men that are engaged and bound to be married but die prior to this and after some time rise from their graves as werewolves.[60] Those who died with an undefined social status may also be ascribed to the category of the so-called unclean or restless dead similar to those who died from unnatural causes, were cursed or those who had not been baptised.[61]

The same explanation can probably be applied to the fate of unborn children that die together with the pregnant mother and become werewolves or they can both face the same fate.[62] A matching destiny awaited those involved in the breaking of some kind of a taboo. For example, a child born out of an incestuous relationship will also turn into a werewolf.[63] The same would happen to a child conceived during a woman's monthly period or before the period of 40 days after childbirth was over.[64] The so-called unclean dead – those who were cursed, killed, committed suicide or died far away from home, such as soldiers, and did not receive proper burials or confession before death – were also bound to the above-mentioned experience.[65]

It seems that the breaking of moral, social, customary, or religious rules caused the alteration individuals underwent after death. Crossing different culturally imposed borders also affects one's identity and sets the commencements on the margins of society. As stated by Radin, changes into werewolves (vampires) are perceived as a punishment that affects the individual person as well as the community i.e., threatened by the activities of the former.[66] The above-mentioned and other similar life histories were believed to affect the dead bodies and what further happened to them once they were dead and buried, not accepted by the earth, not decomposing, but instead, blowing up like a bellows and/or manifesting themselves in different physical forms.

It should be pointed out, however, that not following the rules, customs and practices of complex burial rituals could also influence one's incarnation into a werewolf regardless of the morality of the person during their lifetime. It was important to meet the terms of different mortuary practices that involved treatments of the dead body from the time it was still in the house until it was finally placed into the grave.

Dead bodies and customary practices

Generally, dead bodies undergo different practices. The data at hand show that dead bodies were especially manipulated and interacted with in order to prevent the unwanted and feared changes of a human being into a werewolf. Besides honouring the deceased, the dead body was guarded because of the belief that it carried the potential to transform into a werewolf.[67] The notion that the dead body would undergo such changes if an animal crossed or ran over it or passed under the catafalque was quite common; animals were kept away from the corpse.[68] However, certain objects were thought to be able to affect the space around the body keeping it unharmed even if the animal managed to cross it. For example, for such means Muslims in some parts of Bosnia placed a knife in/on the floor next to the head of the dead person.[69]

Not only did the motion of animals around the deceased have to be avoided, but so did certain actions undertaken by humans, which could otherwise affect the undesired transformation of the body. Objects were forbidden to be handed over or tumbled across the dead body.[70] In some places the orientation of the way one should move around the body was important as well, as one had to take care not to pass near the head of the corpse but closer to the feet.[71] Emotions and the way they were expressed are also significant. As Tomo Dragičević noted it was forbidden to cry over a dead body – if tears were to drop on the body they would affect its change.[72] Besides the preservation of the integrity of the dead body and the space around it through the above-mentioned ways, certain actions involving physical interaction with the body were (however) welcomed and thought to be preventive.

One such example was the placement of different iron objects on the body itself. The specificity of the objects involved that underlines their apothropaic character is the material they were made of. Iron knives, part of a brooch or an iron part of a fireplace would be placed on the body while the body was still kept in the house.[73] In addition to iron objects, it seems that the soil was believed to have similar characteristics. According to accounts from Bosnia and Herzegovina a small amount of earth was placed on the body at chest level in order to prevent the dead getting up from the grave.[74] In some places in Bosnia, the placement of small amounts of soil or grass on the chest of the dead body was considered to protect the body from changing if some forbidden activities, such as handling objects over the body, were nevertheless carried out.[75]

Other activities manipulate the body itself while still preserving its integrity or evermore, trying to keep it tightly bounded. In some places in Croatia, the dead body is placed in a shroud and sewed in tightly so that the deceased would not be able to get out of the grave.[76] In a number of Croatian narratives about werewolves tying the deceased's legs together seem to be one of the most important things that would prevent them from coming back from their graves.[77]

The accounts that follow speak of manipulations of the body that actually affect its integrity and wholeness. A number of accounts speak of marking the body by cutting or piercing it, leaving the skin opened. Usually, the body was cut or pierced while the corps was kept in the house, but sometimes such actions were done during the placement of the body into the coffin or while lowering the deceased into the grave. It was believed that opening the skin in such a way would prevent the body from being blown up into a bellows once buried.[78] In other words, it would not be able to take the shape in which werewolves tend to appear. Generally, huge nails would be stabbed into ones belly or nape or a cutting mark by an iron object made on one's toe or ear.[79] An example from Serbia speaks about a spindle being heated (set on fire) and then used to pierce a person's belly button and afterwards placing a small stone in the hole.[80]

Aside from iron objects, nails, needles, axes or different blades, the body could also be probed with a hawthorn stick. In some places, buckthorn is also considered to have protective properties, for example, in Bosnia and Herzegovina where a buckthorn or a hawthorn stick was placed under the corpse's tongue to protect the body from transforming.[81] Generally regarded as strong protection from werewolves, a hawthorn stick or small pieces of it would be stung into the belly, the bellybutton, one's hand or elsewhere.[82] Another widespread practice was to cut the blood vessels located below one's knees as prevention.[83] Some accounts from Boka region in Montenegro specifically point to the disarticulation of certain body parts such as the nose or a finger by cutting them off or even more severe activities such as cutting the person into pieces and burying them in this manner[84]. In some cases, multiple actions that caused an effect on the dead body were thought to be necessary. The bodies of individuals considered to be evil during their lifetime such as male witches were, in order to prevent their post mortal transformations into werewolves, stabbed in the stomach by a hawthorn stick, small pieces of it were stuck under their nails and the blood vessels under the knees were cut as well.[85]

Apart from affecting the dead body itself, the location of its burial, how and with what, was also important. As was noted in Serbia, a dead, non-baptised child that was thought to be prone to easily turning into a werewolf had to be buried among the hawthorn bushes.[86] Sometimes a hawthorn stick was placed inside the grave next to the body.[87] It was also considered useful to turn the dead body upside-down so that it would lie in the grave with the back oriented upwards. An interesting account was noted in Croatia at the beginning of the nineteenth century. A gravedigger from Opatija had to answer to authorities for stabbing five axel pins into the body of a man that was believed to be a witch during his lifetime and finally placing his body on his belly into the grave.[88]

The examples above highlight the ways in which dead bodies are handled in order to communicate specific ideas about society and its members. Motivated by cultural norms, as well as fear itself, members of the living society interact with the dead body in order to influence and prevent the changes that are considered unwanted. Being aware of their actions (motions), emotions and conscious activities through different mortuary rituals, customs and other practices, the living tend to change the reality of the dead bodies, affect their afterlives along with the future of the living community as well. This is achieved by taking care of the integrity of the dead body as well as making it incomplete, disintegrated, disarticulated or inserted with specific objects.

Conclusion

Examined narratives and practices evolving around werewolves in some South Slavic folk traditions highlight notions about the transformational potential of the human form through death, images of werewolves as malevolent supernatural beings and the treatment of dead bodies in order to prevent the possible change of a dead human being into a werewolf. Ideas about the body that does not decompose but only changes its form speak about a transformation of a deceased human into a specific being with its own bodily characteristics. The usual corporeal form of a human being is believed to change into a form that, at least externally, resembles the human shape, or is just a bellows or even an animal. Even more, such corporeality is fluid and can modify from one mentioned form to another.

These changes affect human's ontological positions and social identities and link individuals of particular social or ritual status with the status of malevolent and threatening supernatural beings. It has already been suggested that, in the supernaturally oriented mode of experience of early modern Europe, the boundaries between oneself and the natural world were seen as more permeable.[89] Exploring different notions about the body in traditional rural Finnish society, Stark noted that the human body in rural Finland was conceived as porous and easily invaded by harmful forces from the outside world as the boundaries of the embodied self were depicted as 'opening up' to the external environment.[90] In order to deepen the understanding of such a 'body schema' and the ways in which such bodies were invaded from the outside, she suggested going beyond the Western notions of the body as a physical instrument or an object which is stable, concrete and bounded, and think instead of *a body in flux and motion* defined through what might be called energies and forces, even though such terms might not have been used by the traditional cultures' inhabitants.[91] Keeping this in mind, it is not difficult to consider the fact that due to the liminal potential of death itself a dead human body is even more open to different types of influences from the outside world, boundary transgressions and changes. The dead body is especially easily affected in certain stages of mortuary practices that precede the burial. At that point, the dead body is not part of the living culture anymore but is affected by it through different practices that aim at conquering death itself, mastering the boundaries of the human identity and the human body, and preventing the undesirable changes. Being unable to decompose and disintegrate in the ground is another liminal condition attributed to the once buried body placing it somewhere between culture and nature itself, leaving it even more uncertain and subject to changes.

The body image of werewolves speaks of their otherness as supernatural beings. Their abject bodily characteristics locate werewolves outside the social order as they also reflect the post-mortal embodiment of individuals that transgressed social and cultural boundaries. At the same time, through their specific reality (activity) these supernatural beings continue to threaten the social order itself.

In addition to death and the immorality of the individual it is the animal, or more precisely, its movement around the dead body that affects the destabilisation of the human identity, which in this case puts the animal in opposition with the human as the other. Animality manifests itself as an inherent part of the werewolves' being as well. However, the unfavourable relation of humans to supernatural beings indicates a fear of changes brought by death and gives the theriomorphic symbol negative connotations. The vast majority of animals, whose shape a werewolf can appropriate, are domestic animals that live in everyday close contact with humans. Such a fact might indicate an awareness of the ubiquitous liminal potential of animals as well as of the possibility that the human body, finding itself in the state of death becomes so open to the natural world that it can even appropriate an animal form, primarily of those animals that are close to humans in their everyday lives.

Notes

1. Aleksandar V. Gura, Elena E. Levkievskaya, 'Volkolak', in N.I. Tolstoy (ed.) *Ethnoliguistic dictionary Slavjanskie drevnosti* I (Moskva, 1995), 418–420. According to another theory also mentioned in the dictionary the etymology of the word *volkolak* could be also derived from the baltoslavic *dlak(i)as 'bear'. It is interesting to note that according to some Zakarpattian beliefs a werewolf can spend one month as a wolf and the one as a bear.
2. Vice Vuković, 'Kako čeljade može postati vukom', *Zbornik za narodni život i običaje Južnih Slavena* 28/1 (1931), 239; Stjepan Banović, 'Manji prinosi. Pričanje o vuku u Vrličkoj krajini i u Makarskom primorju', *Zbornik za narodni život i običaje Južnih Slavena* 29/1 (1933), 235–237; Maja Bošković Stulli, *Istarske narodne priče* (Zagreb, 1959), 218.
3. See Chapter 2 of this book.
4. Irena Benyovsky, 'Vampiri u dubrovačkim selima 18. stoljeća', *Otium* 4, 1/2 (1996), 118–130.
5. Benyovsky, 'Vampiri u dubrovačkim selima 18. stoljeća', 119; Matthew Bunson, *The Vampire Encyclopedia* (New York, 1993), 280.
6. Stevan Dučić, 'Život i običaji plemena Kuča', *Srpski etnografski zbornik* 48 (1931), 292; Natalija Blagojević, 'Vuk u narodnim običajima i vjerovanjima u užičkom kraju', *Užički zbornik* 7 (1978), 375–404.
7. 'Vukodlak', *Riječnik hrvatskog ili srpskog jezika* XXI (Zagreb, 1971–1972), 622–623.

8. Tihomir Đorđević, 'Vampir i druga bića u našem narodnom verovanju i predanju', *Srpski etnografski zbornik* 30 (1953), 147–282.
9. Luka Šešo, *Nadnaravna bića u kontekstu etnologijskih istraživanja tradicijskih vjerovanja u dalmatinskom zaleđu danas. Doktorska disertacija* (Zagreb, 2010), 27.
10. The same has already been noticed by Bošković-Stulli and Šešo. Maja Bošković-Stulli, 'Usmene pripovijetke i predaje i otoka Brača' *Narodna umjetnost* 11/12, 1 (1975), 5–157; Šešo, *Nadnaravna bića u kontekstu etnologijskih istraživanja tradicijskih vjerovanja u dalmatinskom zaleđu danas. Doktorska disertacija,* 26.
11. Damjan J. Ovsec, *Slovanska mitologija in verovanje* (Ljubljana, 1991), 416.
12. Gábor Klaniczay, 'The Decline of Witches and the Rise of Vampires under the Eighteenth-Century Habsburg Monarchy', in: *The Uses of Supernatural Power. The Transformation of Popular Religion in Medieval and Early Modern Europe* (Princeton, 1990), 178.
13. Benyovsky, 'Vampiri u dubrovačkim selima 18. stoljeća', 119.
14. Bošković-Stulli, 'Usmene pripovjetke i predaje s otoka Brača', 157; Šešo, *Nadnaravna bića u kontekstu etnologijskih istraživanja tradicijskih vjerovanja u dalmatinskom zaleđu danas. Doktorska disertacija,* 28–29.
15. Šešo, *Nadnaravna bića u kontekstu etnologijskih istraživanja tradicijskih vjerovanja u dalmatinskom zaleđu danas. Doktorska disertacija.*
16. Benyovsky, 'Vampiri u dubrovačkim selima 18. stoljeća', 118–130; L. Čoralić et al. 'Vampiri u korčulanskom selu Žrnovo: tragom jednog arhivskog spisa iz 18. stoljeća', *Acta Medico Historica Adriatica* 9(1) (2011), 33–46.
17. Pieter Plas, 'Vukovi i smrt: Tanatološko značenje vuka u tradicijskoj kulturi zapadnojužnoslavenskog područja', *Narodna umjetnost* 47(2) (2010), 77–95.
18. Đorđević, 'Vampir i druga bića u našem narodnom verovanju i predanju', 147–282.
19. Dagmar Burkhart, *Kulturraum Balkan. Studien zur Volkskunde und Literatur Südosteuropas* (Berlin/Hamburg, 1989); Ana Radin, *Motiv vampira u mitu i književnosti* (Beograd, 1996).
20. Dušan Bandić, *Tabu u tradicionalnoj kulturi Srba* (Beograd, 1980), 118.
21. Animistic conceptions imply that the death of a human being is followed by the decomposition of the body and the departure of the soul which continues to live by itself. Such notions about the soul may, according to Suzana Marjanić, also be referred to as the monistic soul. According to Marjanić, the concept of the dyadic soul which distinguishes the *bodily soul* and the *free soul* and the concept of the monistic soul, where the *bodily soul* becomes *free* only after death are dominant notions in the folkloristic, ethnological and anthropological discourse devoted to Southern Slavic folk traditions. Suzana Marjanić, 'Južnoslavenske folklorne koncepcije drugotvorenja duše i zoopsihonavigacije/zoometempsihoze', *Kodovi slovenskih kultura* 9 (2004.), 208–248; Đorđević, 'Vampir i druga bića u našem narodnom verovanju i predanju', 164; Radoslav Vešović, *Pleme Vasojevići*, 426. see Đorđević, 'Vampir i druga bića u našem narodnom verovanju i predanju', 164, Dučić, 'Život i običaji plemena Kuča', 288. According to some accounts it is the breath of the devil that blows up the body. Šešo, *Nadnaravna bića u kontekstu etnologijskih istraživanja tradicijskih vjerovanja u dalmatinskom zaleđu danas. Doktorska disertacija*, 30.
22. Bandić, *Tabu u tradicionalnoj kulturi Srba*, 106.
23. Åsa M. Larsson, *Breaking and Making Bodies and Pots. Material and Ritual Practices in Sweden in the Third Millennium BC* (Uppsala, 2009), 297.
24. Nicole Sault (ed.) *Many Mirrors: Body Image and Social Relations* (New Brunswick, 1994), 13–14.

25. Laura Stark, *The Magical Self. Body, Society and the Supernatural in Early Modern Rural Finland* (Helsinki, 2006), 153.
26. Liv Nilsson Stutz, 'Embodied Rituals & Ritualized Bodies. Tracing Ritual Practices in Late Mesolithic Burials', *Acta Archaeologica Lundensia,* series 8(46) (2003), 95; Chris Shilling, *The Body and Social Theory* (London, 2003), 155.
27. Dušan Borić, 'Body Metamorphosis and Animality: Volatile Bodies and Boulder Artworks from Lepenski Vir', *Cambridge Archaeological Journal* 15(1) (2005), 51.
28. Borić, 'Body Metamorphosis and Animality: Volatile Bodies and Boulder Artworks from Lepenski Vir', 51.
29. Borić, 'Body Metamorphosis and Animality: Volatile Bodies and Boulder Artworks from Lepenski Vir', 52.
30. Borić, 'Body Metamorphosis and Animality: Volatile Bodies and Boulder Artworks from Lepenski Vir', 52.
31. Larsson, *Breaking and Making Bodies and Pots. Material and Ritual Practices in Sweden in the Third Millennium BC,* 264, 297.
32. Larsson, *Breaking and Making Bodies and Pots. Material and Ritual Practices in Sweden in the Third Millennium BC,* 393; George Lakoff, Mark Johnson, *Philosophy in the Flesh. The Embodied Mind and Its Challenge to Western Thought* (New York, 1999), 555.
33. Josip Lovretić, *Otok: narodni život i običaji* (Vinkovci, 1990), 537.
34. Lovretić, *Otok: narodni život i običaji,* 537.
35. Julia Kristeva, *Powers of Horror. An Essay on Abjection* (New York, 1982) 2,3; Deborah Caslav Covino, *Amending the Abject body: Aesthetic Makeovers in Medicine and Culture* (Albany, 2004), 17.
36. Šešo, *Nadnaravna bića u kontekstu etnologijskih istraživanja tradicijskih vjerovanja u dalmatinskom zaleđu danas,* 27; Ivan Milčetić, 'Vjera u osobita bića. 1. Vukodlak i krsnik. b) Krk i Kastav u Istri', *Zbornik za narodni život i običaje Južnih Slavena* 1 (1896), 224–226.
37. Stjepan Banović, 'Vukodlaci', *Zbornik za narodni život i običaje Južnih Slavena* 26(2) (1928), 347–357; Tomo Dragičević, 'Narodne praznovjerice. Nastavak', *Glasnik zemaljskog muzeja u Bosni i Hercegovini* 20 (1908), 449–466.
38. Frano Ivanišević, 'Poljica: narodni život i običaji', *Zbornik za narodni život i običaje Južnih Slavena* 10(2) (1905), 181–307.
39. Banović, 'Vukodlaci', 348–349.
40. Lovretić, *Otok: narodni život i običaji,* 543. Milčetić, 'Vjera u osobita bića. 1. Vukodlak i krsnik. b) Krk i Kastav u Istri', 225; Stanoje M. Mijatović, 'Narodna medicina Srba seljaka u Levču i Temniču' Srpski etnografski zbornik 13 (1909), 259–482.
41. Lovretić, *Otok,* 543; Vladimir Ardalić, 'Vukodlak. (Bukovica u Dalmaciji)' *Zbornik za narodni život i običaje Južnih Slavena,* 13(1) (1908), 148–154.
42. Lovretić, *Otok: narodni život i običaji,* 538; Ivan Filakovac, 'Vjerovańa. Retkovci u Slavoniji' *Zbornik za narodni život i običaje Južnih Slavena* 10(1) (1905), 144–149.
43. Lovretić, *Otok: Narodni život i običaji,* 543; Milčetić, 'Vjera u osobita bića. 1. Vukodlak i krsnik. b) Krk i Kastav u Istri', 225.
44. Stevan Mačaj, 'Crnorečki okrug', *Glasnik srpskog učenog društva* (1892), 1–186.
45. For example: Ardalić, 'Vukodlak. (Bukovica u Dalmaciji)', 150; Lovretić, *Otok: narodni život i običaji,* 546; Emilijan Lilek, 'Etnološki pabirci po Bosni i Hercegovini', Glasnik zemaljskog muzeja 11 (1899), 699–721; Radin, *Motiv vampira u mitu i književnosti,* 44.

46. Filakovac, 'Vjerovańa. Retkovci u Slavoniji', 146; Josip Milićević, 'Etnološka i folklorna građa Daruvarskog kraja. Terenski zapisi iz 1966. i 1967. godine', *Rukopisna zbirka Instituta za narodnu umjetnost* 758 (1970), 38.
47. Lovretić, *Otok: narodni život i običaji*, 546.
48. Clothes appear to be an important motive in European narratives about the lycanthropic character of a living man that were investigated by Lecouteux. Claude Lecouteux, *Witches, Werewolves and Fairies. Shapeshifters and Astral Doublers in the Middle Ages* (Rochester, Vermont, 2003), 113–118.
49. Ardalić, 'Vukodlak. (Bukovica u Dalmaciji)', 150.
50. Lovretić, *Otok: narodni život i običaji*, 538.
51. Lovretić, *Otok: narodni život i običaji*, 537.
52. Jadranka Grbić, 'Dekodiranje ovozemaljskih čina: vjerovanja o životinjama u hrvatskoj etnografiji' in: S. Marjanić and A. Zaradija Kiš (eds) *Kulturni bestijarij* (Zagreb, 2007), 217–229.
53. Ardalić, 'Vukodlak. (Bukovica u Dalmaciji)', 150; Lovretić, *Otok: narodni život i običaji*, 548.
54. Milčetić, 'Vjera u osobita bića. 1. Vukodlak i krsnik. b) Krk i Kastav u Istri', 226; Lovretić, *Otok: narodni život i običaji*, 545.
55. Ivanišević, 'Poljica: narodni život i običaji', 245; Pavel A. Rovinskij, *Černogorija v eja prošlom i nastojaščem 2* (1901), 526; Dučić, 'Život i običaji plemena Kuča', 288; Dragičević, 'Narodne praznovjerice. Nastavak', 460.
56. Ardalić, 'Vukodlak. (Bukovica u Dalmaciji)', 148.
57. Lovretić, *Otok: narodni život i običaji*, 537.
58. Rovinskij, *Černogorija v eja prošlom i nastojaščem 2*, 526.
59. Banović, 'Vukodlaci', *Zbornik za narodni život i običaje Južnih Slavena* 26(2) (1928), 347–357.
60. Lovretić, *Otok: narodni život i običaji*, 540, 542, 549–51.
61. Jadranka Grbić, 'Predodžbe o životu i svijetu' in: J. Čapo Žmegač et al. (eds) *Etnografija: svagdan i blagdan hrvatskog puka* (Zagreb, 1998), 298.
62. Josip Lalić, 'Odgovor na neka pitanja' *Arkiv za povjestnicu jugoslavensku* II, 411–422; Gjuro Deželić, 'Odgovor na pitanja stavljena po historičkom društvu', Arkiv za povjestnicu jugoslavensku 7 (1863), 199–232.
63. Dragičević, 'Narodne praznovjerice', 460.
64. Osman Đikić, *Zora* (Mostar, 1899), 68.
65. Mijatović, 'Narodna medicina Srba seljaka u Levču i Temniču', 446; Lovretić, *Otok: narodni život i običaji*, 543; Radin, *Motiv vampira u mitu i književnosti*, 44.
66. Radin, *Motiv vampira u mitu i književnosti*, 44.
67. Grbić, 'Predodžbe o životu i svijetu', 296–336.
68. For example: Ljubinko Radenković, *Simbolika sveta u narodnoj magiji Južnih Slavena* (Niš, 1996), 120; Milenko S. Filipović, 'Život i običaji narodni u Visočkoj nahiji' *Srpski etnografski zbornik* 61 (1949), 176; Mijatović, 'Narodna medicina u Srba seljaka u Levču i Temniću', 116; 'Vampir' in: Špiro Kulišić et al. (eds) *Srpski mitološki rječnik* (Beograd, 70), 50; Lovretić, 'Otok. Narodni život i običaji', 26–34. Luka Lukić, 'Varoš. Narodni život i običaji'. *Zbornik za narodni život i običaje Južnih Slavena* 25(2) (1924), 25–349.
69. Emilijan Lilek, 'Vjerske starine iz Bosne i Hercegovine' *Glasnik zemaljskog muzeja* 6 (1894), 141–161.
70. Marinko Stanojević, 'Zaglavak. Antropogeografska proučavanja' *Srpski etnografski zbornik* 20 (1913), 3–89.
71. Lilek, 'Vjerske starine iz Bosne i Hercegovine', 144.

72. Dragičević, 'Narodne praznovjerice', 460.
73. Đikić, *Zora* (Mostar, 1899), 69; Dučić, 'Život i običaji plemena Kuča', 248; Andrija Jovičević, 'Zeta i Lješkopolje' *Srpski etnografski zbornik* 38 (1926), 538.
74. Lilek, 'Etnološki pabirci po Bosni i Hercegovini', 703.
75. Savo Semiz, 'Nekoliko narodnih gatanja' *Bosanska vila* (1901), 185.
76. Ivanišević, 'Poljica: narodni život i običaji', 246.
77. Lovretić, *Otok: narodni život i običaji*, 540–542, 542–543.
78. Dorđević, 'Vampir i druga bića u našem narodnom verovanju i predanju', 41; Milan Đ. Miličević, 'Život Srba seljaka', *Srpski etnografski zbornik* 1 (1894), 342.
79. Miličević, 'Život Srba seljaka' 342; Stanojević, 'Zaglavak. Antropogeografska proučavanja', 37; Aleksandar Vasiljević, *Bosanska vila* (1897), 229.
80. S. Tanović, *Srpski etnografski zbornik* 50, 12, 259, see Đorđević 'Vampir u našem narodnom verovanju i predanju', 189.
81. Lilek, 'Etnološki pabirci po Bosni i Hercegovini', 703.
82. A. Vasiljević, Bosanska vila (1897), 229; Stanojević, 'Zaglavak. Antropogeografska proučavanja', 37.
83. Milčetić, 'Vjera u osobita bića. 1. Vukodlak i krsnik. b) Krk i Kastav u Istri, 225; Ardalić, 'Vukodlak. (Bukovica u Dalmaciji)', 148, 149; S. Banović, 'Vjerovańa (Zaostrog u Dlamaciji)', *Zbornik za narodni život i običaje* 23, 185–214.
84. Sava Nakičenović, 'Boka. Antropogeografska proučavanja', *Srpski etnografski zbornik* 20 (1913), 189–625, 346.
85. Stanojević, 'Zaglavak. Antropogeografska proučavanja', 37; Dučić, 'Život i običaji plemena Kuča', 288.
86. Atanasije Petrović, 'Crni glog u narodnom vjerovanju', *Glasnik etnografskog muzeja u Beogradu* 13, 132–137.
87. Mitar S. Vlahović, *Bratstvo* 30, 133, see Đorđević, 'Vampir u našem narodnom verovanju i predanju', 190.
88. Vatroslav Jagić, *Archiv für slaviche Philologie BD* 6 (1882), 619.
89. Charles Taylor, *Sources of the Self. The Making of Modern Identity* (Cambridge, 1985).
90. Stark, *The Magical Self. Body, Society and the Supernatural in Early Modern Rural Finland*, 254.
91. Stark, *The Magical Self. Body, Society and the Supernatural in Early Modern Rural Finland*, 254.

Index